ANTHROPOLOGICAL PAPERS OF
THE UNIVERSITY OF ARIZONA
NUMBER 84

El Fin del Mundo

A Clovis Site in Sonora, Mexico

EDITED BY

Vance T. Holliday, Guadalupe Sánchez, and Ismael Sánchez-Morales

WITH CONTRIBUTIONS BY

Joaquín Arroyo-Cabrales
Jordon Bright
James K. Feathers
Edmund P. Gaines
Thanairi Gamez
Gregory W.L. Hodgins
Susan M. Mentzer
Carmen Isela Ortega-Rosas
Manuel R. Palacios-Fest
Kayla B. Worthey
Kristen Wroth

THE UNIVERSITY OF
ARIZONA PRESS
TUCSON

The University of Arizona Press
www.uapress.arizona.edu

Printed in the United States of America.

29 28 27 26 25 24 6 5 4 3 2 1

ISBN-13: 978-0-8165-5299-3 (paper)
978-0-8165-5300-6 (ebook)

Editing and indexing by Linda Gregonis.
InDesign layout by Douglas Goewey.

Library of Congress Cataloging-in-Publication Data
Names: Holliday, Vance T., editor. | Sánchez, Guadalupe, editor. | Sánchez-Morales, Ismael, 1985– editor.
Title: El Fin del Mundo : a Clovis site in Sonora, Mexico / edited by Vance T. Holliday, Guadalupe Sánchez, Ismael Sánchez-Morales.
Other titles: Anthropological papers of the University of Arizona ; v. 84.
Description: Tucson : University of Arizona Press, 2024. | Series: Anthropological papers of the University of Arizona ; volume 84 | Includes bibliographical references and index.
Identifiers: LCCN 2023048389 (print) | LCCN 2023048390 (ebook) | ISBN 9780816552993 (paperback) | ISBN 9780816553006 (ebook)
Subjects: LCSH: Clovis culture—Mexico—Fin del Mundo Site. | Antiquities, Prehistoric—Mexico. | Fin del Mundo Site (Mexico)
Classification: LCC F1219.1.S65 F56 2024 (print) | LCC F1219.1.S65 (ebook) | DDC 972/.1701—dc23/eng/20231213
LC record available at https://lccn.loc.gov/2023048389
LC ebook record available at https://lccn.loc.gov/2023048390

♾ This paper meets the requirements of ANSI/NISO Z39.48-1992 (Permanence of Paper)

Cover: A Clovis point recovered from Upper Bonebed in Locus 1 and the three late Pleistocene proboscideans found at El Fin del Mundo: L-R, mastodon (*Mammut americanum*), mammoth (*Mammuthus columbi*), and gomphothere (*Cuvieronius* sp.). Point photograph by Ismael Sánchez-Morales. Sculptures and sculpture photography by Sergio de la Rosa.

ABOUT THE AUTHORS

JOAQUÍN ARROYO-CABRALES (Ph.D., Texas Tech University, 1994) is Senior Scientist at the Mexican National Institute of Anthropology and History (INAH) and is currently head of the Archaeozoological Lab and in charge of the Paleontological Collection. He is a member of the Mexican National Researchers System. His research focuses on Late Quaternary vertebrates and their contribution to understanding the paleoenvironments in which humans survived in the Americas.

JORDON BRIGHT (Ph.D., University of Arizona, 2017) is a research scientist in the School of Earth and Sustainability at Northern Arizona University (NAU). He specializes in using ostracode faunas and stable isotope data to reconstruct Pliocene to Holocene paleolake and wetland environments in the arid western United States and northern Mexico. Currently he manages the Amino Acid Geochronology Laboratory at NAU.

JAMES K. FEATHERS (Ph.D., University of Washington, 1990) is Research Professor Emeritus at the University of Washington and was the director of the Luminescence Dating Laboratory in the Anthropology Department there from 1993 to 2023, when the lab closed. During that time, he processed nearly 4,000 luminescence dates on sediments, rocks, and ceramics, including a large number of sediments from Late Pleistocene-Early Holocene dates at archaeological sites from North and South America. He is also interested in luminescence dating of archaeological structural remains and the evolution of ceramic technology.

EDMUND P. GAINES (M.A., University of Arizona, 2006) specializes in remote fieldwork and prospecting for Pleistocene archaeological sites. He led the field surveys that identified Fin del Mundo in Sonora, as well as McDonald Creek, Upward Sun River, and more than a dozen other Pleistocene-aged sites in interior Alaska. He currently is the head of Brice Environmental Services' cultural resource division in Fairbanks, leading historic and prehistoric projects across Alaska and the Pacific.

THANAIRI GAMEZ (Master in Geology, Universidad de Sonora, 2022) has worked since 2015 as a technician on many projects related to paleoecology, archaeology, and paleoclimatology for the State University of Sonora. As a postgrad student she has published one article in an international journal.

GREGORY W.L. HODGINS (D.Phil. University of Oxford 1999) is a researcher at the University of Arizona Accelerator Mass Spectrometry Laboratory and director from 2015 to 2022. He works on archaeological and forensic applications of radiocarbon dating and specializes in bone dating and the development of new applications for radiocarbon measurement.

VANCE T. HOLLIDAY (Ph.D., University of Colorado, 1982) was on the Geography faculty at the University of Wisconsin-Madison from 1986 to 2002 and the Anthropology and Geosciences faculty at the University of Arizona from 2002 to 2022 when he retired. He is Executive Director of the Argonaut Archaeological Research Fund, which is devoted to exploring the early peopling of the greater Southwest. His interests include Paleoindian archaeology and geoarchaeology as well as Quaternary soils and paleoenvironments, and Paleolithic geoarchaeology of eastern Europe.

SUSAN M. MENTZER (Ph.D., University of Arizona, 2011) is a researcher in the Senckenberg Centre for Human Evolution and Palaeoenvironment, located at the University of Tübingen in Germany. Her geoarchaeological research focuses on using micromorphology and related microanalyses to reconstruct the formation processes and post-depositional diagenesis of archaeological deposits. She has particular interests in hunter-gatherer lifeways, organization of living space and ancient fire use, and has worked on cave, rockshelter, and open air sites around the world.

CARMEN ISELA ORTEGA-ROSAS (Ph.D., University Paul Cezanne, France, 2007) is a member of the Mexican National System of Researchers. She has studied the paleoecology and paleoclimatology of northern Mexico for 20 years using proxies such as pollen, fossils, geochemistry, and paleomagnetism. Her research has led to more than 15 articles in international journals and books. She is an adviser to the government of the state of Sonora on issues related to climate change and has participated in the creation of the State Climate Change Plan for Sonora.

MANUEL R. PALACIOS-FEST (Ph.D., University of Arizona, 1994) is a geoscientist specializing in micropaleontology and paleoecology. He is the owner of Terra Nostra Earth Sciences Research, LLC, a firm dedicated to studying environmental change using ostracodes, mollusks, and the microflora present in aquatic systems. His experience includes research across the southern United States, from Texas to California and abroad.

GUADALUPE SÁNCHEZ (Ph.D., University of Arizona, 2010) is at the National Institute of Anthropology and History (INAH) and a member of the Mexican National System of Researchers. She has studied the geoarchaeology and lithic technology of sites in northern Mexico together with hunter-gather prehistory, paleoethnobotany, and paleoecology.. Her research has led to more than 50 articles in international journals and books. Her 2016 book *Los Primeros Mexicanos: Late Pleistocene/Early Holocene People of Sonora* (University of Arizona Anthropological Papers No.76), received honorable mention for Best Archaeological Investigation in Mexico.

ISMAEL SÁNCHEZ-MORALES (Ph.D., University of Arizona, 2023) is the curator of Anthropology at the Arizona Museum of Natural History. He is an archaeologist specializing in the study of human-landscape interactions among archaeological hunter-gatherer societies through the analysis of lithic technologies. His research focuses on the Paleoindian and Archaic occupations of northwest Mexico and the American Southwest and on the Middle Stone Age of the Maghreb.

KAYLA B. WORTHEY (M.A., University of Arizona, 2017) is a PhD candidate in the University of Arizona School of Anthropology, a zooarchaeologist, and a stable isotope geochemist. Her research has focused on the study of Pleistocene paleoenvironments at archaeological sites in northern Mexico, Morocco, and Turkey.

KRISTEN WROTH (Ph.D., Boston University, 2018) is the Chemistry Lab Manager and Chemical Hygiene Officer for Earlham College. She specializes in the combination of micromorphology and plant microfossil analysis. Her work focuses on human-landscape interactions and paleoenvironmental reconstruction from a variety of locations and time periods and has recently centered on Middle Paleolithic sites in Europe and the Middle Stone Age of South Africa.

To La Familia Placencia

Frontispiece: View northeast down the main site arroyo and across the north end of El Fin del Mundo and the alluvial fan to the Arroyo Carrizo in the distance. The white coating is the Strata 5/C carbonate across the distal fan surfaces. Loci 1, 2, 3, and 4 are within but near the distal edge of the Strata 5/C area with Locus 1 just right of the arroyo (black plastic is visible). Loci 3 and 4 are farther right, and Locus 2 is to the left of the arroyo. Photograph by Henry D. Wallace, courtesy Archaeology Southwest.

Contents

TABLES

FIGURES

Acknowledgments

This work was funded by the Argonaut Archaeological Research Fund (University of Arizona Foundation; V. T. Holliday, Director) established by Joe and Ruth Cramer of Denver, Colorado; the National Geographic Society; Instituto Nacional de Antropología e Historia, Mexico City; and Archaeology Southwest (formerly the Center for Desert Archaeology), Tucson.

Our field crews were directed by Edmund "Ned" Gaines (seasons 1 and 2; in 2007 and 2008), Natalia Martínez-Tagüeña (seasons 3 and 4; in 2010 and 2012), Ismael Sánchez-Morales (seasons 5, 6 and 7; in 2014, 2018, and 2020), and Hugo García-Ferrusca (during the second part of season 7 in 2020). Mike Brack (Desert Archaeology) set up the grid and mapped the site. William Doelle (Desert Archaeology and Archaeology Southwest), arranged for the mapping and provided other logistical support.

Figures for this volume were prepared by Jim Abbott (Scigraphics), Ismael Sánchez-Morales, and Paul Neville (University of New Mexico). Several were based on maps prepared by Ned Gaines and Mike Brack. We are pleased to include aerial photography by Henry Wallace (Desert Archaeology), and the spectacular image of late Pleistocene proboscideans of Mexico by artist Sergio de la Rosa (http://smilodon.com.mx).

Greg Hodgins and Andrew Kowler were instrumental in the collection, processing, and analysis of radiocarbon-dating samples. John Carpenter and Elisa Villalpando shared their extensive knowledge of Sonoran archaeology and provided moral and logistical support throughout this project. Dennis and Martha Fenwick provided support for some of the zooarchaeological research.

We recognize the late Manuel Robles for pioneering Clovis research in Sonora. Gustavo Placencia first told us about the bones on his ranch and, together with his son Chantín, was a gracious host throughout our fieldwork and allowed us to use his water and firewood. Jano Valdez, cowboy at the ranch, was always very helpful and would often pay us a visit at the site. We are very grateful to our dear friends and cooks Nora Granillo 'Doña Chichí' and Edilia Reyna, and to Adolfo León and Javier Saavedra, who worked with us at the site and were our local guides in this area of the Sonoran Desert and at the town of Félix Gómez 'El Dipo.'

Donald Grayson, Vance Haynes, Greg Hodgins, Steve Kuhn, Gary Morgan, Richard Hulbert, Spencer Lucas, David Meltzer, and Mike Pasenko provided valuable insights and suggestions regarding the archaeology, geology, and paleontology of the site.

T. J. Ferguson, previous Anthropological Series editor was the first to encourage our submission to the University of Arizona Press. We appreciate the cheerful, enthusiastic, and competent help from Allyson Carter (Senior Acquisitions Editor) and Alana Enriquez (Editorial Assistant) at the Press and Linda Gregonis (University of Arizona Anthropological Series Editor), in all stages of editing and production. Two anonymous reviewers provided helpful comments that improved the manuscript. Diane Holliday provided technical editing and checked references. Brendan Fenerty calibrated the radiocarbon dates.

Our sincere thanks and gratitude to our cheerful and hard-working field, lab and kitchen crews who made this project so memorable and enjoyable. Thanks also go to grant reviewers who disparaged our work at the site (it was "a mess") and did not think that it was worth funding. They inspired us!

CHAPTER ONE

Background to the Research at El Fin del Mundo

Vance T. Holliday, Guadalupe Sánchez, Ismael Sánchez-Morales,
and Edmund P. Gaines

The Clovis occupation of North America is the oldest generally accepted and well documented archaeological assemblage on the continent, dating to ~13,000 cal yr B.P. (Meltzer 2021). The distinctive Clovis points have been reported from throughout most of the lower 48 United States, parts of Canada, as well as Mexico, throughout Central America, and possibly in Venezuela (Smith. Smallwood, and DeWitt 2015; Pearson 2017). Clovis is classically associated with mammoth, although only about 12 firm Clovis/mammoth associations are known (Grayson and Meltzer, 2015). Associations of Clovis and other late Pleistocene megafauna are more rare, consisting of mastodon and bison (Grayson and Meltzer 2015). In this volume we provide a full report on the site of El Fin del Mundo, the first documented Clovis association with gomphothere (*Cuvieronius*). The site is in Sonora, Mexico (Figure 1.1), making it the northernmost dated late Pleistocene gomphothere and the youngest in North America. It is the first documented intact, buried Clovis site outside of the United States and the first in situ Paleoindian site identified in northwestern Mexico.[1] The site also includes a Clovis activity area on the "upland" surface (described in Chapter 2) that rises gradually from the area with the buried features. In addition, a paleontological bonebed below the Clovis level includes a rare association of mastodon, mammoth, and *Cuvieronius* sp. The site also provides a paleoenvironmental record rare for the region spanning the Late Glacial Maximum (LGM), Bølling-Allerød Chron, Younger Dryas Chron, and the early Holocene. These archaeological, geological, paleontological, and paleoenvironmental records are presented and synthesized here.

Some of the first buried Clovis sites reported from North America came from the southwestern United States at the Naco and Lehner sites in the San Pedro Valley of Arizona (Figure 1.1; Haury 1953; Haury, Sayles, and Wasley 1959). Indeed, they were the first well-documented Clovis sites to be reported after the finds from the Clovis type site (Blackwater Draw Locality 1) in New Mexico (Cotter 1937, 1938; Sellards 1952).[2] The three sites in New Mexico and Arizona helped define the notion of Clovis as "big game hunters." This was further reinforced by additional discoveries of mammoth in the San Pedro Valley, including more Clovis-mammoth associations. The most extensive, complex (in terms of activity areas), and best known of the group was the last one excavated, the Murray Springs site (Haynes and Huckell 2007), which is about 250 km (~155 mi) northeast of El Fin del Mundo. The San Pedro Valley revealed the highest concentration of buried, intact Clovis sites in North America and perhaps the densest concentration of mammoth kills in the world (Ballenger 2015). A ~55 km (~34 mi) reach of the upper San Pedro River Valley revealed a minimum of four Clovis-mammoth

[1] The Wally's Beach site in Canada was interpreted as a Clovis horse and camel kill (Kooyman, Newman, and others 2001; Kooyman, Hills, McNeil, and Tolman 2006; Kooyman, Hills, Tolman, and McNeil 2012) but the relationship of the Clovis artifacts and faunal remains was unclear, and the artifacts appeared to be redeposited. The site is a kill but with no clear Clovis association (Waters and others 2015). Debert, Canada, is an Eastern Fluted site with artifacts similar to Clovis (MacDonald 1985) but with some technological differences; it is later than classic Clovis (Miller, Holliday, and Bright 2013).

[2] The Dent site, Colorado (Figgins 1933), and the Miami site, Texas (Sellards- 1938), are Clovis mammoth kills that were reported before Clovis was recognized as a distinct artifact style different from and older than Folsom (Sellards 1952; Wormington 1957).

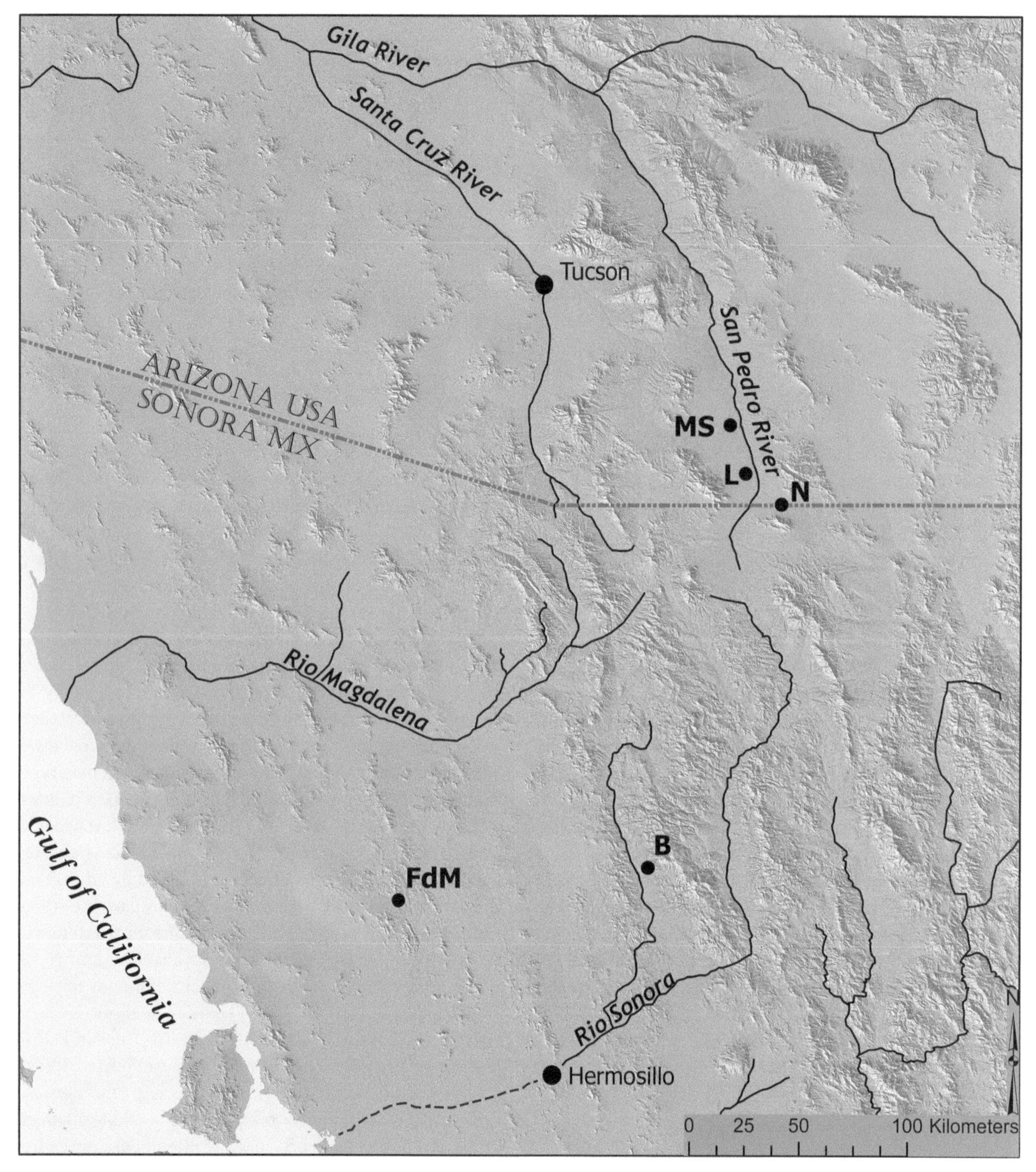

Figure 1.1. Northern Sonora and southern Arizona with key cities and drainages and the locations of El Fin del Mundo (FDM) and El Bajío (B) in Sonora and three key Clovis sites in the upper San Pedro River Valley of southern Arizona (N = Naco; L = Lehner, and MS = Murray Springs). The Naco area also includes the Leikem and Navarette sites. Prepared by Paul Neville.

sites, and possibly as many as six (Figure 1.1; Haynes and Huckell 2007; Ballenger 2015; Holliday, Haynes, and Huckell in press). Besides their archaeological records, these sites provided important clues regarding the environment of the final millennia of the Pleistocene in the Southwest based on geology, paleontology, and paleobotany. For example, alluvial systems evolved into palustrine settings, while plant communities became more xeric (Haury 1953; Haury, Sayles, and Wasley. 1959; Haynes and Huckell 2007; Mehringer and Haynes 1965; Ballenger and others 2011).

Given the proximity of the San Pedro Valley sites to the international border (Figure 1.1), an obvious question was what sort of Clovis and Terminal Pleistocene[3] record was preserved in northern Sonora. The first clue to answering this question was a report by Mexican archaeologists on El Bajío, a surface Clovis site north of Hermosillo (Figure 1.1; (Robles 1974; Robles and Manzo 1972). In 2002, with the establishment of the Argonaut Archaeological Research Fund at the University of Arizona (Vance Holliday, director) and in cooperation with the Instituto Nacional de Antropología e Historia (INAH) in Hermosillo, Sonora, the two senior authors planned a research effort focused on Sonora.

The *Proyecto Geoarqueología y tecnología lítica de los sitios Paleoindios de Sonora* was established and led by Guadalupe Sánchez. We initiated a systematic survey looking for Clovis and other Paleoindian sites in the headwaters of the San Pedro River in Sonora (Gaines 2006) and in the Hermosillo/lower Río Sonora region. This work revealed the presence of 3 Clovis site complexes and 21 localities with isolated or unmapped finds of fluted points similar to Clovis (Gaines, Sánchez, and Holliday, 2009). Subsequent research documented Paleoindian materials from throughout Sonora, but no sites with intact buried deposits were reported until the discovery of El Fin del Mundo in 2007 (Sánchez 2016).

In June 1997, during a visit to the municipal museum in Carbó, Sonora, north of Hermosillo, Guadalupe Sánchez and Vance Haynes observed an unfossilized mammoth femur and rib. The bones were recovered more than 30 years earlier on a remote ranch in the municipio of Pitiquito. The owner, Gustavo Placencia, invited the group to his ranch. They had to decline his generous offer due to the time (four hours one way) involved and lack of access to a suitable field vehicle required in the rainy season. The principal objective of the Spring, 2007 field season of the *Proyecto* was to visit all the known localities in northern Sonora where Paleoindian artifacts and/or remains of late Pleistocene megafauna were reported. A priority on this list was the remote ranch in the municipio of Pitiquito.

On February 5, 2007, Guadalupe Sánchez, along with Edmund Gaines and Alberto "Beto" Peña, led by Alejandro "Jano" Valdez, the ranch cowboy, visited the locality that produced the bones on display in the Carbó museum. The exposure was an "island" of sediment in the middle of an arroyo system. Two bone layers were observed, exposed in the profiles around the island (Figure 1.2). The size of the bone and presence of tusk fragments (Figure 1.3) indicated that both layers contained the remains of Pleistocene megafauna. The first artifact found was a yellow chert uniface (#45980) that had recently fallen from the exposed upper bone layer, confirming that the bonebed was archaeological. Shortly thereafter, a large rhyolite Clovis-style biface (#46021) was discovered about 3 meters from the island exposure, followed by discovery of the middle portion of a quartz crystal biface (#46022) next to the north wall exposure (Chapter 4). The team knew they had found a potentially important archaeological and paleontological site and named it "El Fin del Mundo" (the end of the world) on the basis of a comment made when the team first arrived at the site. On a return trip two days later, a complete Clovis point of white chert (#46023; Figure 1.4) was found about 28 meters to the south of the island, confirming the team's suspicions that they had a new Paleoindian site.

As a result of this discovery, we organized a joint Mexican-American program through INAH for the archaeological investigations (under the direction of Guadalupe Sánchez) and the University of Arizona for geoarchaeological research (under the direction of Vance Holliday). We also brought in Joaquín Arroyo-Cabrales (INAH) to work on vertebrate paleontology and zooarchaeology. Recognition of an archaeological record in the context of a lacustrine and paludal sequence at the site resulted in involvement of paleobiological specialists to reconstruct the paleoenvironmental history preserved at the site. Seven field seasons of excavation were completed as of the winter of 2020. Following each field season, a report was prepared in Spanish describing the excavations and analysis of the artifacts. These reports were submitted to the Consejo de Arqueología of the Coordinación Nacional de Arqueología, Instituto Nacional de Antropología e Historia in Mexico City and they are housed in the archives of the Coordinación Nacional de Arqueología, INAH.

In the first season we made a number of significant discoveries that placed the site in the context of other

[3] Terminal Pleistocene is an informal term referring to the final millennia of the Pleistocene following the Last Glacial Maximum, including the Bølling-Allerød and Younger Dryas chrons (~19k – 11.7k cal yr B.P.).

Figure 1.2. Discovery of El Fin del Mundo in February 2007. View of the north end of the south wall of Locus 1. The trowel on the ground just beyond the tape measure is the location of a uniface (#46294); found directly below an overhang, above and left of the end of the tape. The overhang is Stratum 4 (white due to precipitation of calcium carbonate in Stratum 5 and wash-over by carbonate over Stratum 4). The overhang had the negative impression of the artifact and exposed some bone (Upper Bonebed). Photograph by Edmund P. Gaines.

Figure 1.3. Farther west along the same exposure, Alberto Peña examines a tusk fragment at the Stratum 2/3 contact (Lower Bonebed). Note burrowing in Stratum 4 at the top of the photograph. Photograph by Guadalupe Sánchez.

Figure 1.4. The first Clovis point (#46023) from the site, found on the surface by Alberto Pena ~28m south of the south wall exposure of Locus 1. Photograph by Edmund P. Gaines.

Paleoindian and paleoenvironmental research in North America. The two bonebeds were separated by alluvial deposits. The Upper Bonebed was buried within alluvium and beneath lake and wetland deposits. On the upland surface to the south and southwest of the island was an extensive activity area with Clovis and later artifacts. Of particular significance, El Fin del Mundo was the first Paleoindian site that we found that was not collected by artifact hunters. In addition, a primary source of white quartz and quartz crystal was identified about 7 km (~4 mi) west of the site.

El Fin del Mundo appeared to represent a locality where Clovis groups lived and hunted and butchered Late Pleistocene mammals. This site was the first Pleistocene megafauna hunting-butchering site discovered in Mexico since 1957 (Sánchez 2016). The stratigraphic sequence was also very similar to the terminal Pleistocene-early Holocene stratigraphy of the Clovis, New Mexico, and Lubbock Lake, Texas, Paleoindian sites on the High Plains of the southcentral United States, with lake beds resting on alluvium and megafauna remains throughout (Holliday 1997). El Fin del Mundo therefore held the possibility of reconstructing Clovis and early post-Clovis environments and address broader questions regarding regional environmental characteristics before, during, and after the Younger Dryas Chronozone (YDC) (Eren 2008; Haynes 2008; Holliday and Meltzer 2010; Meltzer and Holliday 2010; Strauss and Goebel 2011; Palacios-Fest and Holliday 2017).

The association of Clovis artifacts with gomphothere and our dating of that feature were the focus of a paper by Sánchez and others (2014). The site is also summarized in several synthetic papers (Sánchez 2016; Sánchez and Carpenter 2012; Sánchez and others 2015), but we have a considerable body of data on the artifact assemblages, geology, zooarchaeology, and paleobiology. Those data are the focus of this volume.

SETTING

El Fin del Mundo is in the Basin and Range landscape of the Sonoran Desert, approximately 100 km (~62 mi) northwest of Hermosillo, in the Mexican state of Sonora (~300 km or ~185 mi southwest of Tucson, Arizona; Figure 1.1). The site is at an elevation of approximately 630 meters above mean sea level. The site area covers about 3 km^2 and is exposed at the toe of an ancient alluvial fan that drains to the northeast. The area around the site is an eroded landscape crossed by a series of arroyos incised into the fan and draining into the Arroyo Carrizo (Frontispiece, Figure 1.5). The Carrizo is an axial drainage of the intermontane basin where the site is located; it flows to the southeast and joins the Río Bacoachi. The Río Bacoachi flows to the south and then to the southwest and on into the Gulf of California.

The site is along the distal margin of a relatively intact remnant of the larger fan system. The arroyo system that cuts through the site and drains into the Arroyo Carrizo also left several islands of fan deposits (Frontispiece, Figure 1.6). Only one (Locus 1) is composed of Terminal Pleistocene-Early Holocene alluvial, lacustrine, and palustrine sediments. The other islands consist of older fan alluvium, a superimposed soil, and localized younger tufas. These are identical to the stratigraphy exposed in the uneroded fan deposits. Locus 1, with the buried archaeological bonebed, and the surrounding arroyo landscape, with surface artifacts, covers an area of approximately 100m by 100m. The upland activity area is in an arc about 200m to 1000m south and southwest of Locus 1, on the fan surface that rises from approximately 2 to12 meters above the surface of Locus 1 (Figure 1.6).

METHODS

Fieldwork included excavation and survey. Each area of the site that became the focus of archaeological, paleontological, or geological investigation was formally identified as a "locus." To date, 25 loci have been recognized (Figure 1.6). Excavations focused on Locus 1 (Figures 1.6 and 1.7), the island with the exposed, in situ bonebed. Test excavations were conducted in archaeological Loci 5, 8, 19, and 23. These loci were deemed significant because either lithic scatters or isolated Clovis artifacts were found on the surface or because auger testing revealed potentially undisturbed stratigraphic deposits. No buried archaeological contexts were found during these investigations. Intensive archaeological survey focused on upland "camp" Loci 5, 8, 9, 10, 14, 17, 19, 21, 22, 23, and 24, where scatters of lithic materials and fire features (i.e., hearths and fire cracked rocks) were observed on the surface. Systematic survey was also conducted in Locus 25, which contains a rhyolite outcrop that was a primary source of lithic raw material. Large blocks of this toolstone were observed along with manufacturing debris. Additional survey was conducted in the arroyos and stream beds in the surrounding area of the site, with the objective of identifying locally available lithic raw materials from secondary sources. Most of the loci found during surface survey along with 2, 3, 4,

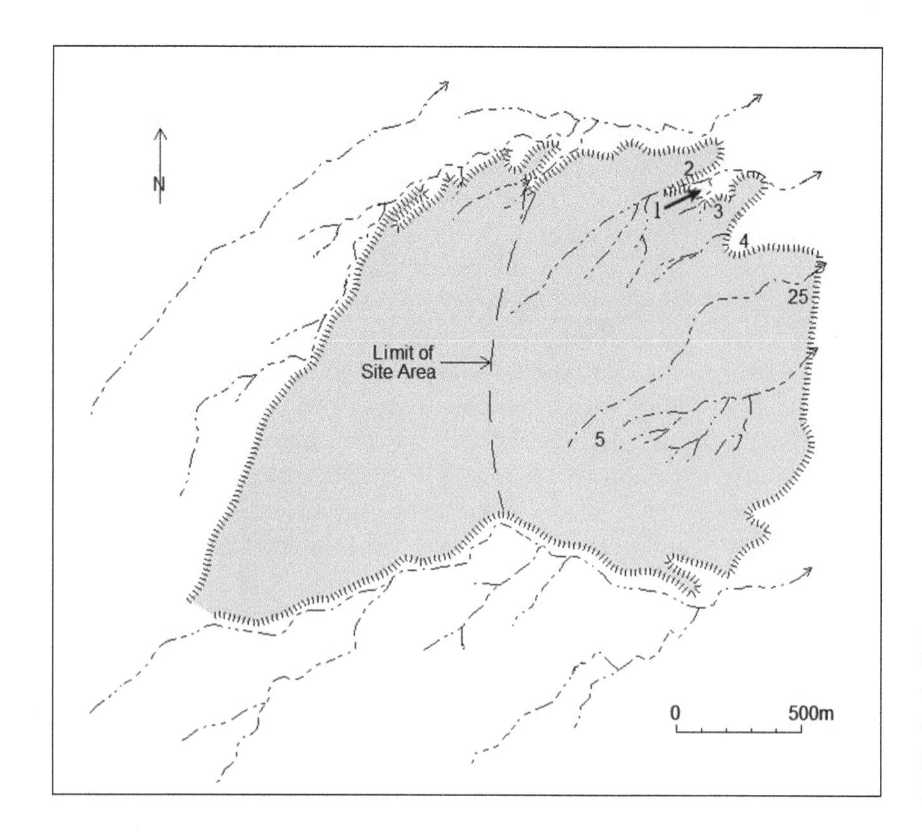

Figure 1.5. The alluvial fan in the Arroyo Carrizo drainage (to the upper right) with El Fin del Mundo toward the distal end. Key site loci are identified by number (1, between 2 and 3, 4, 5, 25). Prepared by Jim Abbott.

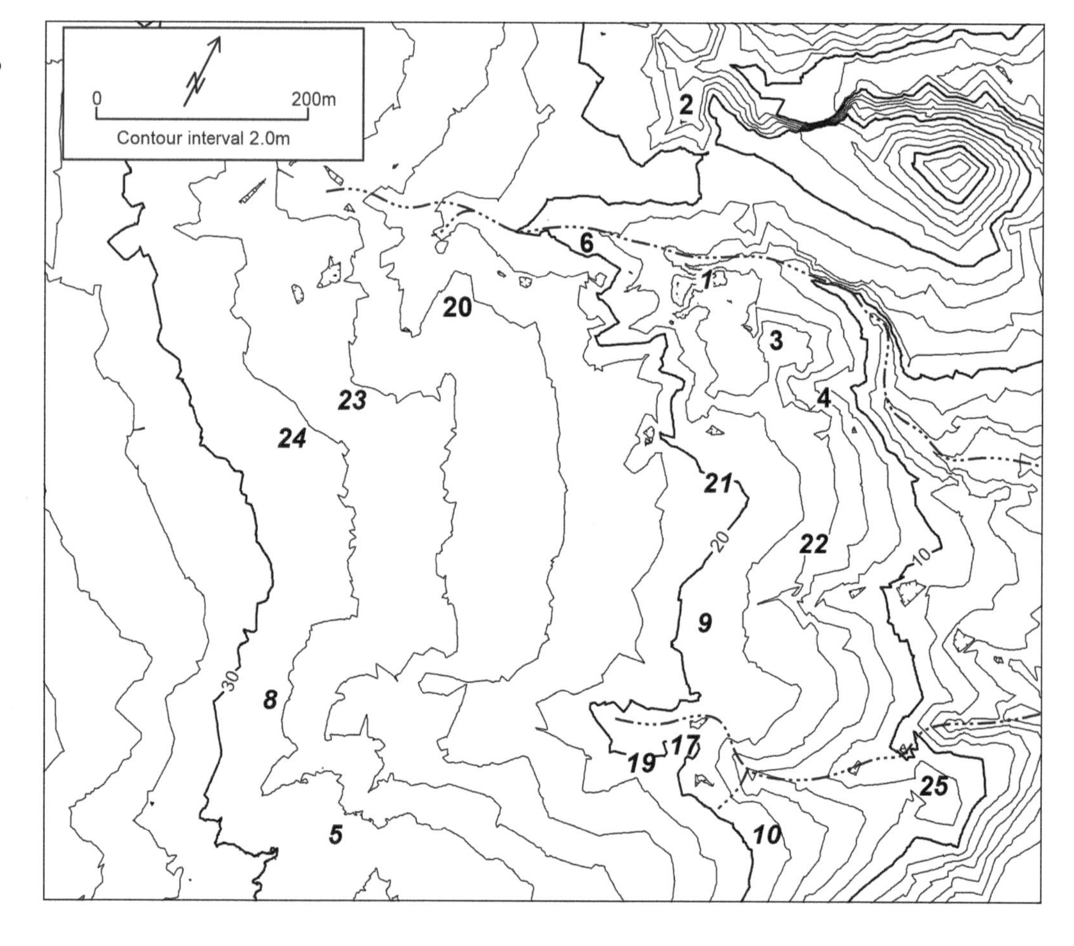

Figure 1.6. Topographic map of El Fin del Mundo with archaeological (italics) and geological loci. Prepared by Jim Abbott based on a map by Mike Brack with support from Desert Archaeology.

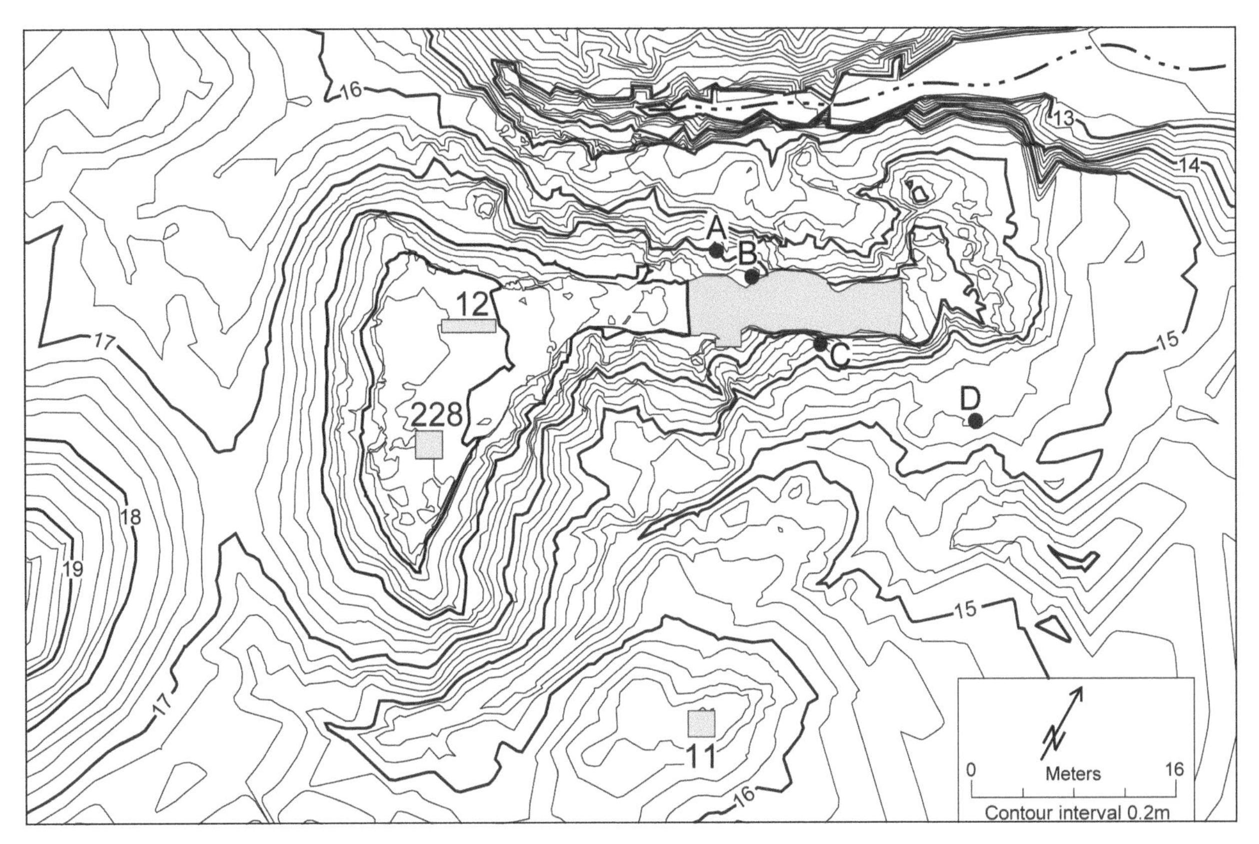

Figure 1.7. Topographic map of Locus 1 and vicinity showing the main excavations on the eastern end of the Locus 1 island, other areas of test excavations, and surface finds. Test excavations: 12 and 228 are tests through Strata 4 and 5 to Stratum 3; 11 is a test block in the area where Clovis point #46023 was found on the second visit to the site. Other surface finds include (A) a quartz crystal biface (#46022), (B) a quartz crystal Clovis point (#58342), (C) a yellow chert uniface (#46294), and (D) a rhyolite biface (#46021). Prepared by Jim Abbott based on a map by Mike Brack with support from Desert Archaeology.

and 6 were also the focus of geological or paleontological investigation.

The excavations and survey were tied to a site datum. A metric grid system was established to maintain horizontal and vertical control for excavations in the Locus 1 island, for testing elsewhere, and for recording surface artifacts recovered during survey. The grid was set using a high-precision Sokia™ Real Time Kinematic (RTK) Global Positioning System (GPS). This system utilizes a pole-mounted GPS receiver unit in concert with a stationary reference (base) station. The grid was established at the same orientation as the island at 329.3° (i.e., 30.7 degrees west of north; Figure 1.7). All elevation measurements were taken in reference to an arbitrary elevation set for the site datum and designated in meters below datum (MBD). To facilitate fieldwork, several sub-datums were set around the site. The entire site area was mapped using the same GPS and total station, yielding a topographic map with 20-cm contour intervals (Figures 1.6 and 1.7). All in situ artifacts and bone recovered during excavation and survey were plotted horizontally with subcentimeter precision in reference to northing (N) and easting (E) coordinates. Each 1-m by 1-m unit was excavated in 50-cm by 50-cm squares and vertically in 5-cm levels or to the top of the next stratigraphic break; whichever was encountered first. All excavated sediment was screened through ¼-inch and ½-cm mesh. In the 2008 season, most of the excavated matrix was water screened, but this was discontinued in subsequent seasons owing to logistical difficulties in getting water to the site.

When necessary, fossil material recovered via excavation was stabilized using a 1:10 mixture of Resistol™ (akin to Elmer's glue) and water. This was applied with either a paintbrush or an aerated sprayer along with ample water to ensure maximum penetration of the bone. In some cases, however, identifiable elements were removed without adding this material to keep them free from contamination that would affect radiocarbon analysis.

In some cases, the remains were encased in polyurethane foam to remove them in sound condition. This was accomplished by first applying wet tissue entirely around the bone surface. As the tissue dried it formed a protective casing that prevented the foam from sticking to the bones and facilitated removal in the laboratory. The remains were then encased in polyurethane foam that forms a 10- to 20-cm thick hardened jacket that holds them together and protects them during removal. Cardinal direction, unit number, and grid coordinate information were recorded on the polyurethane casing prior to removal. When possible, individual elements were jacketed separately. It was, however, necessary to group large concentrations of multiple bones together in a single polyurethane jacket.

Samples for laboratory analyses were largely collected from Strata 3, 4, and 5 in Locus 1. Those for phytoliths (Chapter 7), pollen (Chapter 8), ostracodes (Chapter 10), and most diatoms (Chapter 9) were collected along the north side of Locus 1 at the intersection of the 150E grid line with 546.90N (Profile 08-1; Chapters 2 and 4). The sampling was in contiguous 5-cm segments. Additional samples for diatomite microstratigraphy were collected in contiguous 2-cm intervals from profile 10-4, which was approximately 10 meters east of 8-1, where the diatomite was thicker and microstratified (Chapters 2 and 4). Samples for micromorphology were collected from Profile 8-1 and from other exposures in the site (Chapters 2 and 3). Radiocarbon samples were collected from a number of exposures in the site (Chapter 2), but most were recovered during excavations and systematic examination of the walls of Locus 1.

Archaeological survey consisted of GPS-enabled systematic parallel pedestrian transects. In areas closer to known archaeology or those deemed to have a higher probability of containing artifacts, fossils, or preserved stratigraphic exposures, transect intervals were closed to 5-m spacing. In areas farther from known archaeology or of lower probability, transect intervals were opened to 10-m spacing. No shovel testing or other subsurface testing techniques were utilized during survey. Excavations outside of Locus 1 involved controlled test excavations of 1-m by 1-m or 1-m by 2-m units in most cases. While all archaeological and paleontological materials encountered during excavation were collected for analysis and curation, collection of surface artifacts encountered during survey was restricted to diagnostic or suspected Paleoindian materials and Archaic projectile points at the discretion of the survey team leader.

THIS VOLUME

The rest of this monograph presents the large body of data available for El Fin del Mundo and some of our interpretations. The geologic setting, stratigraphic context, soil and sediment micromorphology, and geochronology of the site are presented in Chapters 2 and 3. These sections build on the basic stratigraphic and chronologic sequence described by Sánchez and others (2014). The following chapter (4) is a discussion of the Upper Bonebed in Locus 1, which contained the remains of two gomphotheres and an associated Clovis lithic assemblage along with the Lower Bonebed, which appears to be paleontological. This section also builds on the description and discussion provided by Sánchez and others (2014). The next chapter (5) deals with the Clovis stone tool assemblage from the site and comparisons with other Clovis assemblages in the region. This chapter is based on Ismael Sánchez-Morales's master's research project and expands on previously published descriptions of the Clovis lithic component from the site (Sánchez-Morales 2018; Sánchez-Morales, Sánchez, and Holliday 2022). Chapter 6 is an inventory and discussion of the vertebrate paleontology and zooarchaeology of the Clovis bonebed. The next four chapters present data and summaries on pollen (Chapter 7), phytoliths (Chapter 8), diatoms (Chapter 9), and ostracodes (Chapter 10). We close with our synthesis of the terminal Pleistocene-early Holocene record preserved at El Fin del Mundo (Chapter 11).

CHAPTER TWO

Stratigraphy and Geochronology

Vance T. Holliday, Susan M. Mentzer, Guadalupe Sánchez,
and Edmund P. Gaines

El Fin del Mundo is one of the few documented Terminal Pleistocene-Early Holocene stratigraphic sequences in Sonora and northwestern Mexico and is the only one with a documented in situ Paleoindian archaeological assemblage. This chapter discusses the geologic record at El Fin del Mundo with several specific goals. The first is to present the stratigraphy and geochronology of the site area to reconstruct its formation processes and landscape evolution. Sánchez and others (2014) report on the archaeology and stratigraphy specific to the Clovis bonebed but not the surficial upland component. This chapter integrates all the Fin del Mundo geological information, including micromorphology (Chapter 3) with the archaeological (Chapters 4, 5) and paleontological records (Chapter 6). Related discussions of pollen (Chapter 7), phytoliths (Chapter 8), diatoms (Chapter 9), and ostracodes (Chapter 10) recovered from the site aide in reconstructing trends in the local hydrology and plant communities. Finally, the geoarchaeological and paleoenvironmental implications of El Fin del Mundo are integrated into the broader context of archaeological sites and geological localities from across the U.S. Southwest and Northwest Mexico to glean clues about the evolution of the region during the Terminal Pleistocene (Chapter 11).

GEOLOGIC SETTING

El Fin del Mundo is located in a small intermontane basin within a chain of volcanic hills about 100 km (~62 mi) northwest of Hermosillo, Mexico (Figure 1.1). The site area covers ~3 km^2 and is exposed at the toe of an ancient alluvial fan that drains to the northeast (Figure 1.5). The fan is dissected and the erosional exposures exhibit fan alluvium with a well-expressed and presumably ancient soil beneath the fan surface. The area around the site is an eroded landscape crossed by a series of arroyos incised into the fan and draining into the Arroyo Carrizo, which drains into the Río Bacoachi. Like most arroyos in the region, these drainages carry water only intermittently.

Erosion in the site area exposed the late Quaternary stratigraphic record and the buried, in situ component of the archaeological record. Dissection left the local basin fill exposed as stratigraphic sections in the walls of arroyos, head cuts, and along two isolated erosional "islands." The walls of the arroyos and islands are vertical and maintained by a surface caprock of carbonate (discussed below). The bonebeds, artifacts, and their containing strata were preserved in only one of these islands (Locus 1; Figures 1.6 and 1.7). The other island (Locus 3, Figure 1.6) and all other exposures exhibit primary or reworked Late Pleistocene fan alluvium locally capped by carbonate at the surface.

METHODS

Geological investigation focused on stratigraphic exposures in Loci 1, 2, 3, 4, and 6, and all areas of surface archaeological finds (Loci 5, 9, 10, 17, 19, 22, 23, 25; see Figure 1.6). Geoarchaeological analyses included field-based stratigraphic and geomorphic observations and descriptions, integration of micromorphology (Chapter 3), and an array of laboratory analyses of sediment samples. Field descriptions of sediments and soils follow standard criteria (Table 2.1; AGI 1982; Birkeland 1999; Holliday 2004). The laboratory analyses of sediments and soils from El Fin del Mundo include calcium carbonate content (using the Chittick method; Singer and Janitzky 1986)

Table 2.1. El Fin del Mundo Profile Descriptions

			Locus 1, Profile 08-1 (N wall; N end of E150 line)
Strat	***Soil Horizon***	***Depth (mbd)***	***Description***
	Lma/di	10.93–11.03	Surface carbonate "crust" v. hard; ***C14 samples***
5	Ldi/Bw1	11.03–11.18	10YR 7/2d 6/2m; weakly oxidizes; v. hard; common, faint carb films on ped faces; wk pr & mod sbk; clear, smooth; ***C14 samples***
	Ldi/Bw2	11.18–11.33	10YR 7/1.5d 6/2m; weakly oxidized; hard; common, faint carb films on ped faces; wk sbk; clear, smooth; ***C14 samples***
4de	Ldi1	11.33–11.78	Diatomaceous Earth; upper half 2.5Y 6.5/1d 5/1m (could be 10YR, too); lower half 2.5Y 7/1d 6/1m (10YR, also); massive, soft; faint horizontal bedding in lower 10cm; abrupt, smooth to locally irregular; ***C14 samples***
4d	Ldi2	11.78–11.83	Diatomite; 10YR 8/1; locally laminated; locally mixed with the diatomaceous earth; locally mixed with Strat 3 (below); abrupt, wavy; ***C14 samples***
3B	2C1	11.93–12.23	cse SC w/few pebbles; few cobbles; 2.5Y 7/2d 6/3m; fine, str nonpedogenic pr; clear, wavy; ***C14 samples***
3A	2C2	12.23–12.33	cse SC with v common pebbles & cobbles; 2.5Y 7/2d 5.5/3m; massive w/v wk, v coarse pr or partings (nonpedogenic) (these partings seem to be continuation of structure in overlying zone); clear, wavy; ***C14 samples***
2	3C3	12.33–12.58	SC w/rare f pebbles & cse sand; 5Y 8/2d 6.5/3m; fine str nonpedo pr; gradual; ***C14 samples***
	3C4	12.58–13.28+	LcseS; 5Y 8/2d 6.5/3m; large mottles of 2.5YR 4/3.5d 4/4sm in lower half of exposure; indurated; mostly massive w/ sbk partings in upper 20cm
			Locus 1, Profile 10-4 (N wall; N end of E160 line)
Strat	***Soil Horizon***	***Depth (mbd)***	***Description***
4de		11.25–11.83	Diatomaceous earth; lower ***~4cm sampled for diatoms***
4d		11.83–11.88	Diatomite; finely bedded; ***sampled for diatoms in 5mm increments***
3B		11.88+	Alluvium; upper ***~2cm sampled for diatoms***
			Locus 2, Pr 07-6
Strat	***Soil Horizon***	***Thickness, cm***	***Description***
C			Upper massive carbonate; missing across most of Locus 2
		10	Middle platy carbonate; locally missing across Locus 2
		~18	Lower massive carbonate (thins to south and west)
Big Red	Btkb1	30	10YR 6/3sm (sl redder than this) SL; fine qtz pebbles common in upper ½; wk cse pr & str cse sbk; slightly firm; thin cont clay films on ped faces; prominent vertical plates of carb (up tp 2mm thick) along major ped faces; clear smooth
	Bkb1	20	7.5YR 5/3sm, cse pebbly S; wk sbk; firm; common films & thr carb on ped faces; clear smooth
	Btkb2	30	7.5YR 4/4m L; wk cse pr & str cse sbk; thin cont clay films on ped faces; distinct common carb films & thr on ped faces; very firm (local ledge forming unit along exposure); abrupt smooth boundary

Table 2.1. (continued)

Strat	*Soil Horizon*	*Thickness, cm*	*Description*
			cse platy carb; abrupt; ***C14 samples***
		~13	Carbonate below Big Red is locally more strongly expressed platy in some sections, esp. upper 10 cm; but more massive to blocky in other.
2		40	Whiter than 10YR 8/1d; massive carb that weathers to str med pr; clear smooth
		~25	2.5Y 6/2 sm; str pr clay; clear smooth
			10YR 5/2sm; massive clay weathering to wk cse pr; locally weakly platy

Locus 3, Profile 07-5

Strat	*Soil Horizon*	*Thickness, cm*	*Description*
C		2–3	Hard carbonate crust; 10YR 8/2-7/2m; clear, smooth; missing across most of Locus 3
		10	Upper Massive Carbonate 1; 10YR 5/2sm; hard; massive; clear, irregular bndy; missing across most of Locus 3; ***C14 samples***
		20	Upper Massive Carbonate 2; almost yellow enough to be 2.5Y 7.5/2sm; weathers into long "alcoves," missing across most of Locus 3
		15	Middle Platy Carbonate; almost yellow enough to be 2.5Y 7.5/2sm; missing across much of Locus 3
		15	Lower Massive Carbonate; almost yellow enough to be 2.5Y 7.5/2sm; common across Locus 3; ***C14 samples***
		up 10–15	Carbonate w/locally common bodies of organic matter or Mn-ox; where absent the lower carbonate is just thicker; abrupt smooth
		8	Interbedded lenses of carbonate, brown sandy clay, and pink-red (2.5 YR 6/4-5/4sm) sand; locally common bodies of organic matter (as above); abrupt, smooth
Big Red	Btk1b	14	5YR 4/4m SCL w/medium sand; common fine siliceous pebbles throughout; str cse pr & str cse sbk; cont, thick clay films on ped faces; common carb films & thr; clear, smooth
	Btk2b	20	5YR 4/4m SCL w/medium sand; str cse pr & str cse sbk; cont, thick clay films on ped faces; common carb films & thr; clear, smooth
	Btb	20+	7.5YR 4/4m SCL; wk cse pr & med str sbk; discont thin clay films on ped faces

Locus 4, Profile 07-8

Strat	*Soil Horizon*	*Thickness, cm*	*Description*
C		~20	Weathered remnant of lower massive carb
Big Red	Btkb1	~20	7.5YR 6/3d SL; cse mod pr & case str abk; v common thr & films of carb; abrupt
	Btk2b1	~40	7.5YR 6/3d (locally 5/5d) SL; cse mod pr & case str abk; common thr & films of carb (locally v common); abrupt
1		~20	7.5YR 7/3d (w/domains of 7/4d) hard platy carbonate w/more massive plates 1-2cm thick at top & bottom; thinner plates common in between in calcareous pebbly sand; abrupt
		~35	7.5YR 7/2d calcareous, blocky pebbly sand (the blocks hint at coarse platiness)
		12–15	Plates of carbonate (ave 1cm thick) w/intercalated layers of calcareous pebbly sand 7.5YR 6.5/2d; Mn-ox staining locally follows plates; ***C14 samples***
		~35	7.5YR 6/2d (gray) calcareous pebbly sand; cse blocky to cse prismatic; medium pebbles locally common

(continued)

Table 2.1. (continued)

Strat	Soil Horizon	Thickness, cm	Description
Locus 5, Unit 20			
Strat	*Soil Horizon*	*Thickness, cm*	*Description*
5	Bw	≤20	Cse LS, 7.5YR 6/3d; v wk sbk; abrupt
Big Red	Bt	~10	F pebbly SCL, redder than 7.5YR 5/4d; wk med pr & wk med bk; thick continuous clay films; hard; clear, smooth
	Bk1	~10	Common thr & films of carb; abrupt
	Bk2	~10+	Common thr & films of carb; abrupt
Locus 5, Profile 10-2			
Strat	*Soil Horizon*	*Thickness, cm*	*Description*
Big Red	Btk	~15	F pebbly SC, redder than 7.5YR 5/8m, 5/6d; cse med pr & mod med sbk; cont thin clay films; thin continuous clay films; v hard; clear, smooth
	Btb	~45	F pebbly SC (pebbles & cse S less common), redder than 7.5YR 5/8d; mod cse pr & str med sbk; cont thin clay films; few faint carb thr & films in upper 20cm
Locus 19, Bucket Auger			
Strat	*Soil Horizon*	*Thickness, cm*	*Description*
5	Lma/di	~20	Surface carbonate "crust" 10YR 6.5/2d; v. hard; common carb threads; pitted; str eff
4de	Ldi	~10	10YR 7/2d; hard; rare carb films on ped faces; str eff
	Ldi	~10	10YR 5.5/2d; hard; v rare carb films on ped faces; str eff
	Ldi1	~10	10YR 5.5/2d; sl hard; str eff
	Ldi2	~10	10YR 5/3d; soft; str eff
	Ldi	~10	10YR 4.5/2d (darker gray); soft; str eff
	Ldi	~20	10YR 4/1.5d (darker gray); soft; mod eff
	Ldi	~25	10YR 4/1d (v dk gray); soft; mod-wk eff
3 or 2		~20	Olive gray pebbly SC
			Conglomeritic bedrock

Note: Loci 2, 3, 4, 5, and 19 are outside of the excavation area and were not tied to the site elevation datum.

Key:

Colors = Munsell d = dry; sm = slightly moist; m = moist.

Textures: S = Sand; SL = Sandy Loam; SC = Sandy Clay; SCL = Sandy Clay Loam; f = fine.

Structure: sbk = subangular blocky; abk = angular blocky; pr = prismatic; wk = weak; mod = moderate; str = strong; f = fine; med = medium; cse = coarse.

Miscellaneous descriptors: bndy = boundary; carb = carbonate; cont = continuous; disc = discontinuous; Mn-ox = manganese oxides; sl = slightly; eff = effervescent (reaction to HCl).

Ldi and Lma are horizonation nomenclature terms for lacustrine marl and lacustrine diatomite/ diatomaceous earth following Soil Survey Staff (2017).
The term "marl" is used here for Stratum C following a generalized definition of fine-grained lacustrine or palustrine carbonate after Neuendorf et al (2005).

Table 2.2. Laboratory Data for Calcium Carbonate Content of Soils and Sediments through Strata 5, 4, and Upper 3B, Profile 8-1, Locus 1

Stratum	*Depth Below Datum, Meters*	*CaCO3 %*
5	11.08–11.03	54.2
	10.98–11.03	53.0
	11.03–11.08	46.2
	11.08–11.13	47.0
	11.13–11.18	43.4
	11.18–11.23	41.2
	11.23–11.28	33.2
	11.28–11.33	33.2
	11.33–11.38	22.3
4de	11.38–11.43	1.7
	11.43–11.48	2.1
	11.48–11.53	1.5
	11.53–11.58	0.9
	11.58–11.63	1.3
	11.63–11.68	1.0
	11.68–11.73	1.3
	11.73–11.78	1.9
4d	11.78–11.83	1.7
3/4	11.83–11.88	0.8
3B	11.88–11.93	0.1
	11.93–11.98	0.3

Note: Calcium carbonate content determined using the Chittick method.

and micromorphology (following methods described in Chapter 3; Table 2.2) along with radiocarbon dating and recovery of paleoecological indicators (Chapters 7–10).

Radiocarbon dating provides numerical age control for the site. Dates were obtained on samples of diatomaceous earth, charred plant fragments, and succinid shells. (Succinids are terrestrial gastropods that live in proximity to standing water or moist micro-environments.) Holliday and others (2014) detail the laboratory protocols for the radiocarbon dating. Succinid shells provided the most reliable age control because they do not fractionate carbonate and cannot be contaminated (Pigati and others 2004, 2010). Andrew Kowler processed shells at the Arizona AMS Radiocarbon facility using the extraction line of Jay Quade. Organic matter samples were submitted for conventional or AMS (accelerator mass spectrometry) radiocarbon dating at the University of Arizona's Environmental Isotope Laboratory ("A" numbers) and AMS Radiocarbon facility ("AA" numbers), respectively (Table 2.3). All samples of charcoal and charred plant fragments were processed in the Arizona AMS Radiocarbon facility. Attempts at dating bone and teeth from the Upper (Clovis) and Lower Bonebeds were unsuccessful, as described and discussed by Hodgins in Appendix A. To summarize, samples obtained from gomphothere bone in the Upper Bonebed did not yield collagen and therefore attempts at direct radiocarbon dating of bone were abandoned. Further dating efforts focused on gomphothere teeth from the Upper Bonebed and mastodon teeth from the Lower Bonebed. Collagen was generated from these samples, but most of the resulting dates are significantly younger than the known ranges for Clovis archaeology, gomphotheres, and mastodons. Optically stimulated luminescence (OSL) was also attempted to date the sediment that enclosed the Clovis bonebed, but the results were unsatisfactory, as discussed in Appendix B. Radiocarbon dates were calibrated with CALIB http://calib.org/calib/ (Stuiver and others 2021) using the calibration datasets from Reimer and others (2020).

TERMINOLOGY

Five main lithostratigraphic units were identified at El Fin del Mundo: Strata 1, 2, 3, 4, and 5 (numbered from bottom to top: Figure 2.1). Elsewhere across the site strata C and S were recognized. Stratum 1 is the primary clastic deposit present in all exposures across the site beyond the Locus 1 island. Stratum 2 is also a clastic deposit exposed both in and beyond Locus 1. Stratum 1 was recognized and formally differentiated from Stratum 2 following the work published by Sánchez and others (2014), largely based on dating. The sequence of Strata 2-3-4-5 is preserved only in the Locus 1 island (Figure 2.2). Stratum 5 caps the sequence at Locus 1 and its resistance to weathering helped preserve the underlying sediment (Figure 2.3). Strata 3, 4, and 5 are inset into Strata 1 and 2. Two other strata are locally present on the uplands, separated from the Locus 1 sequence by erosion, geomorphic setting, and chronology and are isolated from one another. They are identified by lithology: "Carbonate" (Stratum C) and "Surface Sand" (Stratum S). Stratum C is further subdivided based on stratigraphic position and visible structure: Lower Massive, Middle Platy, and Upper Massive.

Diatomite and diatomaceous earth are significant components of the stratigraphic record at El Fin del Mundo, comprising Stratum 4. However, scientific literature offers a variety of approaches to defining the terms "diatomite" and

Table 2.3. Radiocarbon Dates from El Fin del Mundo

Stratum[a]	*Context*	*Lab Number*[b]	*Sample Type*	*Fraction*	*^{14}C yr B.P.*	±	*Cal yr B.P. range (2σ)*			*δ13C (‰)*
							median	+	-	
				Locus 1 Excavation Area						
5	Mixed	AA83955	Nacreous shell	NA	4,840	40	**5,560**	**5,605**	**5,475**	0.7
5	**Upper**	**AA88885[c]**	**Shell (Succinid)**	**NA**	**8,870**	**60**	**9,988**	**196**	**262**	-6.2
5	Mid-to-upper	AA81350[c]	Shell (Succinid)[f]	NA	8,030	50	8,884	192	234	***
5	Mid-to-upper	AA88886[c]	Shell (Succinid)[f]	NA	8,470	50	9,490	51	160	-6.3
4de	Lower	A14837	Organic-rich sediment	A	8,375	110	9,353	186	320	-21.8
4de	Lower	A14836	Organic-rich sediment	A	9,465	100	10,749	396	310	-17.1
4de	Basal contact	A14850	Organic-rich sediment	A	9,030	75	10,191	184	287	-24.3
4de	Basal contact	AA80085B[d]	Charred plant fragments	B	9,290	290	10,527	720	845	-13.2
4de	Basal contact	AA80671B[d]	Charred plant fragments	B	9,560	120	10,898	306	327	-22.9
4de	**Basal contact**	**AA80084A**	**Charred plant fragments**	**A**	**9,715**	**64**	**11,127**	**119**	**334**	-13.2
4d	Top	AA80078A	Organic-rich sediment	A	6,921	58	7,754	171	131	-24.6
4d	Upper	AA80076A	Organic-rich sediment	A	7,570	93	8,370	175	187	-25.1
4d	Base	A14896A	Organic-rich sediment	A	6,185	115	7,070	335	316	-17.9
3B	**Upper**	**AA100181A**	**Charred plant fragments**	**A**	**11,550**	**60**	**13,413**	**160**	**106**	-24.2
3B	Upper	AA100182B[d]	Charred plant fragments	B	11,880	200	13,764	517	454	-21.8
3B	Upper	AA114925	Charred plant fragments		16,920	270				-16.3
3B	Upper	AA90708	Shell (Succinid)[f]		17,220	70	20,778	151	213	***
3B	Mid-to-upper	AA83272B[d]	Charred plant fragments	B	11,040	580	12,879	1,356	1,678	-17.8
3B	Lower	AA90707[e]	Shell (Gyraulus)		12,130	10	14,048	33	164	***
3A/3B	Lower	AA94051	Shell (Succinid)[f]		22,490	90	26,785	313	344	***
3A/3B	Base	AA83958	Charred plant fragments		16,130	240				-18.5
3A	**Upper**	**AA94052**	**Shell (Succinid)**		**12,890**	**50**	**15,409**	**184**	**168**	***
3A	Lower	AA83959	Charred plant fragments		21,200	1600				***
2	**Upper**	**AA94053**	**Shell (Succinid)**		**21,530**	**80**	**25,851**	**119**	**126**	***

Table 2.3. (continued)

Stratum[a]	*Context*	*Lab Number*[b]	*Sample Type*	*Fraction*	*^{14}C yr B.P.*	±	*Cal yr B.P. range (2σ)* median	+	-	*δ13C (‰)*
				SE of Locus 1						
5?	(18-1)	AA111506	Soil		5710	26	6,490	113	85	-17.0
				Locus 2						
2		**AA81351**	**Shell (Succinid)**		**27,950**	**150**	**31,863**	**937**	**396**	***
		AA112590	**Shell (Succinid)**		**24,560**	**170**	**28,827**	**344**	**399**	-4.2
				Locus 3						
C U	**Upper massive**	**AA94057**[c]	**Shell (Succinid)**	**NA**	**8,000**	**40**	**8,869**	**137**	**218**	***
C L	**Lower massive Upper half**	**AA81042**[c]	**Shell (Succinid)**	**NA**	**15,060**	**50**	**18,513**	**119**	**290**	***
C L	**Lower massive Lower half**	**AA81043**[c]	**Shell (Succinid)**	**NA**	**16,010**	**80**	**19,324**	**199**	**204**	***
				Locus 4						
1	**top**	**AA81044**	**Shell (Succinid)**	**NA**	**35,990**	**50**	**41,093**	**177**	**194**	***
				Locus 19						
4	47-50cm bs	A15749	Organic-rich sediment	Bulk	7115	90	7,933	199	235	-20.5
4	50-53cm bs	A15750	Organic-rich sediment	Bulk	7085	+75/-70	7,900	165	125	-21.2

Notes: **Samples in bold** are those used for the site chronology. Modified from Holliday and others 2014:Table S8.

Key: a. +Geologic member: Stratum 4, de = diatomaceous earth, d = diatomite; Stratum C, U = Upper Massive, L = Lower Massive.

b. "A" numbers were determined in the conventional radiocarbon laboratory (now defunct) in the Department of Geosciences, University of Arizona; "AA" numbers were determined in the Arizona Accelerator Mass Spectrometry Laboratory (now defunct), University of Arizona.

Suffix indicates A or B fraction of organic carbon.

c. Age calculated on the basis of sample-specific extraction blanks (see Holliday and others 2014); all other dates reflect the long-term (~1 yr) average.

d. Dates of B-fraction from 100% base-soluble samples.

e. AA90707 Modern groundwater dated 1180±30 ^{14}C yrs B.P. (AA92972) was used to "correct" this date 12,190±40 ^{14}C yrs B.P. in Holliday and others (2014).

f. weathered and considered redeposited.

***An estimated δ^{13}C value (-5‰) was used to adjust measured ^{14}C activity for biologic fractionation.

"diatomaceous earth." The *Glossary of Geology, 5th Edition,* defines diatomite as a light-colored, typically white, soft, low-density, friable siliceous sedimentary rock, consisting chiefly of opaline frustules of diatoms (Neuendorf, Mehl, and Jackson 2005:178). "Diatomaceous earth" is defined as a synonym for diatomite (Neuendorf, Mehl, and Jackson 2005:177). Zahajská and others (2020) propose that sediment with more than 50 percent of sediment weight comprised of diatom SiO_2 and having high (>70%) porosity is "diatomaceous ooze" if unconsolidated and "diatomite" if consolidated. They define "diatomaceous sediment" as less than 50 percent of sediment weight comprised of diatom SiO_2. They reject the term "diatomaceous earth" because of its ambiguous colloquial use in nonscientific literature. The term previously had widespread usage among geologists and paleontologists working with late Pleistocene deposits on the Southern High Plains (Sellards and Evans 1960; Sellards 1952:53–55; Green 1962; Haynes 1975, 1995; Stafford 1981; Holliday 1995, 1997).

The primary problem is the inconsistent use of the term in scientific literature rather than colloquial use in nonscientific literature. Patrick (1938:22), in a study of diatoms from the Clovis site, refers to a zone of "almost pure diatomite" from a zone otherwise known as the "Blue

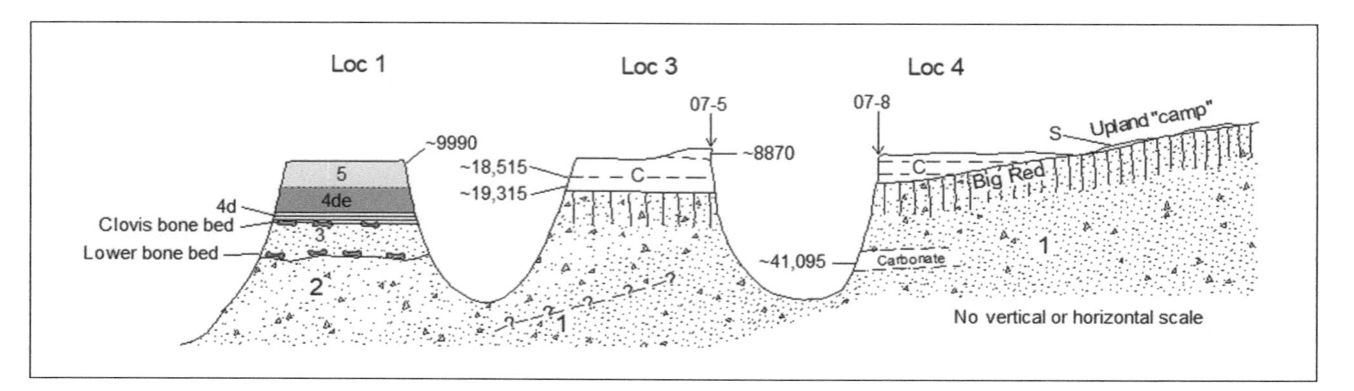

Figure 2.1. Schematic geologic cross section at El Fin del Mundo connecting Loci 1, 3, 4, and the upland camp. Dates are calibrated ^{14}C ages. Samples for all dates on Stratum C in Locus 3 were collected in section 07-5. Prepared by Jim Abbott.

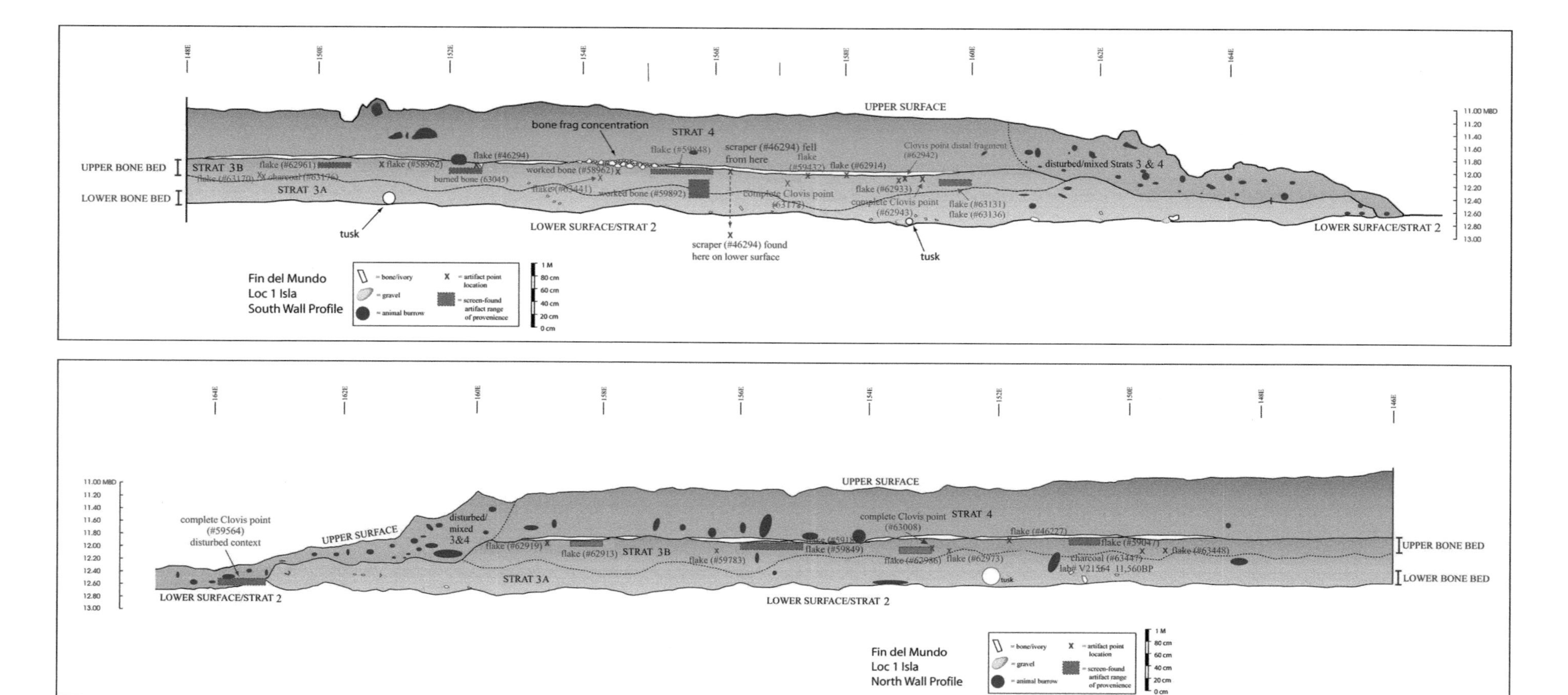

Figure 2.2. Stratigraphic profiles of the north (A) and south (B) walls of the east end of Locus 1 prior to excavation. Prepared by Ismael Sánchez-Morales.

Figure 2.3. Stratigraphic section along the south wall of Locus 1, Units 1 and 2, January 2008. Stratigraphic nomenclature is indicated as well as key calibrated ^{14}C ages (Table 2.3), although the radiocarbon samples are not from this section. Vertical desiccation cracks are apparent in Stratum 3B. The contrast between the darker gray Stratum 4de and lighter gray Stratum 5 shows well due to significantly higher carbonate content compared to Stratum 4de. Photograph by Vance Holliday.

Sand." Sellards (1952:28–30, 53–55), working at Lubbock Lake and the Clovis site ("Blackwater No. 1 Locality") uses "diatomite" interchangeably with "diatomaceous earth" (aka the "Blue Sand") for the Clovis site, but for Lubbock Lake, uses just "diatomite," although in a stratigraphic description (prepared by Glen Evans) for Lubbock Lake, "diatomite" is a variety of diatomaceous earth. Hester, Lundelius, and Fryxell (1972), in a synthesis of the work at the Clovis site, appears to use "diatomite" as a general term for what others refer to as the Blue Sand or diatomaceous earth. Similarly, Haynes and Agogino (1966) equate the terms "diatomite" and "diatomaceous earth." Haynes (1975, 1995) differentiates "diatomite" from "diatomaceous sand and silt", but the diatomite is not necessarily a white or light gray deposit of mostly diatom frustules. Stafford (1981) differentiates "diatomite" and "diatomaceous earth" without elaboration.

The terms used here follow Holliday (1995, 1997). Diatomite is a pure or nearly pure deposit of diatom frustules and is white or light gray. Diatomaceous earth is a mix of diatoms with clastic material and usually some organic matter. The color is medium to dark gray. At El Fin del Mundo and on the High Plains, the purity of diatomite is well over 50 percent and that of diatomaceous earth is well under 50 percent.

STRATIGRAPHY, SEDIMENTS, AND SOILS

Erosion by the modern arroyo system exposed the site and contributed to its discovery, but also removed surrounding contemporaneous deposits. The stratigraphic record and thus the record of landscape evolution at El Fin del Mundo is missing over much of the site area. This erosion removed at least some, if not large parts, of the site area in Locus 1, such that the original extent of the occupation zone is unknown. The extensive erosion via deep arroyo cutting prevents physical tracing of Strata 2, 3, 4, and 5 beyond the

Locus 1 island, and Stratum C and S beyond the uplands, but these strata (or lateral equivalents) may have been present over a much larger area in the past. Other isolated pockets and islands of sediment similar in composition to Strata 3 and 4 are also present in different Loci, but these cannot be directly connected to the strata present in Locus 1. Radiocarbon dating, lithology, and archaeological stratigraphy are the only means of stratigraphic correlation.

Stratum 1

Stratum 1 was not recognized by Sánchez and others (2014). It was differentiated from Stratum 2 in Loci 2 and 4 based on dating and subtle differences in soil development. Stratum 1 rests unconformably on the local bedrock, which is a cemented very-coarse conglomerate with clasts of volcanic rock, possibly the Baucarit Formation of Bartolini, Shafiqullah, and Morales (1994). Stratum 1, based on limited age control, underlies most of the site area and is the principal surface deposit of the fan under the upland camping area. The stratum contains primary or reworked alluvial deposits comprising the toe of the fan into which the Locus 1 site stratigraphy is inset. Local volcanic rock was a source for sediment in the fan alluvium based on presence of volcanic rock fragments and igneous minerals observed in the micromorphological samples. Weathering of this conglomerate likely provided much of the clastic material in Stratum 1. In thicker sections (those uneroded on the uplands), the lower half of Stratum 1 is deep red to purple in color and composed of a massive gravelly, sandy clay. The upper half in thicker exposed sections and most of the stratum in eroded thinner sections is a mixed or mottled pale olive gray and red, massive pebbly sandy clay (Loci 3, 4, 5; Table 2.1). In most exposures, the stratum also exhibits carbonate lenses. In the thickest and apparently most complete field sections (Locus 4), the carbonate consists of thin (~1-3 cm thick) subhorizontal lenses that merge or bifurcate in some places (Table 2.1; Figure 2.6). These lenses appear to follow bedding. This characteristic along with visible merging/bifurcating is indicative of subsurface groundwater (i.e., secondary) carbonate. A well-expressed soil ("Big Red"), discussed below, is present in upper Stratum 1 where erosion was minimal (Loci 3, 4, 5, and others on the south and southeast areas of the site).

Stratum 2

Stratum 2 is differentiated from Stratum 1 in the Locus 1 island based on a distinct erosional disconformity. Stratum 2 is not present in most loci. It is best expressed in Locus 2 (Figures 1.5, 1.6, 2.2). On the landscape around Locus 1, sediments derived from weathering and erosion of the island itself buried most of the Stratum 2 exposures. Dating indicates that alluvium in Locus 2 is also Stratum 2 (Figure 2.5). Stratum 2 in the Locus 1 island and in Locus 2 is up to 2m thick and is composed of pebbly sandy clay that fines upward into a sandy clay (Locus 2, Table 2.1). Like Stratum 1, most of Stratum 2 in the Locus 1 island is a pale, olive gray color with a few hints of redder oxidation zones. These colors represent alteration of the primary sediment; probably starting with oxidation (possibly due to weak soil formation), followed by reduction. This color variation is typical of a fluctuating water table. In settings with minimal erosion (Locus 2), a facies of the "Big Red" soil is preserved in upper Stratum 2.

Stratum 2 likely formed due to alluvial fan deposition, with a sediment source that was similar to that of Stratum 1. It is not exposed in most loci but comprises the surface of all of Locus 2 and extends from there to Locus 1. The clear erosional disconformity between Strata 1 and 2 in Locus 1 and the dating illustrate cyclical alluvial deposition and erosion typical of a fan.

Soil Formation in Upper Strata 1 and 2

"Big Red" is an informal term for the distinct, well-developed soil formed in the Strata 1 and 2 alluvium that comprises the uplands. Ibarra-Arzave and others (2020) recognize the soil only in Stratum 2, based on the stratigraphy defined by Sánchez and others (2014), published before the two strata were differentiated. The soil exhibits a strongly developed soil profile (Bt horizon with reddish hues and thick clay coatings visible in hand samples; Table 2.1, Loci 2, 3, 4, 5; Figures 2.4, 2.5, 2.6, 2.7; Chapter 3). Big Red is the surface soil across most of the upland area of the Clovis camp and the fan surface beyond the site except where it has been locally buried by Stratum C or S. The Paleoindian (and later) occupation debris is on top of the soil. Stratum 1 is the surface deposit and soil parent material beneath the Clovis and Archaic camp area east and southeast of Locus 1 (e.g., Loci 3, 4, 5) whereas Stratum 2 is the surface deposit and soil parent material in Locus 2, west of Locus 1.

Stratum 1, upper Stratum 2, and Big Red are not present in Locus 1 due to Terminal Pleistocene channel incision. Micromorphology of a lower Stratum 2 sample from Locus 1 expresses birefringence fabrics like that in the possible Bk horizon of Big Red sampled in Locus 3 (Chapter 3), which suggests that Big Red may have been present in this location before its truncation.

Figure 2.4. Profile 07-5 exposed on the south side of the Locus 3 island. At right by the tape is Stratum 1, with the Big Red soil and its Bt horizon exhibiting well-expressed rubefication and prismatic structure. At the left on top of the soil is Stratum C with the Lower Massive carbonate (Lm), the Middle Platy carbonate (unlabeled), and Upper Massive carbonate (Um). Radiocarbon dates on Lower Massive are ~19,325 cal yrs (lower LM; AA81043) and 18,515 cal yrs (upper LM; AA81042), and the Upper Massive is ~8870 cal yrs (AA94057) (Table 2.3). Photograph by Vance Holliday.

Figure 2.5. Stratigraphic exposure on the west side of Locus 2, profile 07-6. A well-expressed (strong structure) Bt horizon of the Big Red soil in Stratum 2 is evident. It is capped by a thin (few centimeters) remnant of Stratum C, Lower Massive carbonate. Radiocarbon dates on Succinid shells collected behind the "boulders" at left dated to ~28,825 cal yrs (AA112590) and ~31,865 cal yrs (AA81351) (Table 2.3). Photograph by Vance Holliday.

Figure 2.6. Stratigraphic exposure on the west wall of Locus 4, profile 07-8, illustrating the Big Red soil in Stratum 1. Carbonate lenses are apparent at the base of the soil (forming the ledge), in zone with the date (AA81044), and elsewhere through the section. Photograph by Vance Holliday.

The red soil is well-expressed but varies somewhat in horizon morphology. This is shown where the soil formed in Stratum 1 east and southeast of Locus 1 moving up slope (i.e., over ~700 meters, from the soil exposure in Locus 3 to and across the Locus 5 camp; Figures 2.6 and 2.7). In Locus 3, it is a sandy clay loam with a Bt horizon, 5YR hues, and possibly a weakly expressed Bk horizon based on micromorphology (Chapter 3). In Locus 5 it is sandier with hues between 5YR and 7.5YR and a Bt horizon with very thick, laminated clay coatings observed in thin section (Chapter 3). These variations represent a pedofacies and could represent sedimentological variation of the same layer (within Stratum 1) across the surface of the fan or they could represent variation in the age of the soil due to differences in timing of alluvial deposition across the surface of the fan. That is, there were additional cycles of erosion and deposition after Stratum 1, such as Stratum 2, in the eastern areas of the site. Big Red is somewhat weaker in expression in Locus 2 with 7.5YR hues and thinner clay film development, but it is also a welded buried soil, suggesting local cycles of erosion and sedimentation.[4] The parent material is also younger than Stratum 1.

In Locus 4, a carbonate within Stratum 1 beneath the Big Red soil was interpreted in the field as a possible K horizon. Micromorphological analyses indicate that it is not pedogenic but instead this carbonate is a primary deposit containing bedded clastic inclusions and lenses of intact plants, some of which were replaced with secondary silica. These observations suggest that there were localized wetlands within the fan at the time of the deposition of Stratum 1. Furthermore, the presence of diatoms in Stratum 2 (Chapter 3) also suggests that some of the fan materials were sourced from either eroded wetland deposits on the fan or that stream flow was perennial and supported diatoms that were later eroded and redeposited. This latter

[4] This general morphological distinction was noted by Ibarra-Arzave and others (2020), in their "Loc 2-3" and "Big Red" sections. Both sections are on the uplands roughly near Localities 2 and 5, respectively (their Figure 1), but exact locations are unknown.

Figure 2.7. The Locus 5 uplands of El Fin del Mundo. (Top) View west across an open area of Locus 5. A small test pit exposes Stratum 1 and the Big Red soil (highlighted by strong prismatic structure). (Bottom) Excavation unit exposing upper Stratum 1 with an eroded surface and a discontinuous lens of Stratum S on top. Photographs by Vance Holliday.

Figure 2.8. The karren-like weathering in Stratum 5 carbonate on the surface of the Locus 1 island. The view is west across the west end of the island. In the center is the Unit 12 test pit excavated through Strata 5 (lighter gray) and 4de (darker gray) to the contact with Stratum 3 on the floor of the pit (note the absence of diatomite). The weathering is expressed as sharp, jagged peaks in the area away from the pit. Photograph by Vance Holliday.

scenario is supported by the presence of the weathered and reworked succinid shells in Stratum 2.

Carbonates on Top of Strata 1 and 2

In several localities close to Locus 1 (exposures around Locus 2, the Locus 3 island and the arroyo walls around it, and Locus 1), exposures of Strata 1 and 2 exhibit a capping carbonate that is up to 60cm thick (Frontispiece; Figures 2.4 and 2.5). The carbonate, Stratum C, extends to the west, south, southeast, and east of Locus 1 for up to ~200m. It thins out against Big Red to the southeast. To the west and east, the carbonate is present on top of remnants of Big Red (on Stratum 1 in Locus 4, Stratum 1 or 2 in Locus 3, and Stratum 2 in Locus 2) but its original extent is unknown owing to erosion. Throughout the site, the upper surface of the carbonate is weathered by fracturing and dissolution, producing microfeatures similar to karren on limestone (Figure 2.8).[5] The thickest and apparently most complete section of the carbonate is in Locus 3 (profile 07-5; Table 2.1; Figure 2.4). The Lower Massive carbonate is the least weathered and most horizontally extensive layer, found to the east of Locus 3 and south of Loci 1 and 3. Remnants of the overlying Middle Platy layer are locally common away from Locus 3. The Upper Massive carbonate is found only in Locus 3, although a time equivalent may be present across the surface of Locus 1 (Stratum 5). In the Locus 3 exposures there are no obvious weathering zones visible in the field within the three carbonate layers nor at the top

[5] Karren is defined as "channels or furrows caused by solution on limestone" (Neuendorf, Mehl, and Jackson 2005:348).

of the Lower and Middle zones. This initially suggested essentially continuous carbonate precipitation.

The carbonate is a relatively soft, low-density material, probably because it does not engulf older sediments (except at its base), which is more typical of pedogenic or groundwater carbonates. These characteristics along with inclusions of both aquatic and terrestrial gastropods visible in the field throughout the carbonate suggest that at least the Middle Platy and Upper Massive zones were formed as a result of seep or spring activity.

Micromorphological analyses of samples of the Big Red soil and overlying carbonate layers in Locus 3 generally confirm these hypotheses (Chapter 3). There, the Big Red soil contains redoximorphic features, including gleying, which impacted the color of the clay component. The contact between the Big Red soil and the overlying Lower Massive carbonate layer is sharp, although sand-sized materials that are similar in composition and texture to those that are present within the Big Red soil, as well as Stratum 3 in Locus 1, are present as inclusions. Microscopic redoximorphic features including manganese oxides are present. The microstructure and morphology of the Lower Massive carbonate suggest that it is a groundwater deposit and not pedogenic in origin. Presence of terrestrial or semi-aquatic gastropods suggest that the water was derived from seeps or springs. Soil pedofeatures that would indicate precipitation of carbonates within a calcic horizon are absent, although the presence of channel voids and possible infilled insect burrows indicate a degree of subaerial exposure during formation. The Middle Platy carbonate contains fewer inclusions of sand-sized materials as well as fewer secondary redoximorphic features relative to the underlying Lower Massive carbonate. Abundant succinid and other shells and intact diatoms are present within a matrix of microcrystalline calcite. Infilled chamber voids, and channels containing mineralized remnants of roots indicate subaerial exposure. The Middle Platy zone thus appears to be a paludal carbonate. The Upper Massive carbonate also contains shell fragments and diatoms but notably it contains phytoliths, some of which are in anatomical connection, and exhibits more abundant structural features suggestive of bioturbation and surface weathering following deposition. Although it appeared massive in the field, discrete depositional units are visible in thin section with their boundaries marked by evidence of plant root activity and increased surface weathering. Reworked fragments of calcareous material contain phytoliths and diatoms as inclusions. These intraclasts formed elsewhere within the seep or spring system and were redeposited here.

Stratum 3

Strata 3, 4, and 5 are present only in Locus 1 (Figures 2.1–2.3, 2.9) indicating that they filled a channel of unknown width and length cut into Stratum 2. Based on the minimum distances to exposures of Stratum 2 around Locus 1, the paleo-channel was less than 100m wide. Its length is unknown. No indication of the paleo-channel is preserved along the arroyo upstream or downstream of the Locus 1 island. The arroyo walls away from Locus 1 expose only old fan alluvium (Strata 1 or 2) locally capped with deposits of Stratum C. The course of the paleochannel therefore must have essentially followed the present course of the arroyo.

Stratum 3 is divided into two units, with the base composed of a pebbly (locally cobbly) sandy clay that is 30 to 40 cm thick (3A) overlain by a sandy clay that is 25 to 30 cm thick (3B) with prominent vertical cracks due to the high clay content (Figure 2.3). Stratum 3A rests unconformably on the eroded surface of Stratum 2. The pebbles and cobbles in 3A, particularly along the 2/3A contact, are angular to subrounded. The 3A/3B contact is flat and usually distinct, highlighted by the vertical cracking in 3B. In some exposures, however, especially after weathering, the contact was not clear and the designation "3A/3B" was used. Stratum 3 is pale olive in color throughout.

Differences in composition of the clay, as well as the degree of weak, post-depositional alteration (possible weak soil formation in 3B) are observed between these two strata in the micromorphological samples (Chapter 3). The pebbly to cobbly character of some components of Stratum 3 and the appearance of cut-and-fill sequences within both Strata 3A and 3B indicate cyclical alluvial aggradation. The coarser nature of 3A is indicative of more energetic stream flow in contrast to finer-grained 3B, which suggests lower energy aggradation. Further, micromorphology (Chapter 3) showed a thin lens of diatoms just below the upper contact of 3B, indicating a brief interval of standing water in the final stages of alluviation.

Bone is common throughout Stratum 3 both in the field and as part of the sand fraction observed in the micromorphological samples (Chapters 3, 4, 6). Scattered pieces of bone including fragments and complete but disarticulated elements from a variety of late Pleistocene mammals, comprising the Lower Bonebed are concentrated on and just above the 2/3A contact along with concentrations of pebbles and cobbles (Chapter 6). These associations of individual bone and bone fragments from random species along with stream gravel on top of the that contact suggest that there was alluvial re-deposition of the bone. The Upper Bonebed, the focus of the excavations, is in uppermost

Figure 2.9. The west wall of the Locus 1 excavations in January 2008. Profile 08-1 is just out of view to the right. The white laminated lens of diatomite shows well near the base of the section. Faint bedding is apparent in the darker gray diatomaceous earth (4de). The upper half of the exposure is the lighter gray sediment of Stratum 5. Local surface erosion destroyed or obscured upper Stratum 5. Photograph by Vance Holliday.

Stratum 3B (Chapters 4, 6). That bone, including some articulated elements, is from two gomphotheres, each largely concentrated in two distinct piles (Chapters 4, 6).

Some of the larger bones in this bed protruded above Stratum 3 and were buried directly by Stratum 4d (Chapter 4). Similarities in weathering characteristics of bone buried within 3B and bone exposed above 3B and encased in diatomite (Chapter 6) suggests (1) a brief interval of 3B alluviation following the Clovis activity and then (2) a rapid shift (in a few years) to a lacustrine depositional environment. Cruz-y-Cruz and others (2016:153) refer to "pedosediments" in upper 3B, but there is no evidence of significant subaerial weathering or pedogenesis..

The sedimentology and geomorphic setting of Stratum 3 suggest deposition by spring-fed flow. The relatively poorly sorted character of sediments, especially in 3A, and localized cut-and-fill cycles both suggest variable though likely perennial discharge throughout deposition. The poor rounding also suggests short transport distance. Those characteristics and the lack of evidence for a channel upstream from Locus 1 suggest a nearby but substantial source of water. That and the evidence from the carbonates for seeps in the immediate area (a complete section of Stratum C in Locus 3 is <100 m from Locus 1; Figure 1.6) all suggest that Stratum 3 was deposited by spring-fed waters. The water was in a small stream that flowed down a channel incised into the older alluvial fan.

Stratum 4

Stratum 4 is up to 1m thick and rests conformably on Stratum 3 (Figures 2.3, 2.9). The Stratum 3-4 contact dips down very gently to the northeast (Figure 2.2). Most of Stratum 4 is composed of diatomaceous earth (informally "4de") but locally at the base of Stratum 4 is a discontinuous layer of white to light gray diatomite (informally "4d") up to 10cm thick (Locus 1, Table 2.1). It is found only where Stratum 4 is thickest, that is, where the top of Stratum 3 is lowest. It thins out upslope along the 3-4 contact (Figure 2.2). Most of the rest of Stratum 4 is gray, silty diatomaceous earth. Micromorphology, phytoliths, and diatoms demonstrate that the siliceous component is a mixture of diatoms and phytoliths (Chapters 3, 8, and 9).

The diatomite rests conformably on Stratum 3. At a microscopic scale, the main contact between Strata 3 and 4 is abrupt. However, several millimeters of finer sediment containing fine sand- and silt-sized materials along with the thin zone of diatoms overlay the poorly sorted, coarse, dominantly igneous sands of upper Stratum 3 (but including sand-sized fragments of bone). These fine sediments at the very top of Stratum 3 are in turn overlain by the nearly pure diatomite of lower Stratum 4. In places, the contact is locally mixed due to bioturbation by insects that moved sediment upwards from Stratum 3 into Stratum 4 within mm-scale burrows (Chapter 3). In addition, upper Stratum 3 and lower Stratum 4 differ markedly in porosity. Stratum 3 contains abundant channels and cracks, while Stratum 4 is massive and contains few voids (Figure 2.3; Chapter 3). These differences suggest that upper Stratum 3 supported plant growth and may have been briefly weathered prior to the deposition of Stratum 4. Where bones in the Upper Bonebed were exposed above Stratum 3, they were encased within the diatomite (Chapter 4). Minimal differences in weathering of the bone below and above the contact suggest that the bones were not exposed to a significant duration of surface weathering processes (Chapter 6), and that the change in depositional environment was rapid.

Where the diatomite is thickest it is finely laminated (Profile 10-4; Chapters 3, 9). Where it thins as the surface of Stratum 3 rises (Figure 2.2), the diatomite appears massive in the field, but thin sections show that fine bedding persists (Chapter 3). The bedding consists of strongly laminated, horizontally aligned lenses of intact diatoms, and thin lenses of fine sand and silt, all faintly visible in the field (Chapter 3). The presence of the thin lens of diatoms in upper 3B and the fine clastic sediments in the diatomite suggest a relatively gradual transition from flowing to standing water. The lack of fragmentation of the diatoms indicates that water energies were low. In the field the contact between the diatomite and over lying diatomaceous earth appears erosional, but micromorphology (Chapter 3) does not support that interpretation. The apparent erosional truncation may be due to lenses of diatomite being discontinuous just prior the transition to diatomaceous earth deposition.

The diatomaceous earth exhibits faint bedding in the field, including thin lenses of diatomite, particularly in the lower half of the layer in fresh exposures in the excavation area (Figure 2.9). Microscopically, the diatomaceous earth is rich in diatoms, but also contains lenses of degraded organic material, phytoliths, and fine sand, producing some of the weakly expressed bedding noted previously (Chapter 3). Post-depositional features include the formation of channel voids from plants, as evidenced by preserved root tissues and other bioturbation features. Because the diatomite thins out to the southwest, the diatomaceous earth rests unconformably on Stratum 3 across the southwest half of the Locus 1 island (Figures 2.2, 2.8). In that part of the site, test pits that were excavated through the diatomaceous earth into upper Stratum 3 yielded no bone or stone.

At Locus 19, ~500m south of Locus 1 (Figure 1.6), a small pocket of homogeneous diatomaceous earth (a lithological equivalent to Stratum 4de; Table 2.1) is inset into Stratum 1 and the Big Red soil. Test excavations here revealed a deposit about 1 meter thick, resting on the fan alluvium. The testing revealed no diatomite, bone, or stone.

Field observations (following Holliday 1995) indicate that the diatomite represents standing-water conditions, and the diatomaceous earth represents more marshy conditions with organic-matter production in a wetland, but possibly including some ponding based on the weak stratification. As the diatomaceous earth aggraded, the local hydrologic environment became more alkaline, culminating in deposition of carbonate (Stratum 5). The data from phytoliths (Chapter 8) and diatoms (Chapter 9) confirm these general interpretations. The diatom data also show that 4de formed as a result seasonal formation of ponds or marshes.

Stratum 5

Stratum 5 is a subdivision of what was originally defined as Stratum 4 by Sánchez and others (2014). It is the calcareous upper half of Stratum 4 in that paper. The calcium carbonate was originally interpreted as simply a post-depositional alteration of the diatomaceous earth. But the dramatic increase in $CaCO_3$ (Table 2.2; Figures 2.3, 2.9, 2.10) and the micromorphology (Chapter 3) revealed that the calcareous zone has a more complex origin. In contrast to Stratum 4, Stratum 5 includes in situ precipitates and a variety of exogenous clastic materials. Stratum 5 is up to approximately 40 cm thick and is silty but very hard, dense, calcareous, and diatomaceous (Tables 2.2). The density and hardness of this layer likely protected Locus 1 from more rapid erosion.

Stratum 5 is composed of silt-sized fragments of volcanic glass, diatoms, phytoliths, clay minerals, and

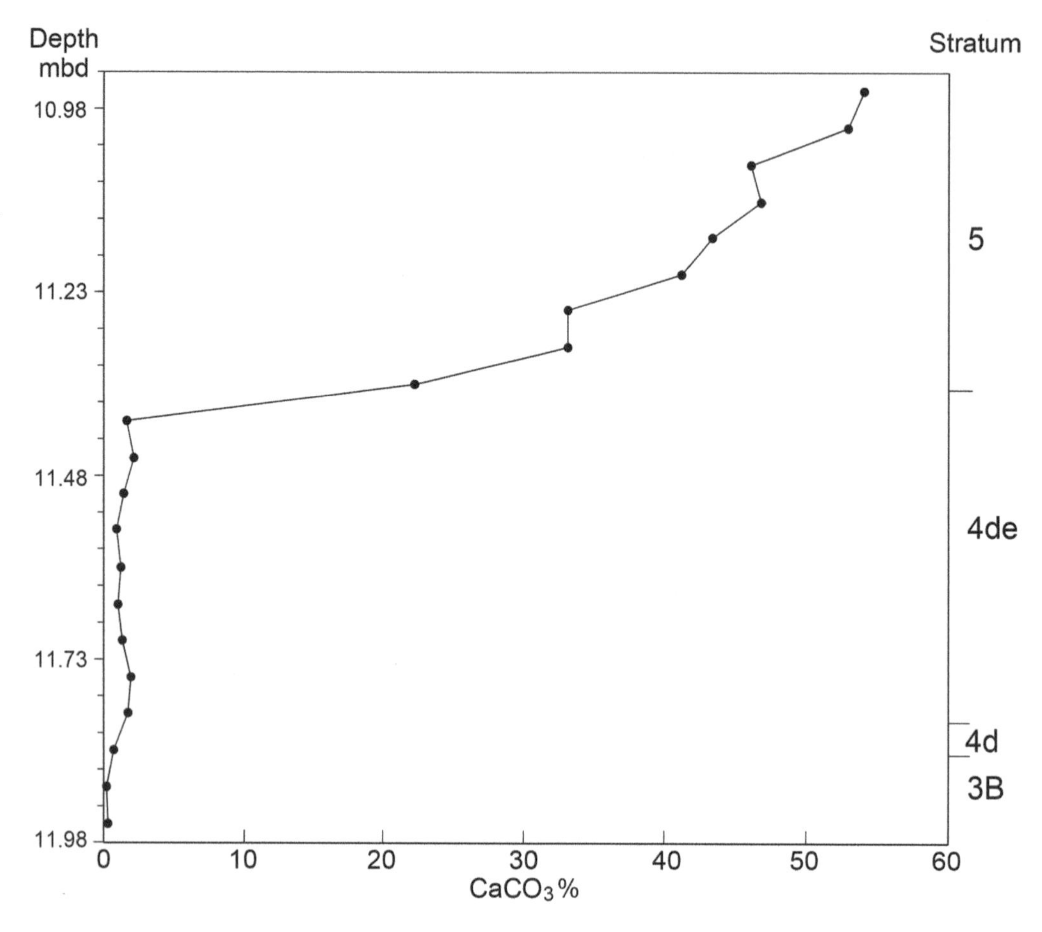

Figure 2.10. Plot of calcium carbonate content (determined by the Chittick method) through Strata 3B, 4, and 5 at El Fin del Mundo. Prepared by Jim Abbott.

microcrystalline calcite. The coarse fraction is dominated by fragments of shell and calcareous intraclasts (semi-consolidated calcareous sediment sourced from elsewhere within the wetland system). Bioturbation features are present and along with phytoliths become more abundant towards the top of the stratum. The carbonate in Stratum 5 therefore appears to have two main sources. Microscopic observations (Chapter 3) identified both a distinctive microcrystalline calcite matrix and the aforementioned coarse inclusions of rounded carbonate clasts. Notably, in thin section the phytoliths that are present within the calcareous inclusions (intraclasts) are different from those of the surrounding matrix. Inside the intraclasts, the phytoliths are well-preserved, articulated, and, in places, laminated. This observation is corroborated by the analyses of the phytoliths extracted from the bulk sediment (Chapter 8), which within this stratum consisted of two different populations: one well-preserved and articulated, and the other more weathered.

Phytoliths (Chapter 8) and diatoms (Chapter 9) both indicate that Stratum 5 represents a wetland roughly like the conditions that produced Stratum 4de. The diatom populations further indicate that Stratum 5 was deposited by increasingly more alkaline water compared to water that fed Strata 4d and 4de (Chapter 9). The phytolith data reveals a shift in the relative abundance of phytoliths compared to diatoms, with the former becoming more abundant and also more weathered towards the top of the stratum. Wroth (Chapter 8) contrasted the relative abundance of phytoliths and diatoms, while Palacios-Fest (Chapter 9) measured the diatom abundance relative to other sediment. However, the depositional environment of Stratum 5 also differed from that of 4de, with input of carbonate clasts and precipitation of crystalline carbonate matrix, producing a marl-like deposit.

Stratum 5 was subjected to locally significant post-depositional weathering. The surface of the layer exhibits distinct karren-like morphology (Figure 2.8). It is also visible in microscopic observations of channel voids, which form because of plant roots. In addition, large channel voids and burrows contain rounded pellets of calcareous materials that were translocated downwards from upper Stratum 5, probably by insects. These infillings may post-date all other primary and secondary features in the Locus. Slight reddening of Stratum 5 by subaerial iron oxidation is also indicated by formation of the "Bw" horizon (Table 2.1). Stratum 5 at the east end of Locus 1 was also mixed throughout by burrowing (Chapter 4; Figures 2.2). Based on extant burrows and krotovinas, at least some of this activity was by mammals.

Sheetwash on Strata 1 and 2

Stratum S is a widespread but discontinuous deposit on the uplands on top of the Big Red soil, consisting of several lithologies. It was likely deposited at various times. The most common lithology is a sandy, gravelly alluvial sheet-wash, best exposed and expressed in Locus 5 (Figure 2.7B). The contact with the underlying soil formed on Stratum 1 or 2. Stratum 2 is unconformable. Stratum S locally buries Archaic and Paleoindian artifacts.

Across the upland landscape as a whole, the modern surface soil ranges from a thin, weak soil with Bw horizonation formed in Strata 5 and S, to the well-expressed Bt horizon characteristics of the Big Red soil (Table 2.1).

GEOCHRONOLOGY

Thirty-five radiocarbon ages are available from El Fin del Mundo (Table 2.3). Most of the dates were determined on shell remains from Succinids, a family of the terrestrial gastropods. Other samples for dating were from other shell, organic-rich sediment (diatomaceous earth), and charred plant fragments. Dating snail shells, organic matter, and charcoal from the same stratigraphic zone allows an assessment of the stratigraphic integrity of a deposit and associated archaeology, ultimately revealing the history of its deposition and post-depositional modification, and its paleoenvironmental significance. Sedimentary organic matter is comprised of materials deposited over the course of an unknown time interval and contains fractions that decompose at varying rates. Its ^{14}C age represents the "mean residence time" (Trumbore 2000:47) except for contexts in which older material might have been introduced into the deposit. Further, post-depositional additions of organic matter can cause heterogeneity in the apparent mean residence time. In the main excavation area in Locus 1, shell from Succinids were scattered throughout the Strata 2-3-4 sequence. Some of the shell from Strata 2 and 3 appeared to be redeposited based on their worn, fragmented appearance. This interpretation was confirmed by age reversals in the stratigraphic sequence (Table 2.4). This is likely, given the alluvial origins of the two deposits. Charred plant fragments were also recovered from Stratum 3. They likewise appear to be stratigraphically mixed (Table 2.4), which is logical given the obvious evidence for redeposition of the shell (i.e., through weathering). Thus, three dates on Succinids provide anchor dates for Strata 2, 3, and 4, and a date on *Gyraulus* shell provides an estimate (following Holliday and others 2014; see Table 2.3). Otherwise, the youngest date for Stratum 3B provides a maximum limiting age; that is, the deposit is no older than the youngest date and could be younger.

Four dates are available on Succinids from the alluvium in Strata 1 and 2, with three of the samples collected from below the Big Red soil. A date of ~41,095 cal yr B.P. is available from the gravelly sandy clay of Stratum 1 crosscut by carbonate lenses in Locus 4 (Table 2.1; Figures 1.4, 2.3). Two dates on shell are available from Locus 2, northwest of and isolated from Locus 1 (Figures 1.6, 2.5). The samples, from carbonate lenses within the alluvium (Table 2.1), date to ~24,560 ^{14}C yr B.P. (~28,825 cal yr B.P.), and ~27,950 ^{14}C yr B.P. (~31,865 cal yr B.P.). The similarity of the two assays suggest that they are good estimates for the age of the alluvium in Locus 2. Further, a date of ~21,530 ^{14}C yr B.P. (~25,850 cal yr B.P.) was determined on shell from Stratum 2 alluvium beneath the Locus 1 excavations. Because of the similarities in lithology and rough age equivalence, the dated deposit in Locus 2 is considered to be Stratum 2.

All four age estimates for the alluvium seem reasonable given the degree of development expressed by the Big Red soil above the dated zones. Because the date from Locus 4 is significantly older than the other three dates, the associated alluvium is identified as Stratum 1. Big Red in both Loci 3 and 4 is better expressed (redder with more translocated clay) than in Locus 2 (Table 2.1; also noted by Ibarra-Arzave and others 2020), supporting differentiation of the alluvium based on numerical dating. The dating and soil development suggest the deposits represent several cycles of alluviation, which is in keeping with their origins as part of an alluvial fan. The dates are considered as rough approximations of an upper age limit for the alluvial fan, given their association with both near-surface and distal components of fan alluvium.

Resting on Stratum 1 in Locus 3, the stratified carbonate (Stratum C; Figure 2.4) produced three radiocarbon dates. The area has the most complete sequence of carbonate

Table 2.4. Radiocarbon Dates on Plant Fragments and Shell from Stratum 3, Locus 1

Fdm#	*Lab Number (AA)*	*Stratum*	*Material*	*14C yr B.P. median*	*Depth below Strat 4d, cm[a]*	*Depth, mbd[b]*
63447	100181A	3B	Charred plants	11,550	18	12.06
63176	100182B	3B	Charred plants	11,880	18	12.01
59354	83272	3B	Charred plants	11,040	20	
74956A	114925	3B	Charred plants	16,920	20	12.15
	90708	3B	Succinid shell	17,220	19	11.88
	94051	3B	Succinid shell	22,490	22	11.78
62372	83958	3A/3B	Charred plants	16,130	30	12.26
	90707	3A/3B	Succinid shell	12,130	32	12.01
	94052	3A	Succinid shell	12,890	nd[c]	12.15
62543	83959	2/3A	Charred plants	21,200	nd[c]	12.95

Notes:

a. AA90707=19cm above Stratum 2; AA94052=15cm above Stratum 2; AA83959 was on Stratum 2 contact.

b. mbd = meters below site datum.

c. No data because the 3/4 contact above the sample location is eroded and mixed.

preserved at the site and yielded Succinids for dating. The Lower Massive Carbonate zone dates to ~16,010 ^{14}C yr B.P. (~19,325 cal yr B.P.; AA-81043) and ~15,060 ^{14}C yr B.P. (~18,515 cal yr B.P.; AA-81042), and the Upper Massive layer dates ~8000 ^{14}C yr B.P. (~8870 cal yr B.P.; AA-94097), an age discussed later in this chapter. The morphology (macro- and micro-) of the carbonate sequence is indicative of changes in depositional environment, noted previously, and, therefore, suggests some breaks in the depositional chronology. The Middle Platy zone is likely a paludal rather than a groundwater carbonate (Chapter 3). Although not directly dated, the bracketing dates show that the Middle Platy zone could be in the same age range as the lacustrine and palustrine deposits of Stratum 4 and thus may be lateral facies of Stratum 4.

Stratum 3 is differentiated from Stratum 2 based on an unconformity representing an unknown amount of time and presence of Stratum 3 only in Locus 1. Stratum 3 yielded four shell samples and six samples of discrete, fragmented, charred plant matter (Tables 2.3, 2.4). Shell sample AA-94052 was unweathered and therefore likely in situ. It provides a date of ~12,890 ^{14}C yr B.P. (~15,410 cal yr B.P.) for upper 3A. That date is also the maximum age for Stratum 3B. Among the seven samples from 3B (two shells and five fragments of charred plants), the youngest date, on charred plant matter, is 11,550 ± 60 ^{14}C yr B.P. (~13,415 cal yr B.P.; AA-100181A; Table 2.4) and provides a maximum limiting age for the sediments and the archaeological feature.

The full age range of Stratum 4 is unknown because of lack of dates from the basal diatomite. The base of the diatomite and the rest of Stratum 4 are younger than the youngest date from Stratum 3. The oldest and only reliable date from Stratum 4, determined on charcoal from flecks scattered across the contact at the top of the diatomite is ~9,715 ^{14}C yr B.P. (11,125 cal yr B.P.; AA-80084A; Table 2.3). That date and the date of <13,415 cal yr B.P. for Stratum 3 provide an age bracket for the pond that produced the diatomite.

Samples of organic-rich sediment, charred plant fragments, and Succinids from Strata 4 and 5 provide a range of dates for the diatomaceous earth in Stratum 4 and the carbonate facies of Stratum 5, but only two are considered valid. Charcoal flecks at the base of the 4de facies, resting on the diatomite that date to ~9,715 ^{14}C yr B.P. (11,125 cal yr B.P.), indicate the onset of the wetland conditions that produced the diatomaceous earth. Most of the rest of the dates from Strata 4 and 5 are out of stratigraphic order. Only one other date is from a sample considered to be in place and thus an accurate indicator of the age of the wetland depositional cycle. A whole, unweathered succinid shell from near the top of the section in Stratum 5 dated to ~8870 ^{14}C yr B.P. (~9990 cal yr B.P.; AA-88885; Table 2.3). The timing of the Strata 4 to 5 transition can only be estimated. Assuming a roughly steady rate of accumulation of Strata 4de and 5, that change in the depositional environment is estimated to be ~10,440 cal yr B.P. (Chapter

9). Younger shell from Stratum 5 (samples AA81350 and 88886 in Table 2.3) was recovered but was fragmented and weathered. As documented in thin section, it was likely mixed (Chapter 3). When compared to that of ~9990 cal yr B.P. for unweathered and likely in place shell collected approximately 10 cm below the surface of Stratum 5, the ~9490 and ~8884 cal yr B.P. dates are likely indicative of mixing of younger shell into that stratum as aggradation of the wetland slowed and ended after ~10,000 years B.P. The lithology, stratigraphic position, and dating indicates that the carbonate of Stratum 5 is a weathered equivalent of the Upper Massive facies of the carbonate sequence in Locus 3, dated to ~8870 cal yr B.P. (Table 2.3).

From the disturbed context of Stratum 5 at the east end of Locus 1 (Figure 2.2; Chapter 4), 50 nacreous shell beads, made of marine shell likely from the nearby Gulf of California, from the same necklace were found. The beads were distributed vertically from the surface down to about 45cm due to mixing. One bead was dated by radiocarbon ~5,560 cal yr B.P. (Table 2.3).

Locus 19 yielded bulk organic matter samples from a deposit that is lithologically similar to Stratum 4de. These samples date to the middle Holocene (Table 2.3). Evaluating these dates is difficult given the isolation of the area and absence of the underlying diatomite, fossils, or artifacts. The dates are on organic matter and therefore could be minimal ages. The Succinids from Stratum 4 in Locus 1 provided dates that are generally a little older than the bulk organic matter (Table 2.3). This deposit in Locus 19 may represent a localized pond of diatom-rich sediment that appeared after the Locus 1 pond and wetland disappeared.

Stratum S, resting on the Big Red soil of the uplands produced no material for numerical dating. However, evidence of soil formation in the deposit is minimal or absent, indicating a late Holocene age.

DATING THE CLOVIS OCCUPATION

The age of the Clovis bonebed at El Fin del Mundo was initially determined to be ~11,550 ^{14}C yr B.P. (~13,415 cal yr B.P.), based on a sample reported as charcoal (AA100181) (Sánchez and others 2014). That date and very similar age determinations averaging ~11,565 ^{14}C yr B.P. from the Aubrey site, Texas (Ferring 2001:50–51), were argued to represent the oldest Clovis dates yet available. Microscopic examination of the material dated revealed it to be some sort of plant material, but possibly not charcoal.

Upper 3B produced five additional samples for dating (Table 2.3). Two samples of discrete but fragmented and charred plant matter were recovered from upper 3B via the excavations in proximity to the bonebed and these samples were subsequently dated. Sample AA100182B dates to ~11,880 ^{14}C yr B.P. (reported by Sánchez and others 2014). Sample AA114925 dates to ~16,920 ^{14}C yr B.P. A succinid sample (AA90708) from just below the bonebed dates to ~17,220 ^{14}C yr B.P. (~20,780 cal yr B.P.) and charred plant material from a little deeper (AA83272B) dates to ~11,040 ^{14}C yr B.P. but with a very large standard deviation. This set of dates indicates that all samples of charred plant matter are detrital fragments of plant matter and, like some of the Succinids, provide only a maximum age estimate. The date of 11,550 ^{14}C yr B.P. provides the youngest date for the bonebed and uppermost Stratum 3B and indicates that they are at or younger than that age. Stratum 3B and the Clovis bonebed must be older than the oldest date in Stratum 4 (~9,715 ^{14}C yr B.P.; ~11,125 cal yr B.P.). In light of the groundwater correction, the date on the *Gyraulus* raises the possibility that the Clovis feature is in the range of ~10,950 ^{14}C yr B.P.

ARCHAEOLOGICAL CONTEXT

Archaeological research at El Fin del Mundo included excavations in Locus 1 (Chapter 4) and survey with testing across the uplands within about 1 km of the excavation area (Figure 1.6). The excavations focused on the uppermost bonebed at the Strata 3/4 contact in Locus 1 (Chapter 4). The bonebed was in the topographically slightly lower eastern half of upper Stratum 3/lower 4 on the Locus 1 island (Figure 2.2). Test excavations through Stratum 4 and into upper 3B in the western half of Locus 1 (Units 12, 228; Figures 1.4, 2.9) where the 3-4 contact rises gradually to the west revealed no archaeological materials of any kind. The archaeological feature was entirely within the eastern half of the island on a flat surface within upper Stratum 3B, with a few larger elements also covered by diatomite. The human activity took place on a flat alluvial plain less than 100 meters wide. Upland settings (2 m higher or more) for observation of the alluvial plain were present within about 50 meters to the south and 100 meters to the northwest (Figure 1.6).

The Upper Bonebed covered approximately 40m^2. Two eroded concentrations of bones representing parts of two juvenile proboscideans, *Cuvieronius* sp., were uncovered along with scattered bone fragments and 27 stone artifacts (Chapters 4, 5, 6). Among the artifact assemblage, four Clovis points were found in association with the bone concentrations.

The biggest bones and three flakes in direct contact with the bone were on the contact between Strata 4 and 3B. The other artifacts and bone were between the top of Stratum 3B and 26cm below the contact, mostly within 15cm of the contact. The deeper material likely moved down due to cracking subsequent to formation of the main bone feature. Visible cracks within Stratum 3B were obvious and pervasive in the field but were significantly less abundant in the immediately overlying sediments of Stratum 4 (Figure 2.3).

This distribution of bone and stone clearly indicates a brief phase of subaerial exposure of upper 3B prior to burial, but not enough exposure to initiate weathering of the unburied portions of bone. Further, small fragments of bone and teeth are scattered through the upper 26cm of 3B. These data suggest that smaller items such as bone bits, plant fragments, and stone artifacts moved down as cracks formed while larger bone and bone fragments were not transported. This process could account for some of the dating inconsistencies among the samples of Succinids and charred plant fragments. Butchering activity under muddy conditions in upper 3B also could have contributed to mixing. The similarities in weathering of the bone between those segments buried within 3B and those encased in the diatomite further support an interpretation of a brief interval of time for final deposition of 3B alluvium and onset of lacustrine deposition.

The upland archaeological loci (5, 7–10, 14, 17, 19, 22–25) are scattered across a rough arc 200 to 1000 meters to the east, southeast, south, and southwest of Locus 1 (Figure 1.6). Fan alluvium with the Big Red soil underlie the upland loci. The various archaeological loci identified on the uplands are accumulations of stone artifacts. Clovis and Archaic artifacts rest directly on the Big Red soil. Locus 25 is also an area of lithic raw material consisting of boulders of rhyolite, vesicular basalt, and quartzite with evidence of quarrying (Chapter 5).

DISCUSSION

El Fin del Mundo includes two basic geologic and geomorphic components of archaeological significance. A buried, in situ lowland component (in Locus 1) contained remnants of a buried bonebed with parts of two gomphotheres. Stone tools, including Clovis points, were recovered in association with the bone (Chapters 4, 5). The full extent and complexity of the bonebed cannot be determined, however, due to extensive erosion. The other component of the site is on stable uplands approximately 200m to 1000m south and southwest of Locus 1. Hundreds of artifacts with Paleoindian and younger affinities were mapped and recovered (Chapter 5), all resting on the Big Red soil in Pleistocene fan alluvium.

The geologically oldest component of the site is the upland, composed of the alluvial deposits of Strata 1 and 2 and the Lower Massive carbonate. The alluvium is either primary or redeposited fan deposits. It is late Pleistocene in age. One radiocarbon date puts upper Stratum 1 at ~41,000 cal yr B.P. (in Locus 4, ~200m southeast of Locus 1) and three samples date Stratum 2 ~32,000 to ~29,000 cal yr B.P. (in Locus 2, ~200m north of Locus 1), and <~25,850 cal yr B.P. (Locus 1). The dating in Loci 2 and 4 accords with the degree of development of the regional soil formed at the surface of the alluvium. In both Loci 2 and 4, the soil, informally called Big Red, exhibits a Bt horizon 60 to 80cm thick with 7.5YR to 5YR hues. In this setting, these characteristics suggest tens of thousands of years of landscape stability (Gile, Hawley, and Grossman 1981; Birkeland, 1999; Holliday 2004; Holliday and others 2006).

Resting on top of Stratum 1 or 2 and the Big Red soil in proximity to Locus 1 is a carbonate, Stratum C. This deposit formed from seeps or springs. No obvious spring throats or spring cauldrons or other point sources of water are apparent, suggesting that water seeped out of Stratum 1 upslope of the Locus 1 kill area. This seems logical given the position of the carbonate low on the distal end of a large fan, where groundwater stored in fan alluvium could "leak" out (e.g., Ashley 2001). The presence of diatoms in the middle and upper carbonate suggests that local wetland conditions accompanied the seeps. Where the carbonate is present, a carbonate overprint in the underlying soil is also noted. Radiocarbon dating and lack of weathering within the carbonate sequence indicate that this seep activity was more or less continuous (through time but not necessarily in one place on the landscape) from ~19,300 to <8,800 cal yr B.P.

Stratum 2 is a component of the alluvial fan that contains El Fin del Mundo because both are more extensive than the strata in Locus 1 that are unique to that setting. Stratum 2 is the youngest fan alluvium and is capped by Stratum C. The oldest component of Stratum 2 is several thousand years younger than Stratum C (although Stratum C is not directly dated where it caps Stratum 2). One scenario is that incision of Stratum 2 and formation of the channel with Stratum 3 and younger deposits in Locus 1 is linked to the onset of the seep or spring activity represented by Stratum C. This is a testable hypothesis dependent on recovery of dateable material from Stratum C in Locus 2 during future fieldwork.

The fan surface and Strata 1 and 2 were incised sometime after ~25,850 cal yr B.P. and before ~15,410 cal yr B.P. The channel was then filled with Strata 3, 4, and 5. Stratum 3 includes lower and somewhat coarser 3A, dated to ~15,410 cal yr B.P., and upper, generally finer Stratum 3B, dated <15,410 to ≤13,415 cal yr B.P. This stratigraphic sequence is unique to Locus 1 and completely isolated from all other exposures at the site due to extensive erosion by arroyo formation. The Lower Bonebed in Locus 1 is a paleontological feature in lower 3A alluvium with redeposited remains of gomphothere, mammoth, mastodon, and tapir, among other mammals. The Upper Bonebed is the Clovis archaeological feature in Locus 1 with the remains of gomphothere; it is associated with Strata 3B alluvium and 4d diatomite. The extensive erosion by arroyo cutting limits our reconstruction of the paleo-landscape associated with the two bonebeds, the streams associated with them, and the subsequent wetland of Strata 4 and 5.

The arroyo system that exposed the site and created the Locus 1 island extends about 1 km upstream. This arroyo provided no evidence for a paleo-channel that could have contained or produced the Stratum 3 alluvium. That observation combined with the seep or spring genesis of the carbonate suggests that the flowing water of Stratum 3 and standing water of Stratum 4 were fed by spring waters derived from the immediate surroundings of Locus 1.

The sedimentary characteristics of Strata 3, 4, and 5 in Locus 1 suggest the following scenario for the local environmental evolution: alluviation with brief subaerial weathering (Stratum 3), followed by initiation of standing water conditions (basal Stratum 4 diatomite), and then transition to seasonal ponding or marsh conditions (Stratum 4 diatomaceous earth). The Strata 3 to 4 transition, roughly contemporaneous with the Clovis activity, included brief periods of standing water during the final phases of 3B deposition and some intermittent flowing water during the subsequent lacustrine phase. Stratum 4de evolved into Stratum 5, indicative of a more alkaline wetland, eventually producing a marl-like cap.

One of the more puzzling aspects of the depositional sequence coinciding with the Clovis activity is how an alluvial system evolved into a lacustrine system; that is, what mechanism impounded the water? A low hill north-northwest of Locus 1 could have generated a mass movement that dammed the drainage, or beavers could have constructed a dam. Stream discharge clearly slowed, but if due to damming, the stream flow that produced Stratum 3 would have filled a lake and overtopped a dam. Unfortunately, the extensive erosion that left Strata 3 and 4 preserved only in Locus 1 removed all evidence of the paleo-topography of the time. This issue is further explored in Chapter 11.

The changes in the sedimentary fabric and in abundance of phytoliths and organic material relative to diatoms in Strata 4 and 5 indicate that the wetland environment shifted over time, with an increase in alkalinity and abundance of plants coupled with perhaps more seasonally arid conditions. Diatoms from Stratum 4de document a shift to more alkaline conditions (Chapter 9). Stratum 5 appears to represent a culmination of alkaline conditions with precipitation of carbonate. In addition, Stratum 5 contains abundant carbonate intraclasts, representing clastic input from elsewhere within the wetland system. Carbonate recrystallization in Stratum 5 and concomitant cementation of this facies likely contributed to the preservation of the Locus 1 sequence via "case hardening" of the exposures.

Stratum 5 in Locus 1 is similar to the Upper Massive facies of the Stratum C in Locus 3 based on the presence of the calcareous intraclasts, shell fragments, and micritic matrix (Chapter 3). The carbonates are also both early Holocene in age and represent the final stage of deposition in their respective stratigraphic settings.

A plausible scenario linking the upland and lowland stratigraphy is that the seeps or springs that produced the Stratum C carbonate also fed water into the Strata 3-4-5 lowland/channel. The differences in the carbonate and diatomite sedimentary sequences are thus attributable to lateral facies changes. The dating of the Stratum 5 carbonate in Locus 1 shows that it fits within the age range of upper Stratum C ("Upper Massive" zone) in Locus 3. The bracketing dates on the Middle Platy carbonate show that this zone could be in the same age range as the lacustrine and palustrine deposits of Stratum 4 and thus be facies of it. An early Holocene carbonate shifting laterally to diatomite facies over a distance of ~5 meters at the Lubbock Lake site, Texas (Stafford 1981), illustrates the potential facies relationship of seep or spring carbonates and diatomite or diatomaceous earth. This scenario further suggests that older seep or spring flow fed water that deposited the alluvium of Stratum 3.

The geological scenario is significant archaeologically because it explains the site formation processes in both the lowland and upland settings. Sedimentation through the final millennia of the Pleistocene preserved the

Upper Bonebed while landscape stability on the uplands through the terminal Pleistocene and into the Holocene resulted in the palimpsest of Clovis and Archaic artifacts. The presence of water in the lowlands through the final millennia of the Pleistocene (flowing to standing water conditions) and into the Holocene (seeps and wetlands) also attracted Clovis and Archaic foragers to the area. The carbonate sequence of Stratum C thus represents a shift from conditions with emergent groundwater creating seeps (and possibly springs) to a wetland or marsh fed by the seeps. The broader paleoenvironmental implications of this record are unclear, as discussed in Chapter 11.

The initial publication of El Fin del Mundo (Sánchez and others 2014) presented a preliminary version of the stratigraphy and interpretation of its formation history. After publication, additional field seasons, continued laboratory analyses and new results, as well as published critiques prompted revisions to both the stratigraphic sequence and the formation model. As described in this chapter, the key differences in the stratigraphy between Sánchez and others (2014) and this volume are (1) the subdivision of the alluvial fan deposits into Stratum 1 and Stratum 2 based on dating and (2) the subdivision of Stratum 4 into Stratum 4 and Stratum 5 based on the appearance of primary and reworked carbonates. This volume also provides a slightly revised interpretation of the radiocarbon dating results for Strata 3 and 4de. These revised interpretations are: Stratum 3A dates to ~15,410 cal yr B.P.; Stratum 3B and the Clovis occupation date to ≤13,415 cal yr B.P.; Stratum 4d is <13,415 to >11,125 cal yr B.P.; and Stratum 4de is ≤11,125 to ~10,440 cal yr B.P., and Stratum 5 is ~10,440 to ~9990 cal years B.P. The lithostratigraphic and chronostratigraphic records from El Fin del Mundo unfortunately preclude recognition of the late Bølling-Allerød, Younger Dryas, and earliest Holocene time intervals.

Based on the post-2014 excavation and laboratory data, as well as the revised interpretations, it is now possible to directly comment on published critiques of interpretations and misinterpretations of the stratigraphy, dating, and isotopic characteristics of El Fin del Mundo. Cruz y Cruz and others (2015:143–144) state that "[t]he majority of the artifacts outside of the kill site were found on the surface of a red paleosol...dated to...~13,390 cal yr B.P." citing Sánchez (2010). The upland artifacts were on the red soil, but that soil represents stability and soil formation spanning several tens of thousands of years (Chapter 2). Their paper also includes a photograph (their figure 9a) said to be the "red paleosol" at FDM with carbonate concretions, but that is a mistake. The sign in the pit clearly indicates that the exposure shown is in Locus 1, west of the main excavations. It therefore illustrates the Stratum 4 diatomaceous earth. The carbonate is a Holocene feature (Chapter 2), but no isotopic data are presented. Cruz-y-Cruz and others (2016:153) list isotopic data associated with radiocarbon dates from organic matter from Stratum 4, citing Sánchez and others (2014). None of those dates are accepted as reliable (Sánchez and others 2014: Table S8), however, because they are out of stratigraphic order (Chapter 2).

Ibarra-Arzave and others (2020), link the Big Red soil along with the weathering in what is now identified as Stratum 5 to the San Rafael Paleosol of Cruz y Cruz and others (2014, 2015). The data presented in this volume show that the stratigraphic and lithologic facies relations are more complex. The Big Red soil that formed in Stratum 1 on the eastern and southeastern regions of the site probably developed over the past 40,000 years based on dating in Locus 3, whereas the Big Red soil that formed in Stratum 2 developed over the past 30,000 years, based on dating at Locus 2 (near "Loc 2-3" of Ibarra-Arzave and others 2020). Throughout the site, the surface finds of Clovis archaeology rest on top of the Big Red soil. Thus, the soil began forming long before Clovis time and, where not buried by Stratum S, is probably still forming. The revised formation model presented in this volume shows that there is no relationship between the Big Red soil and any weathering in Stratum 4. The layer formerly considered to be Stratum 4 was interpreted by Sánchez and others (2014) as "subjected to weathering and soil formation," but is now called Stratum 5, and its calcareous composition is interpreted as primary rather than secondary. Post-depositional weathering of Stratum 5 is observed, however, and includes penetration of plant roots downwards and formation of a solidified caprock, possibly due to recrystallization of the primary carbonate and development of a silica-rich outer rind (see Chapter 3). This weathering in Stratum 5 likely overlaps the latest (ongoing) stage of Big Red formation. In conclusion, the proposition by Ibarra-Arzave and others (2020) that all the soils might be linked as facies of a single soil-stratigraphic unit is based on a preliminary interpretation presented in Sánchez and others (2014) that has since been revised. Although a soil is forming on the surface today, including in areas where the Big Red soil is exposed, to call these a single unit would obscure important chronostratigraphic, lithostratigraphic, topostratigraphic, and geoarchaeologic relationships of the soils.

Another misinterpretation of site stratigraphy at El Fin del Mundo concerns the carbonates present on the uplands and their relationship to the stratigraphic sequence in Locus 1. Waters, Stafford, and Carlson (2020:8) state that "springs deposited carbonates (marls) upslope of the site, and these marls are lateral facies of Stratum 2 and 3 sediments." In contrast, Sánchez and others (2014) argue that Strata 2 and 3 were deposited by spring-derived water and argue for the presence of springs or seeps based on the presence of the upland spring-derived carbonates and broad age correlation (carbonates dated at ~8870, ~15,060 and ~16,010 cal yr B.P.; chronostratigraphic correlation). But as documented by Sánchez and others (2014), nowhere at the site is there direct lithostratigraphic evidence of carbonates as lateral facies of Stratum 3. In fact, the revised stratigraphic sequence and formation model presented in this volume proposes that Stratum 4 may be a lateral facies of the Middle Platy member of the Stratum C upland carbonates, while Stratum 5 may be a lateral facies of the Upper Massive member of Stratum C. These links to different remnants of the wetland system across the landscape all post-date the Clovis occupation of the site.

The remaining critique of the interpretations presented by Sánchez and others (2014) is by Waters, Stafford, and Carlson (2020) and concerns the dating of the Clovis occupation. As sites of similar age, both El Fin del Mundo and the Aubrey site dates have been subject to critique. Waters and Stafford (2007) rejected the Aubrey dates because the context was not clear from the original field report. In a subsequent public presentation, however, Ferring (2012) clearly established the context. In an update of their Clovis chronology, which was based on radiocarbon ages with standard deviations of less than 100 years, Waters, Stafford, and Carlson (2020) continue to reject the Aubrey age, as well as the dating at El Fin del Mundo. They raise several issues with the El Fin del Mundo sample, dated 11,550 ± 60 ^{14}C years B.P. (13,413 ± 160 cal years; AA100181A; Table 2.3). The most immediate critiques focused on the material dated, the context of the sample, and the standard deviation. The dated sample and additional samples subsequently recovered came from the same level within Stratum 3 as the Clovis artifacts. Waters, Stafford, and Carlson (2020:8) refer to the dated sample as "dispersed organic matter." This description is incorrect. As clearly stated (Holliday and others 2014:13), the sample was examined under a microscope and observed to consist of charred material with "woody tissue structure."

Waters, Stafford, and Carlson (2020:8) further argue that "a carbonate reservoir effect may have made samples from the site date older than their true age." This is an assertion with no evidence. The sample was fully decalcified. Moreover, they then assert "Consequently, the 11,560 ^{14}C yr B.P. date on organic matter associated with the gomphothere may be 1000 14C years too old." They fail to acknowledge, much less discuss, that such a recalculation would produce a date of 10,560 ^{14}C yr B.P. for the Clovis occupation, far younger that their own minimum age estimate for Clovis (10,820 ± 10 ^{14}C yr B.P.). In any case, the variation in multiple radiocarbon dates on charred plant material from Stratum 3B, clearly redeposited as discussed in this chapter (Table 2.4) results in a re-interpretation of the ~11,550 ^{14}C yr B.P. (~13,413 cal yr B.P.) as a maximum age for the Clovis bonebed (i.e., dating to ≤13,410 cal yr B.P.).

CHAPTER THREE

Soil and Sediment Micromorphology

Susan M. Mentzer

Micromorphology is a well-established technique used to understand the genesis of soils (soil micromorphology) and to observe the composition, and depositional and post-depositional characteristics of archaeological strata and features (archaeological micromorphology). Typical archaeological applications focus on the identification of anthropogenic, biogenic, and geogenic sediment, and the analysis of microstratigraphy related to human activity (e.g., floor sequences, hearths). In North American Paleoindian open-air sites, micromorphological analyses are conducted infrequently and when used, the goals typically align more with soil micromorphology (e.g., Collins and others 1998; Kellogg, Roberts, and Spiess 2003; Luchsinger 2002; Macphail and McAvoy 2008; Miller and Goldberg 2009), although archaeological features are occasionally sampled (Zarzycka and others 2019). Paleoindian sites associated with wetlands have received some attention, with micromorphological analyses focused on understanding the genesis of so-called "Black Mat" layers (Harris-Parks 2016; Holliday, Dello-Russo, and Mentzer 2020). Outside of North America, micromorphological analyses of lacustrine and wetland hunter-gatherer archaeological sites have contributed to understanding the local environment (e.g., Liutkus and Ashley 2003; Mallol 2006). At El Fin del Mundo, micromorphology was applied to study the formation of soils and to make microscopic observations of the composition, fabric, and structure of the main stratigraphic units and their contacts.

Anthropogenic materials were not observed in any of the samples, and therefore the work is focused almost exclusively on reconstructing depositional and post-depositional processes as they relate to changes in the local environment at the site over time.

METHODS

Samples were collected for micromorphological analyses during the 2007-2008 and 2008 field seasons. Oriented blocks were carved directly from the excavation profile or exposed scarp using a knife. In Locus 1, the entire stratigraphic sequence was sampled as a column in order to obtain a permanent record of the site. Additional isolated blocks were collected in Loci 3, 4, and 5 to sample specific features or strata. Sample numbering reflects the site, locality, and number of blocks collected. For example, sample FDM-3-19 was collected from El Fin del Mundo, Locus 3, and was the nineteenth block collected for the overall study.

At the University of Arizona, the blocks were oven-dried and impregnated with a mixture of unpromoted polyester resin diluted with styrene (7:3; Advanced Coating Company, Westminster, Massachusetts) and catalyzed with methyl ethyl ketone peroxide. The hardened blocks were sliced with a rock saw, and selected portions of the sequence were further processed by trimming to 5-cm by 7-cm chips. The chips were mounted on glass slides and ground to a thickness of 30 microns by Ray Lund at Arizona Quality Thin Sections, Tucson. The thin sections were studied at a variety of magnifications and different types of light using stereomicroscopes and petrographic microscopes equipped with plane-polarized light (PPL), cross-polarized light (XPL), darkfield illumination, oblique incident light (OIL), and blue light fluorescence. Micromorphological analyses were conducted at the University of Arizona and the University of Tübingen (Germany), following standardized criteria for general descriptions of soils (Stoops 2021) and microscopic features of terrestrial carbonate samples

Table 3.1. Micromorphological Descriptions of Samples Associated with Stratum 1 and the Big Red Soil

Sample Number	*Voids*	*Fabric*	*Related Distribution, and Description of Coarse and Fine Components*	*Pedofeatures*
FDM-4-20	Channels, and thin zigzag planes extending outward or connecting small vughs, the former interpreted as desiccation cracks	Weak laminated fabric with bedding of the coarse fraction, and cm-scale bands of coarser and finer sands. Calcitic crystallitic *b*-fabric	Porphyric, with poorly sorted subangular sands in a matrix of microcrystalline calcite	Secondary iron oxides associated with channel voids at the base of the sample. Calcite depletion and enrichment hypocoatings on channels at the top of the sample. Some channels at the base of the sample exhibit a first phase of silica coating, with a later phase of calcareous infilling
FDM-3-18	Simple packing voids between sand grains are dominant; vughs and small channels also present	None	Gefuric, with bridges of clay between poorly sorted sand grains	Clay bridges, but otherwise no structural development. Possible iron reduction contributing to color gradation visible in the field
FDM-3-19	Channels and vughs dominant, planes present but less common	Granostriated *b*-fabric	Porphyric, with coarse fraction consisting of poorly sorted sand. The fine component is noncalcareous	Iron oxides. Rhizoliths composed of microcrystalline calcite in channel voids. Variably calcified root hairs and fungal hyphae in other channels
FDM-5-21	Planes and small channels; some channels interconnected by planes; one infilled chamber	Unoriented sample; fabric observations not possible	Close porphyric, with coarse fraction consisting of poorly sorted sand. Lacks basalt fragments. The fine component is noncalcareous	Iron oxides in the matrix and, less commonly, as coatings in voids. Thick, laminated typic clay coatings of channels and planes

(Alonso-Zarza and Wright 2010; Durand and others 2018; Pimental, Wright, and Azevedo 1996). When necessary, samples were compared to published microscopic images of lacustrine and palustrine deposits (e.g., Freytet and Verrecchia 2002) and pedofeatures and facies unique to paleosols (e.g., Federoff, Courty, and Guo 2010).

Elemental mapping of specific microscopic features was conducted directly on the uncovered thin sections using a Bruker M4 Tornado micro-x-ray fluorescence instrument located at the University of Tübingen. The purpose of these analyses was to study the spatial distribution of specific elements within the samples, and to identify the elemental composition of post-depositional components, such as oxide stains and secondary minerals. Elemental distribution maps produced on micromorphological samples are qualitative rather than quantitative (Mentzer 2017). The scans were conducted with full 30W rhodium x-ray tube settings (50 kV, 600 μA), with spot spacing and dwell times ranging from 50 μm/10 ms to 30 μm/100 ms depending on the desired resolution and target element.

RESULTS

All strata were sampled in various locations at El Fin del Mundo, except for Stratum S. Descriptions and interpretation of the samples are presented in three groups: (1) samples from Stratum 1 and the Big Red soil, (2) samples of Stratum C carbonates from Locus 3, and (3) the stratigraphic sequence at Locus 1.

Stratum 1 and the Big Red Soil Samples from Loci 3, 4 and 5

Four blocks were collected to document the formation processes of the Big Red soil and its spatial variability. One block was collected from a carbonate located in Stratum 1 beneath the Big Red soil in order to determine whether it was a subsurface calcic or petrocalcic horizon associated with the soil. Three blocks were collected from the Big Red soil, which was itself formed on Stratum 1 from two different loci. The sampling locations were selected in order to study variation in color of the soil. These are the only samples of Stratum 1 that were analyzed in this study. Some aspects of the samples are summarized in Table 3.1.

In all of the samples, the sand-sized particle size fraction contains abundant feldspar, quartz, and less common biotite, granitic rock fragments, and rounded fragments of weathered basalt. Accessory minerals from the basalt (e.g., pyroxene) are very rare, but present in most samples. The fine sand and silt component frequently contains quartz and volcanic glass.

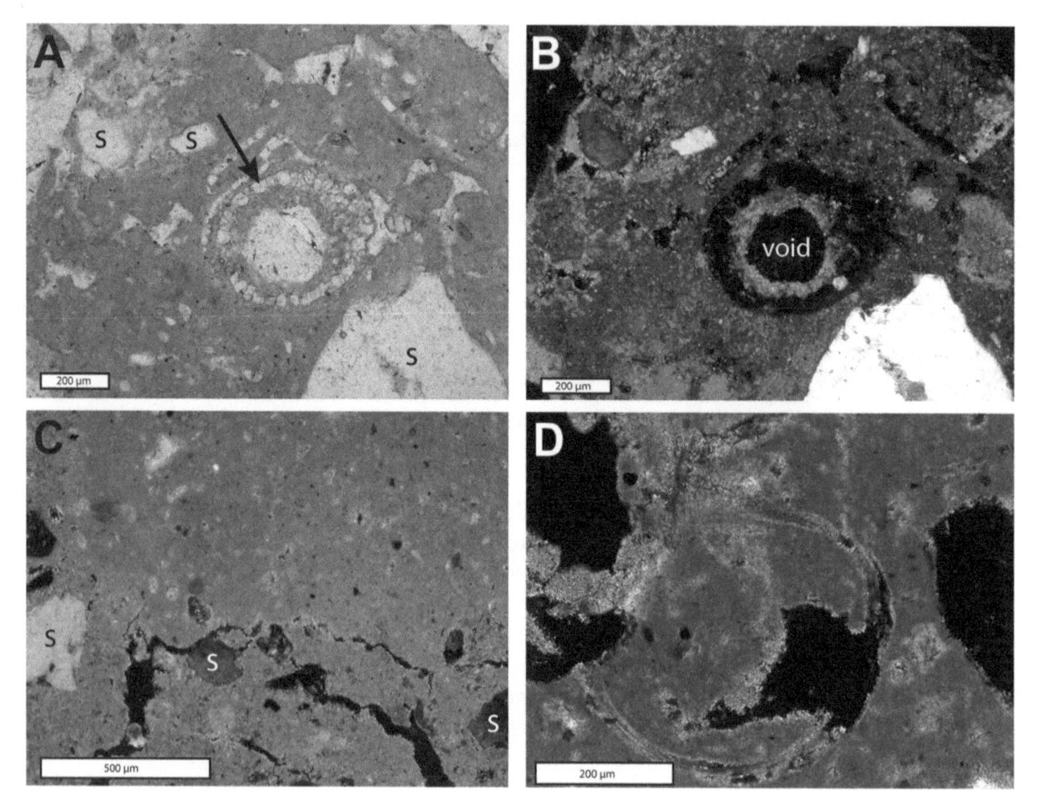

Figure 3.1. Sample FDM-4-20, laminar carbonate beneath the Big Red soil in Locus 4. (A) A cross-section through a lithified root. The arrow indicates the preserved cells of the cortex of the root. Poorly sorted, subangular grains of sand (s) are present in a calcareous matrix. PPL. (B) Same view as (a), XPL. The dark zone shows that the cortex cells have been replaced by silica, while the lighter colored inner zone contains secondary carbonate. The center dark area is a void. (C) Differences in the density of the fine calcareous matrix, with the upper portion of the image exhibiting higher density. XPL. (D) A shell infilled and embedded in a fine calcareous matrix. XPL.

Locus 4

One sample (FDM-4-20) was collected from a laminar carbonate outcrop at Locus 4 (Figures 1.6, 2.6) beneath the Big Red soil. Laminar carbonates are common macro- and microscopic features of calcrete horizons in soils (Durand and others 2018), and here the analyses focused on the formation of the carbonate. Some aspects of the sample can be related to an alluvial mode of deposition. For example, there is an overall laminated fabric (not well-developed) characterized by weak bedding of the coarse material with alternating cm-scale bands of coarser and finer sands. Beds of finer sands are associated with a fine matrix that is slightly less calcareous compared to the beds of coarser sand. The finer beds are also associated with horizontally-aligned rhizolith-like carbonate- and silica-bearing concretions that are pseudomorphic after plant channels. At least three bands of these horizontally aligned plant tissues are present in the sample. In the uppermost band at the top of the thin section, the plant tissues are most abundant and preserve anatomical structures. In some areas, possible calcium oxalate crystals are present within woody plant tissues wherein the original organic material appears to be replaced by silica. Elemental mapping confirms the presence of silicon in these features. One of these features is illustrated in Figure 3.1. Other primary sedimentary components include fragments of shell (Figure 3.1D) and one possible diatom.

Pedofeatures, such as secondary iron oxides and calcite hypocoatings are associated with channel voids. Rock fragments, including biotite grains, are slightly impacted by secondary calcite precipitation. There are some possible orthic calcite nodules, or rounded domains exhibiting differences in the density of carbonate content throughout the sample (Figure 3.1C). Microspar as a recrystallization fabric is rare, but present at the base of the sample. Overall, the sample lacks the types of pedofeatures that are typically associated with subsurface calcic horizon formation. For example, pendants and calcite laminations, such as interbedded micrite and sparite that are typical of calcretes (Durand and others 2018) are absent.

The weak sorting of the coarse components of the sample suggests that primary depositional fabrics are present in Stratum 1 in this locus. The micromorphological observations help to address the question of whether this carbonate could be related to Big Red soil formation (pedogenesis), groundwater, or a wetland system. The morphology of the carbonate as viewed in the field, in particular the presence of laminar horizons, initially suggested a pedogenic rather than groundwater origin. At microscale, carbonate fabrics are dominantly vegetal or biogenic in appearance. In addition, the presence of shell fragments and lenses of plant material replaced by silica and carbonate suggest that at least some of the biogenic materials accumulated on the ground surface and that the local environment was wet and biologically active at the time of deposition. Silicified plant fragments, which form in a variety of settings (Gutiérrez-Castorena and Effland 2010), have been documented in other micromorphological studies of wetland deposits located in proximity to volcanic bedrock, which provides the source of silica (Liutkus and Ashley 2003). The presence of silicified plants in discrete lenses within this sample may indicate both primary depositional plant accumulations and periodic shifts in the water chemistry to a pH that would favor silica preservation over carbonate. Based on these observations, the primary depositional components of this sample could indicate a localized and dynamic wetland during the time of alluvial fan activity. The carbonate itself is secondary and could form as a result of groundwater.

Locus 3

In Locus 3 (Figure 1.6), two samples were collected. Sample FDM-3-18 targets a gradation in color of the Big Red soil that was observed in the field with red (5YR 4/4) present at the base, and pale green (2.5Y 7/2) present at the top. The block spans the contact between the two colors. The sample is composed of densely packed, poorly sorted sand (Figure 3.2). The material is massive and lacks any structural development. The color observed in the field is visible in both the indurated chip and the thin section, with strong red at the base, and greenish color at the top. The high magnification analyses reveal that these two portions of the sample have identical texture, structure, and mineralogy. The dramatic difference in color is due only to the color of the rare concentrations of clays bridging the spaces between sand grains (Figure 3.2B). Elemental maps reveal that the clay in both zones of the sample contains iron. The color difference is therefore likely a result of gleying of the upper portion of the deposit. Overall, the sample is interpreted as a massive alluvial sand due to the dense packing of the grains. Soil development here is very weak and is expressed mostly as clay bridges between sand grains. The color differences observed in the field are likely a result of post-depositional gleying due to a local perched water table.

Sample FDM-3-19 was collected from a different expression of the Big Red soil in Locus 3, which is higher in the stratigraphy compared to sample FDM-3-18. Here, the soil contains more fine material with a fine fraction composed of noncalcareous clay. The calcareous features in this sample (Table 3.1, Figure 3.2C) indicate that it could represent development of a calcic horizon (Bk) in Big Red. The rhizoliths could also form due to plant root interaction with an alkaline perched groundwater table that at greater depth could have also contributed to the gleying observed in sample FDM-3-18. Other aspects of soil formation expressed here is the development of granostriated birefringence fabric (*b*-fabric).

Locus 5

One small block (FDM-5-21) was collected without documented orientation in order to make observations about typical expression of the Big Red soil in Locus 5 (Figure 1.6). Thick, laminated clay coatings (Figure 3.2D) are present around coarse mineral grains, between grains, and within voids. The coatings are fragmented and the aligned *b*-fabrics are locally disrupted. This sample is interpreted as a weathered (exposed) Bt horizon of a soil that is presently undergoing chemical and mechanical break-down, as evidenced by fragmentation of clay coatings and disruption of laminations. The iron staining along voids, as well as in matrix impregnations may be related to weathering of the clays.

Trends in Big Red

The three samples collected from Stratum 1 and the Big Red soil illustrate that there is a range in the color, texture, and degree of soil development within the Big Red. The samples likely come from different horizons of the soil (FDM 3-18 = C horizon, FDM-3-19 = Bk horizon, FDM-5-21 = Bt horizon). Of the three samples, the most advanced soil formation is observed in sample FDM-5-21 and is evidenced by the presence of very thick clay coatings. In Locus 3, the soil is impacted by groundwater, which contained anaerobic zones that caused localized gleying

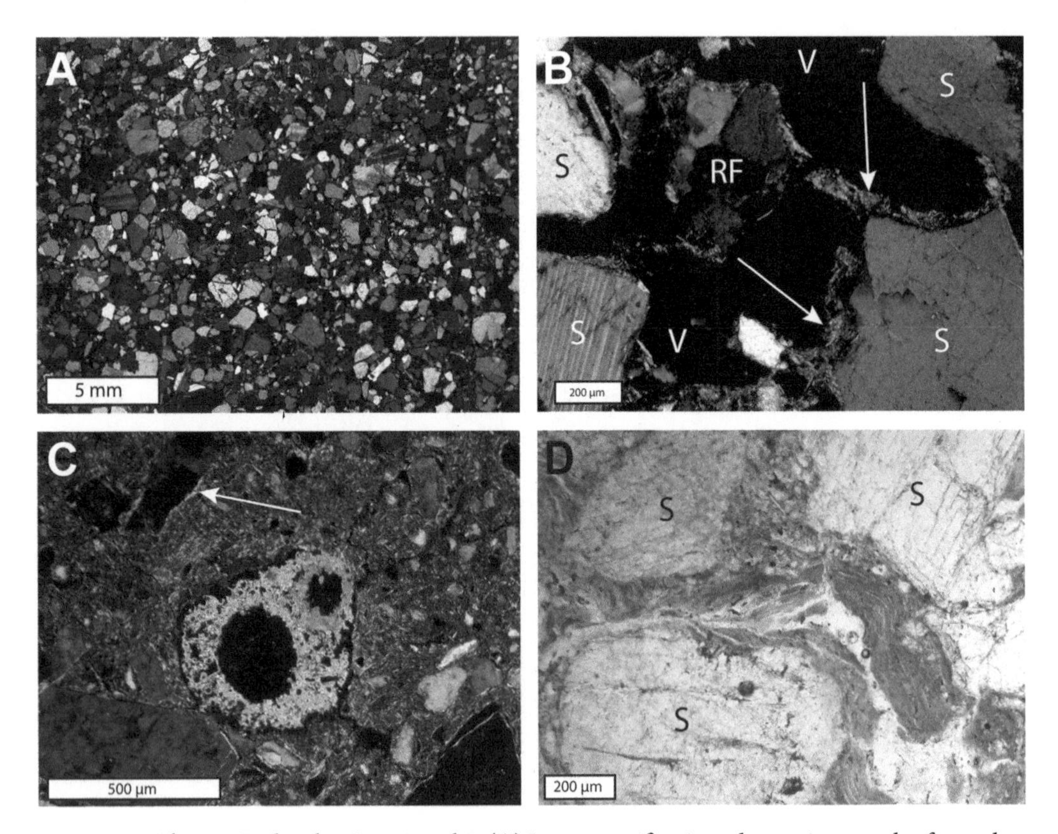

Figure 3.2. The Big Red soil in Loci 3 and 5. (A) Low magnification photomicrograph of sample FDM-3-18 showing the massive structure composed of packed sand grains. XPL. (B) At high magnification, thin coatings of clay (arrow) infill packing voids (v) and form bridges between sand grains (s) and rock fragments (RF). The macroscopic differences in color are due only to the color of these thin concentrations of clay. XPL. (C) Calcic pedofeature within a void in sample FDM-3-18. The surrounding sediment exhibits a porphyric related distribution, and alignment of clays along the edges of coarse sand grains (arrow). XPL. (D) Thick, laminated clay coatings between sand grains (s) are present in the Big Red soil in Locus 5. PPL.

of the soil, while more aerated zones may have contributed to formation of rhizoliths in plant root channels.

In Locus 4, carbonate beneath the Big Red soil was initially hypothesized to be a K horizon of that soil, formed on bedded alluvial sands of Stratum 1. The presence of both secondary carbonate and secondary silica in this sample, as well as primary shell fragments and secondary mineralization of lenses of plant tissues is more consistent; however, with formation due to wetland conditions at the time of sediment deposition. This sample could therefore represent a localized wetland associated with the alluvial fan that formed Strata 1 and 2.

The Carbonates in Locus 3

Three blocks were collected from Stratum C to make observations about the three different facies of carbonate present in Locus 3 above the Big Red soil (Profile 07-5; Figures 1.6, 2.4). All three samples are illustrated in Figure 3.3, with most of the routine description summarized in Table 3.2 and more unique aspects of the microscopic observations noted in the text.

Sample FDM-3-15 was collected from the Lower Massive carbonate. Secondary oxides, confirmed with elemental mapping as containing manganese, are present as dendritic orthic nodules (Figure 3.3A). Clusters or concentrations of an unknown secondary mineral are present throughout the sample. The mineral is yellow brown in plane-polarized light and is highly birefringent under crossed polars. Elemental mapping indicates that it contains only iron, suggesting it is a secondary iron mineral or amorphous iron compound. In addition, nodular calcite forms and masses are present, while other calcareous features of pedogenic origin such as pendants, laminar calcite, and broken mineral grains are absent. Diatoms are exceptionally rare.

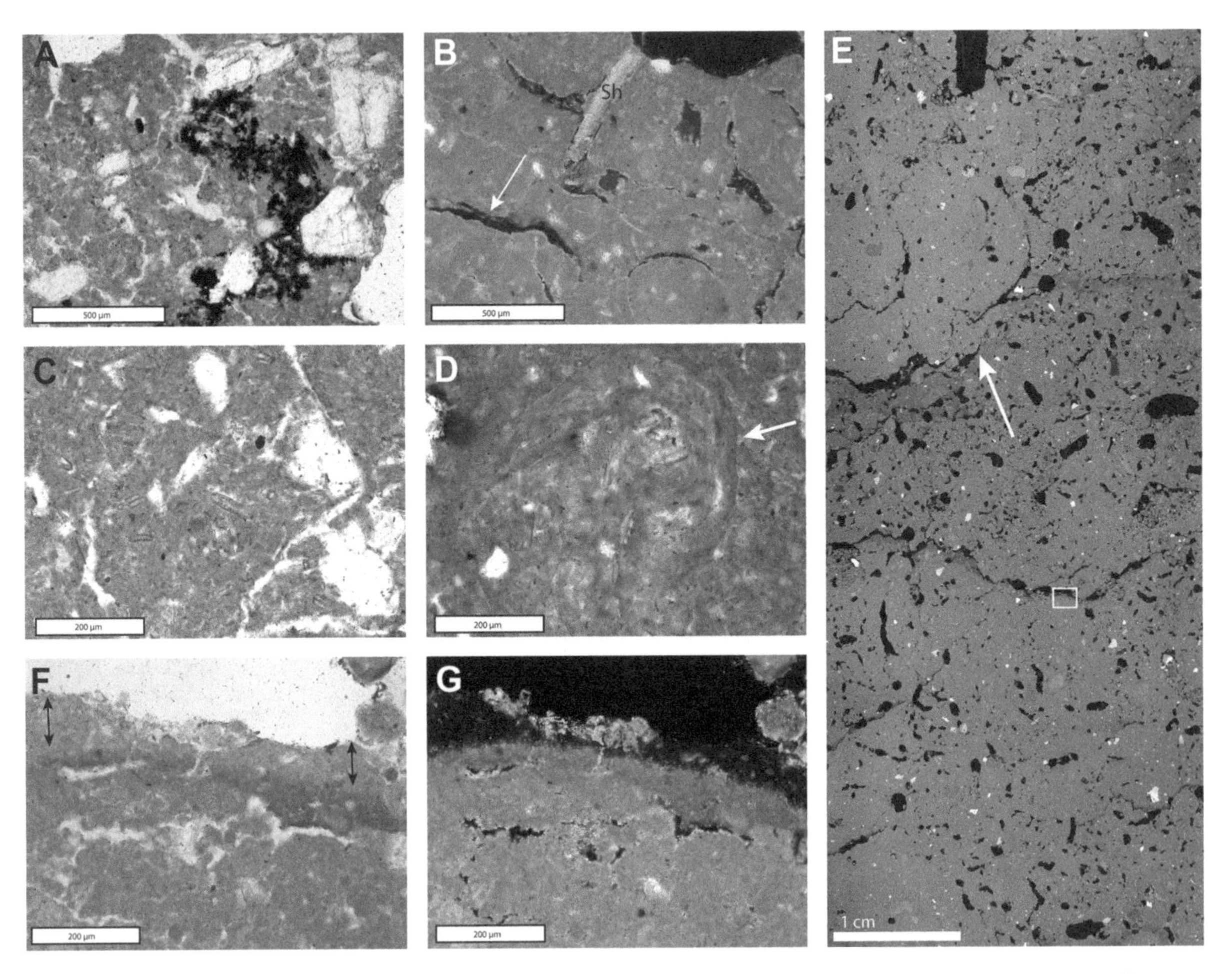

Figure 3.3. Carbonates in Locus 3. (A) Lower Massive carbonate, with dendritic secondary manganese oxides. PPL. (B) Middle Platy carbonate, with horizontal fissures (arrow) that produce a weak platy microstructure, and shell fragments (Sh). XPL. (C) Diatoms embedded in the calcareous fine fraction of the Middle Platy carbonate. PPL. (D) Diatoms in the Upper Massive carbonate, here, concentrated inside a rounded calcareous intraclast, the edge of which is marked by the arrow. PPL. (E) Multiple depositional units are visible within the Upper Massive carbonate, with the contact between the lower and upper units described in the text indicated with the arrow. The slightly lighter color of the upper unit is due to a greater abundance of carbonatc, while the lower unit exhibits greater porosity, mostly in the form of channel voids and zones of spongy microstructure. Secondary silica, located along a horizontal crack is detailed further in (f-g) with the image location indicated by the white box. (F) Secondary silica formed along a crack. The arrow indicates the silica zone. PPL. (G) Same view as (f), XPL. The silica zone is isotropic.

Sample FDM-3-16 was collected from the Middle Platy carbonate. Shell fragments are abundant within this sample (Figure 3.3B). Secondary manganese oxides are present but are less abundant compared to the Lower Massive carbonate. Only one possible large (mm-scale) calcite nodule is present. Diatoms are abundant and many appear to be intact (unfragmented; Figure 3.3C). Diatoms seem to be most abundant in the middle and upper middle zones of the thin section, although true quantification would be difficult.

Sample FDM-3-17 was collected from the Upper Massive carbonate. At a microscale, however, the sample is not massive and there are at least two units distinguished on the basis of composition and microstructure. The two units are defined by a gradual shift in microstructure within the lower unit, as well as differences in the calcareous component of the fine fraction. In the lower unit, calcareous material is more abundant at the base relative to the top. Towards the top of the unit, the fine matrix becomes darker in color, which is possibly due to an increase in clay content.

Table 3.2. Micromorphological Descriptions of Carbonates in Locus 3

Sample Number; Facies	*Voids*	*Fabric*	*Related Distribution, and Description of Coarse and Fine Components*	*Pedofeatures*
FDM 3-15; Lower Massive	Large and small channels and thin zigzag planes, which are more abundant at the base of the sample.	None, but overall fining upwards sequence within Stratum C. Calcitic crystallitic *b*-fabric.	Close porphyric with coarse fraction composed of very poorly sorted angular to subrounded sands. The sand mineralogy is similar to other samples, however fragments of biotite are less abundant. Basalt fragments are also less common and show increased evidence of weathering (rubification of the glassy matrix, secondary clays, and internal oxides). The fine matrix is calcareous and is composed of microcrystalline calcite.	Dendritic orthic nodules containing manganese oxides. An unknown secondary mineral. Calcareous nodules.
FDM 3-16; Middle Platy	Channels and planes, the latter having a dominant horizontal orientation and contributing to a weak platy microstructure. Large chambers (mm-cm diameter) are also present.	None, but overall fining upwards sequence within Stratum C. Calcitic crystallitic *b*-fabric.	Open porphyric, with the coarse fraction composed of very poorly sorted angular to subrounded sands, with mineralogy similar to other samples, but with finer grain size, and therefore fewer fragments of volcanic rock. The fine matrix is highly calcareous and is composed of microcrystalline calcite.	Chambers are infilled with loose, well-rounded aggregates of calcareous material identical to the sample matrix (but possibly slightly decalcified), interpreted as insect excrements. Possible fungal filaments or calcified root hairs and some microspar crystals are also present in voids. Manganese oxides as orthic aggregates and small dendritic nodules, and one carbonate nodule.
FDM 3-17; Upper Massive	Channels at the base, becoming most abundant at the top of the lower unit. Infilled channels at the top. Vughs throughout, but more common in the upper unit.	None, but overall fining upwards sequence within Stratum C. Calcitic crystallitic *b*-fabric.	Open porphyric, with the coarse fraction composed of very poorly sorted angular to subrounded sands, with mineralogy similar to other samples. An additional component of the coarse fraction is sand-sized calcareous intraclasts. The fine matrix is highly calcareous and is composed of microcrystalline calcite mixed with clay. The proportion of carbonate relative to clay varies vertically within the sample.	Secondary silica infilling horizontal planes within the lower unit. Channels and small chambers filled with well-rounded aggregates of calcareous material, interpreted as insect excrements.

Elemental analysis shows an enrichment in silicon at the top of the unit. The microstructure also varies from massive with channels at the base, to a more porous, spongy structure at the top with infilled chamber voids and cemented, rounded, and partially coalescing aggregates of calcareous matrix. Channel voids are also most abundant at the top of the lower unit. Two roughly horizontal cracks are present in the lower unit and extend across the thin section; both are associated with coatings of secondary silica, their composition confirmed by elemental mapping (Figure 3.3E–G). At least two other roughly horizontal but discontinuous cracks are present, with thin coatings of secondary silica on these. The upper unit is more calcareous compared to the lower unit and contains fewer channel voids. The nature of the coarse component is consistent throughout both units. Both units contain fragments of shell that appear diagenetically altered. Both units also contain inclusions of calcareous sediment in secondary position (termed "intraclasts"; see section on Stratum 5 in this Chapter), and the abundance of these is variable. Finally, both units contain diatoms; however, the distribution of diatoms within the sample is variable, with some intraclasts in the upper unit containing very abundant diatoms as inclusions (Figure 3.3D).

Vertical Trends in the Locus 3 Carbonates

Of the three sampled carbonate facies, the coarsest carbonate sample is the Lower Massive carbonate, which is in proximity to the top of Big Red soil formed on

Stratum 1. The three carbonate samples from Locus 3 exhibit fining-upward texture from the base to the top and could represent a sequence of decreasing energy of deposition of coarse material or, towards the end of the sequence, a lack of coarse inputs into a wetland system. The abundance and degree of development of secondary iron and manganese oxide features is strongest at the base of the Locus 3 sequence and decreases towards the top. This pattern could relate to the influence of the water table on the sediment. Unlike the carbonate sampled in Locus 4 beneath the Big Red soil, visible depositional fabric such as bedding or laminations is absent. The coarse material in the carbonate samples is identical in composition to the coarse material from Stratum 1 in the Big Red samples.

Biogenic components in Stratum C are diatoms and shells. Diatoms are most abundant in the upper two carbonate samples from Locus 3. The diatoms are well-preserved, especially in the Middle Platy carbonate. Lack of fragmentation may indicate primary deposition in this case. Diatoms are slightly less abundant in the Upper Massive carbonate, and their distribution is variable within the sample, associated in greatest abundance with the reworked calcareous materials.

The sample from the Lower Massive carbonate is interpreted as a groundwater carbonate, due to the sharp basal contact with the underlying Big Red soil, the presence of redoximorphic features and their degree of development within the carbonate as a whole, and the spatial proximity to gleyed zones of the Big Red soil. Following this interpretation, the gleying of the Big Red soil could be attributed to a perched water table. Unlike the Lower Massive carbonate, the Middle Platy carbonate is interpreted as forming under palustrine conditions with the carbonate impacted by a later phase of subaerial weathering and soil formation, which caused desiccation and growth of plant roots that formed channel voids. Further information on the water composition could be provided by identifying the species of shells and diatoms.

The sample from the Upper Massive carbonate is interpreted as two palustrine carbonate units with a weathered surface at their contact. The surface exhibits evidence of plant activity in the form of increased abundance of root channel voids and a multi-cm zone of increased bioturbation, evidenced by the spongy microstructure. Together, the two units and the surface between them may represent wetland conditions, with a brief period of exposure and weathering, followed by a return to higher moisture conditions. Some calcareous materials within this sample are in secondary position. These intraclasts are composed of palustrine sediment that was fragmented and redeposited. The presence of intraclasts containing diatoms suggests that some of this material is sourced from a location within the wetland of even higher biological productivity. This sample, the upper unit in particular, is most similar to Stratum 5 in Locus 1.

Locus 1, Strata 2, 3, 4 and 5.

A continuous column of sediment was collected from Locus 1 (Profile 08-1; Table 2.1; Figure 2.9) in order to provide a permanent archive of the entire sequence. From the cut blocks, six areas were selected to process into petrographic thin sections. These areas were selected in order to observe general characteristics of the strata and the contacts between them. One additional isolated block was collected from the diatomite at the base of Stratum 4 in order to make microscopic observations about the nature of laminations visible in this deposit. The samples are described and interpreted here individually in the context of the main stratigraphic units (Table 3.3), with description in the text focused on the nature of stratigraphic contacts.

Stratum 2, and the Contact with Stratum 3

Sample FDM-1-11 was collected at the contact between Strata 2 and 3, and therefore it provides an example of the very top of Stratum 2 (Figure 3.4). The contact between the two units is visible in thin section, and it is sharp. The two units have similar compositions, with low to moderate pedogenesis and bioturbation, and are distinguished primarily on the basis of texture (Stratum 3 is coarser), structure, and post-depositional features. The sharp contact between the two units supports the idea that there is an erosional unconformity between them. This is further supported by the elemental mapping, which shows that the fine fraction of Stratum 2 is enriched in iron and magnesium relative to the base of Stratum 3, which suggests that the clay component in each unit may have a different source. The coarse component is very similar to that of the Stratum 1 (Big Red) and Stratum C samples. Overall, this sample, when combined with the observations from samples from other Loci, shows that Strata 1, 2 and 3 are very similar in their composition and mechanism of deposition.

Stratum 3

Three samples span the base, middle, and top of Stratum 3 (FDM-1-11, FDM-1-8, and FDM-1-6). Sample FDM-1-8 from the middle of the unit is very similar to the base, but with a few important differences. First, rounded fragments

Table 3.3. Locus 1 Micromorphological Descriptions

Sample Number; Depth in Meters below Datum	*Stratum*	*Voids*	*Fabric; b-fabric*	*Related Distribution; Description of Textural Elements*	*Pedofeatures*	*Other*
FDM-1-11; 12.51–12.61	2	Channels and planes	No primary depositional fabric. Porostriated, cross-striated and granostriated *b*-fabrics. Granostriations are particularly pronounced around siliceous materials such as volcanic glass and diatoms.	Open porphyric; sand fraction is consistent with the samples from Stratum 1, poorly sorted, with abundant feldspar, quartz and fragments of granite, as well as fragments of basalt and other accessory minerals derived from both rock types. Noncalcareous matrix that is rich in clay.	Some voids contain calcified fungal filaments, needle fiber calcite and/or degraded remnants of organic material. Large channels contain rounded aggregates of sediment similar in composition to the matrix. Overall lack of pedofeatures associated with soil-forming processes (e.g. clay coatings of voids) or groundwater activity (e.g. iron oxides).	Diatoms throughout.
FDM-1-11; 12.41–12.51	3A	Channels, planes, and chambers	No primary depositional fabric. Strongly expressed cross-striated *b*-fabric, moderately expressed porostriated and granostriated *b*-fabrics.	Close porphyric; sand fraction is consistent with the samples from Stratum 1 and 2, poorly sorted, with abundant feldspar, quartz and fragments of granite, as well as fragments of basalt and other accessory minerals derived from both rock types. More coarse sand relative to Stratum 2. Noncalcareous matrix that is rich in clay.	Some voids contain calcified fungal filaments, needle fiber calcite and/or degraded remnants of organic material. Chambers contain rounded aggregates of sediment sourced from the surrounding matrix, or from disrupted calcareous void coatings.	Diatoms rare.
FDM-1-8; 12.05–12.15	3B	Channels, planes, and chambers	No primary depositional fabric. Strongly expressed cross-striated, porostriated and granostriated *b*-fabrics.	Close porphyric.	Some voids contain calcified fungal filaments, needle fiber calcite and/or degraded remnants of organic material. Channels rarely have thin carbonate coatings and hypocoatings, which in some cases have been partially reworked by insects into rounded aggregates. Chambers contain rounded aggregates of sediment sourced from the surrounding matrix, with less common aggregates of calcareous sediment similar to that of Stratum 5, in one case exhibiting internal iron oxide staining.	Diatoms, bone fragments present.
FDM-1-6b; 11.78–11.85	3B/4d	Channels, planes, and chambers	Locally laminated, with one thin lens of strongly aligned diatoms. Fining upwards towards the contact with Stratum 4de.	Close porphyric.	Chamber voids contain rounded aggregates of calcareous sediment similar to that of Stratum 5.	Diatoms increasingly abundant in the upper cm of the stratum; also concentrated in a lens with clay. Bone fragments present.
FDM-1-6a & b; 11.71–11.78	4d/de	Small channels	Strongly laminated within the diatomite, including the presence of thin lenses of fine sand and silt, and horizontal alignment of diatoms. Locally laminated within the diatomaceous earth.	Monic to very close porphyric; sediment composed of diatoms with clay in between.	One large tubular insect burrow is infilled with sediment sourced from Stratum 3 (passage feature, see Figure 6b), with multiple phases of infilling. Smaller channel voids post-date the infilled burrow. A few smaller channels and chambers also contain sediment from Stratum 3. Larger channels contain organic material and/or root hair and fungal filaments which appear to be composed of calcium oxalates, in some places overprinted by calcium carbonate.	Abundant diatoms, organic material becoming more abundant in the diatomaceous earth. Phytoliths in diatomaceous earth.

Table 3.3. (continued)

Sample Number; Depth in Meters below Datum	*Stratum*	*Voids*	*Fabric; b-fabric*	*Related Distribution; Description of Textural Elements*	*Pedofeatures*	*Other*
FDM-1-14; 11.71–11.78	4d/de	Chambers and channels	Laminated within the diatomite, locally laminated within the diatomaceous earth.	Monic; sediment composed of diatoms.	Smaller channel voids contain dense infillings of root hairs or fungal filaments. Some channels contain degraded organic material likely sourced from roots.	Diatoms, organic material, phytoliths.
FDM-1-5; 11.61–11.715	4de	Vertical channels and chambers	Gray zones (lenses) of strongly horizontally aligned diatoms, darker zones with chaotic orientations.	Monic; sediment composed of diatoms and phytoliths. Very rare inclusions of poorly sorted sand.	Vertical channels and chambers contain infillings composed of abundant rounded aggregates of sediment similar to Stratum 5, cemented by fungal filaments or root hairs.	Diatoms present as concentrated lenses (gray zones), or mixed with organic material and phytoliths (brown zones).
FDM-1-4; 11.485–11.61	4de	Vertical channels and chambers	Gray zones (lenses) of strongly horizontally aligned diatoms, darker zones with chaotic orientations.	Monic; sediment composed of diatoms and phytoliths. Very rare inclusions of poorly sorted sand.	Vertical channels and chambers contain infillings composed of abundant rounded aggregates of sediment similar to Stratum 5, cemented by fungal filaments or root hairs. Rare degraded roots in some channels.	Diatoms present as concentrated lenses (gray zones), or mixed with organic material and phytoliths (brown zones).
FDM-1-2; 11.15–11.32	5	Channels and chambers, some thin planes	Calcareous intraclasts have internal vegetal fabric and alignment of components. Calcitic crystallitic *b*-fabric	Porphyric; sand inclusions are exceptionally rare but are present. The fine fraction is composed of silt-sized fragments of volcanic glass, diatoms, phytoliths, clay minerals, and microcrystalline calcite. The coarse fraction is dominated by calcareous intraclasts, and fragments of shell.	Vertical channels and some planes contain thin carbonate coatings and calcified fungal filaments and root hairs. Chambers are infilled with aggregates of sediment that is similar in composition to the matrix, with the packing voids between these infilled by calcified fungal filaments. A few channels contain degraded organic material, or rounded aggregates of organic material.	Diatoms and phytoliths in both the matrix and intraclasts. Shell fragments.
FDM-1-13; 11.15–11.24	5	Chambers, channels, and planes	Calcareous intraclasts have internal vegetal fabric and alignment of components. Calcitic crystallitic *b*-fabric.	Porphyric. The fine fraction is composed of silt-sized fragments of volcanic glass, diatoms, phytoliths, clay minerals, and microcrystalline calcite. The coarse fraction is dominated by calcareous intraclasts, and fragments of shell.	Planes contain thin carbonate coatings and calcified fungal filaments and root hairs. Chambers are infilled with aggregates of sediment that is similar in composition to the matrix, with the packing voids between these infilled by calcified fungal filaments.	Diatoms and phytoliths in both the matrix and intraclasts. Shell fragments.

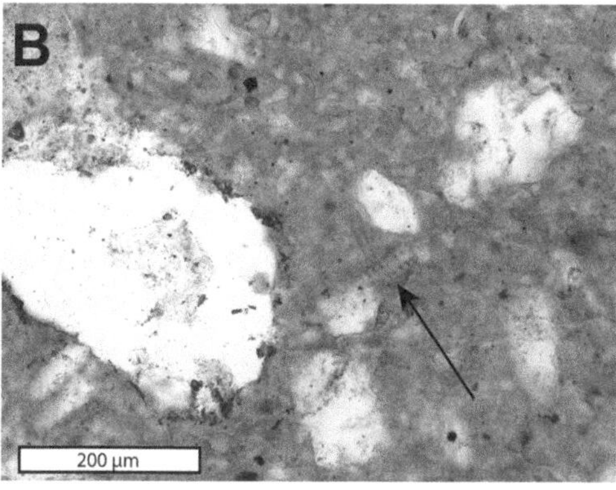

Figure 3.4. Stratum 2 in Locus 1. (A). Fine fraction of the stratum, showing a channel void with poro-striated birefringence fabric in the matrix around it. XPL. (B) Diatoms (arrow) and other siliceous materials in the fine matrix of Stratum 2. PPL.

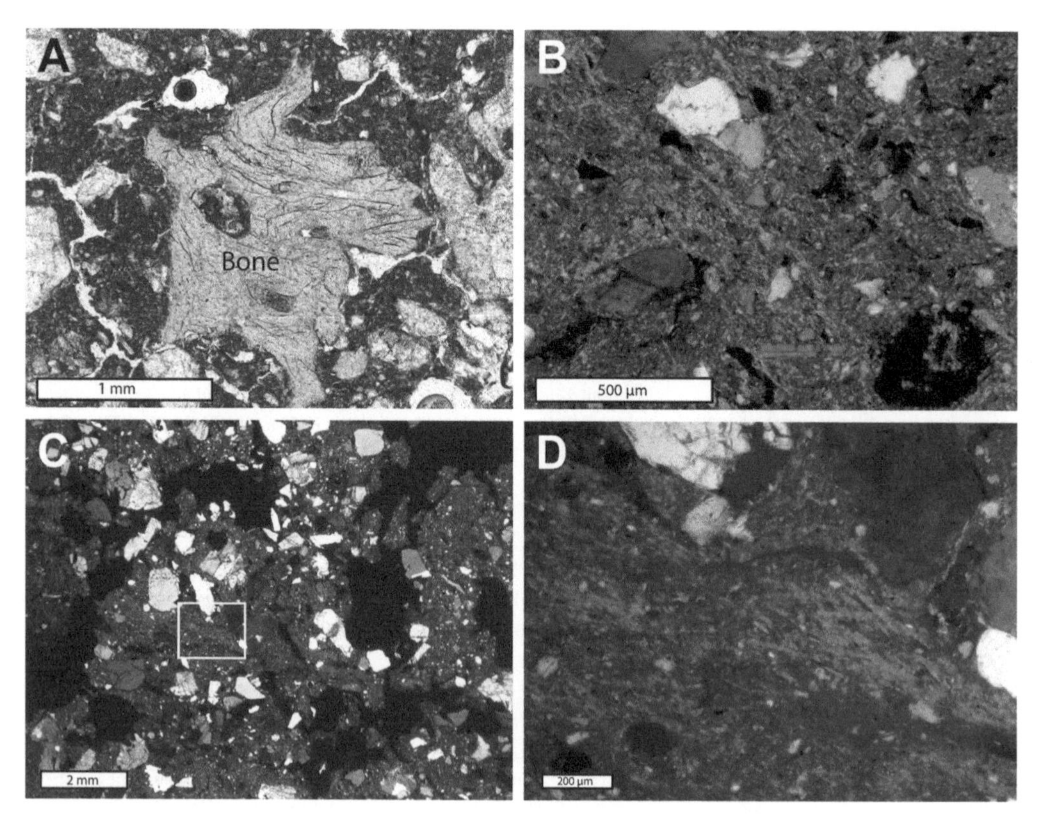

Figure 3.5. Stratum 3 in Locus 1. (A) Bone fragment in the middle portion of Stratum 3. PPL. (B) Well-developed cross-striated birefringence fabric in the middle portion of Stratum 3. XPL. (C). Upper Stratum 3, showing the overall poorly sorted sand in a matrix of clay and silt. The box indicates the area illustrated in (D). XPL.(D). A lens of bedded and laminated clay and diatoms is present just below the contact with Stratum 4. The diatoms are not visible in this image. XPL.

of bone are present (Figure 3.5A). The bones are not burned. Second, some of the rounded aggregates in chamber voids contain material sourced from a calcareous deposit. Third, diatoms are slightly more abundant. Overall, the middle portion of Stratum 3 is an alluvial sediment with primary inputs of bone and diatoms, the latter indicating higher moisture relative to sediment beneath and possibly brief periods of standing water. Weak to moderate pedogenesis is indicated by the *b*-fabric (Figure 3.5B), which is also more developed relative to the base of the stratum.

The Contact between Stratum 3 and Stratum 4

The base of sample FDM-1-6 spans the contact between Strata 3 and 4 (Figure 3.6). The upper part of Stratum 3 is similar in composition and texture to the samples from lower elevations within the sequence. Bone fragments are present in the sand fraction. A thin lens of finer sediment containing horizontally-aligned diatoms and clay is located approximately two cm below the contact with Stratum 4. The contact between the two stratigraphic units is sharp when viewed in the field, and at microscale, there is visually an abrupt shift in composition, based mainly on the appearance of diatoms, but a gradational shift in texture. The upper few millimeters of Stratum 3 contain finer sediment than the majority of the stratum, with a shift towards more abundant fine sand and silt-sized particles. Just above this shift, the nature of the matrix changes abruptly to sediment dominated by diatoms, with inclusions of silty material. The elemental mapping also reveals that the clay minerals, which are rich in aluminum and iron, comprise much of the matrix of upper Stratum 3, and are also present in the basal cm of Stratum 4 (Figure 3.6B). Just above the contact, Stratum 4 also contains inclusions of thin horizontal lenses of sand and silt, and mm-scale laminations are visible as fluctuations in the texture of the inclusions. Lenses of coarser and finer sediment are present. Porosity decreases rapidly across the contact as well. Stratum 3 contains large channels and cracks, as well as infilled

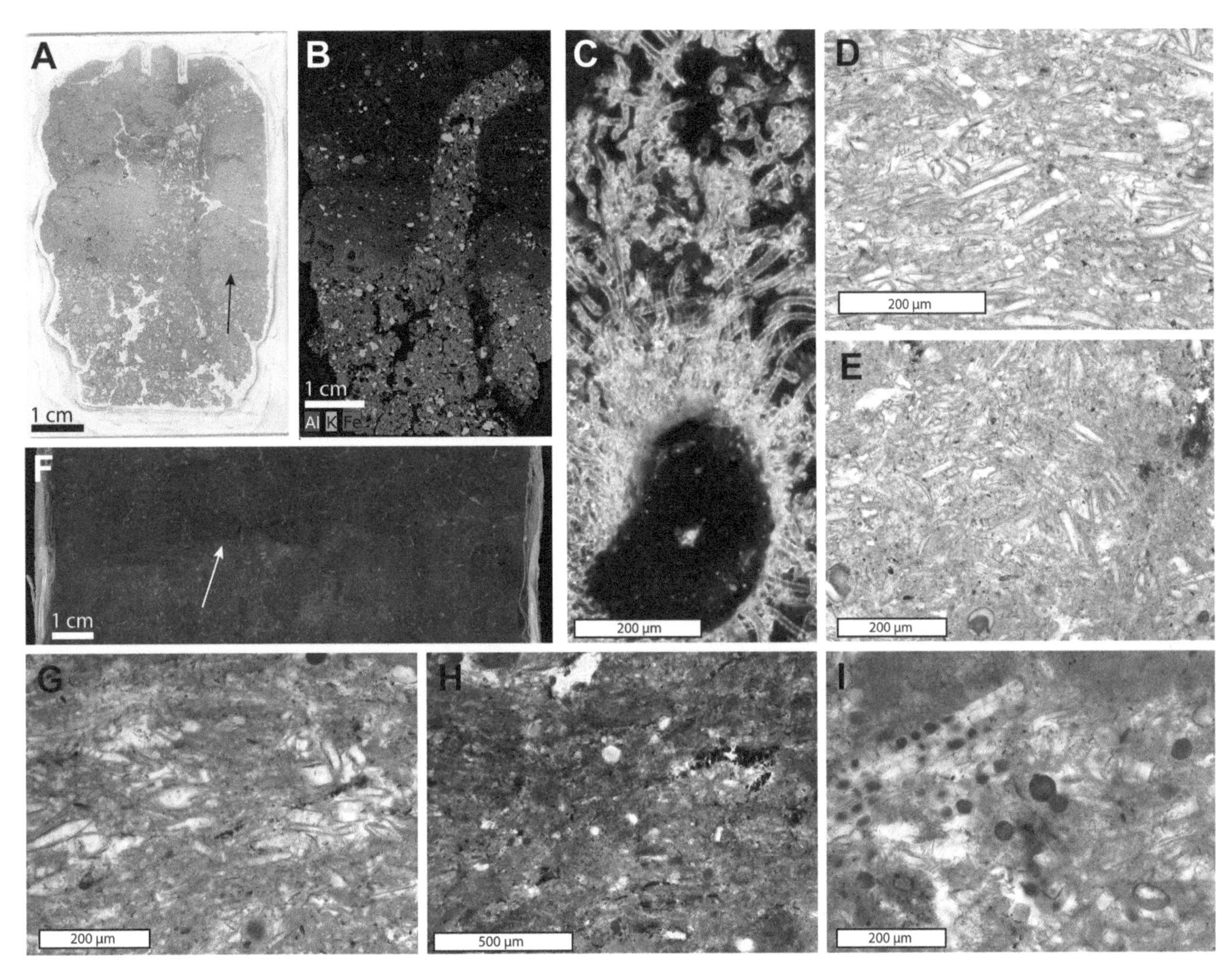

Figure 3.6. The base of Stratum 4. (A) Incident light scan of the thin section FDM-1-6b that spans the contact between Stratum 3 and Stratum 4. The contact is indicated by the arrow. (B) µXRF elemental map of the thin section in (A), showing the distribution of Al, K, and Fe, which are elements that are present in the geogenic fraction, in particular the feldspar, volcanic glass, and clay minerals. Al and Fe, when present together, produce bright pink regions illustrating areas rich in clay minerals, which are present in Stratum 3 as well as the very base of Stratum 4. Where clay minerals are absent at the top of the thin section, the fine fraction is composed only of diatoms. (C) A cropped image of fungal hyphae in a void within Stratum 4. The interference colors indicate that the hyphae are variably composed of calcium oxalates and calcium carbonate, here dominated by calcium oxalates. This composition is widely reported in fungal features in calcareous soils and limestones (see Burford, Hillier, and Gadd, 2006), but the timing of fungal activity relative to deposition of Stratum 4 is unknown. XPL. (D) Strongly laminated diatoms just above the contact. PPL.
(E) Unoriented diatoms from the top of the thin section in (A). PPL. (F) The contact between the diatomite at the base of Stratum 4 and the diatomaceous earth above it is visible in the cut face of block FDM-1-14 (arrow) but is not visible in thin section. Incident light scan of the wet surface of the block. (G) Sample FDM-1-14, which was collected several meters away in an area where dark laminae were visible in the field at the base of Stratum 4. Here, the laminated diatoms at the base are interbedded with organic material. PPL. (H) The top of sample FDM-1-14 contains more abundant organic material as well as phytoliths, in addition to diatoms. PPL. (I) Articulated phytoliths, which indicate that the plant fragments were not significantly disturbed following deposition. PPL.

chambers, while Stratum 4 contains small channels. Larger channels in Stratum 4 contain organic material and/or root hair and fungal filaments that appear to be composed of calcium oxalates (Figure 3.6C), in some places overprinted by calcium carbonate. The diatoms at the base of Stratum 4 just above the contact are very strongly aligned and laminated (Figure 3.6D). The very top of the sediment visible in thin section lacks alignment (Figure 3.6E) but has a similar composition.

The center of the sample is a large, tubular burrow containing material from Stratum 3 that worked upwards into Stratum 4 (visible in Figure 3.6B). The burrow infilling also contains multiple phases of sedimentation, with later burrows cross-cutting it with their own filling and a final phase of cross-cutting by new channels containing roots. Shell fragments are absent in both strata. Rare sand-sized fragments of bone are present at the top of Stratum 3. Bone is not present in Stratum 4.

This sample represents a rather abrupt shift in the nature of sediment deposition from an alluvial setting to standing water. The uppermost alluvial sediment contains bone inclusions and exhibits a shift in grain size towards finer particles, which suggests that the final phase of alluvial activity was lower in energy. A thin lens of diatoms and laminated clay (Figure 3.5D) may suggest a brief phase of standing water. As in the field observations, there were no microscale observations of weathering at the contact between the two units. The visible contact in the field is marked in thin section by an abrupt increase in the abundance of diatoms, but the presence of clay in the matrix of the lowest cm of the stratum indicates that clays present in the water column likely settled at the same time that biological productivity rapidly increased. The diatomite at the base of Stratum 4 is strongly laminated and contains primary depositional fabrics, with the exception of areas that have been disrupted by bioturbation. This bioturbation has also moved older coarse material upwards into the diatomite. There is no evidence of downward movement of material across the contact in the micromorphological sample.

Stratum 4

Two samples were collected from the contact between the diatomite in Stratum 4 (4d) and the overlying diatomaceous earth (4de). The upper part of sample FDM-1-6 (thin section FDM-1-6a) is from the main sampling column (Profile 08-1), while FDM-1-14 was collected from several meters away. The 4d/de contact is visible in the sliced faces of both blocks, where it is sharp, horizontal and marked by a difference in color (Figure 3.6F). However, no obvious changes in sediment are present at the contact in either thin section. Like the diatomite observed in sample FDM-1-6b (see above), the sediment in FDM-1-14 contains several thin lenses of coarser sandy material in a matrix of strongly laminated diatoms (Figure 3.6F). There is also an irregular concentration of clay mixed with phytoliths, marked in the elemental maps as an enrichment in potassium, aluminum, magnesium, and iron (not pictured). Both FDM-1-14 and FDM-1-6a are exceptionally fine in texture and are dominated by diatoms. Other organic material is present interbedded with the diatoms. Moving upwards within the unit, the composition of the siliceous fraction changes. Diatoms become less dense and are mixed with bulliform phytoliths and organic material (Figure 3.6G and H), the latter of which is present in thin lenses.

Two samples were collected from the middle of Stratum 4de, with FDM-1-5 lower in the sequence than FDM-1-4. Both samples are similar in composition but differ slightly in fabric and development of secondary features (voids, void infillings, and microstructure). Compared to samples collected at lower elevation in Stratum 4, the fabric in these samples is more chaotic and looks bioturbated, with zones that appear in hand sample to be lighter gray in color that contain more diatoms with original depositional fabric, and zones appearing more brown in color that contain more organic material, clay, and phytoliths with particles exhibiting random orientation. Organic material and phytoliths also visually appear to be more abundant in FDM-1-4 compared to FDM-1-5. Elemental maps of the samples also revealed that iron is more abundant in FDM-1-4 compared to FDM-1-5.

In the samples from Stratum 4, the basal pure diatomite with primary depositional fabric (4d) is overlain by a more mixed deposit (4de). The mixed deposit is associated with plant phytoliths, possibly indicating the appearance of vegetation within the wetland system in this location. Organic material in the fine fraction also increases in association with phytoliths and sediment mixing that obliterated much of the primary depositional fabric. The upper sediment within these samples may be indicative of shallower water, more plant activity, and more syndepositional mesofaunal activity compared with the basal diatomite. Although the basal diatomite is lacustrine in origin, the overlying diatomaceous earth is interpreted as palustrine. Moving upwards, the mixed biogenic deposit continues to contain abundant diatoms, phytoliths, and organic material. Syndepositional and post-depositional bioturbation features become even more abundant. The sediment composition

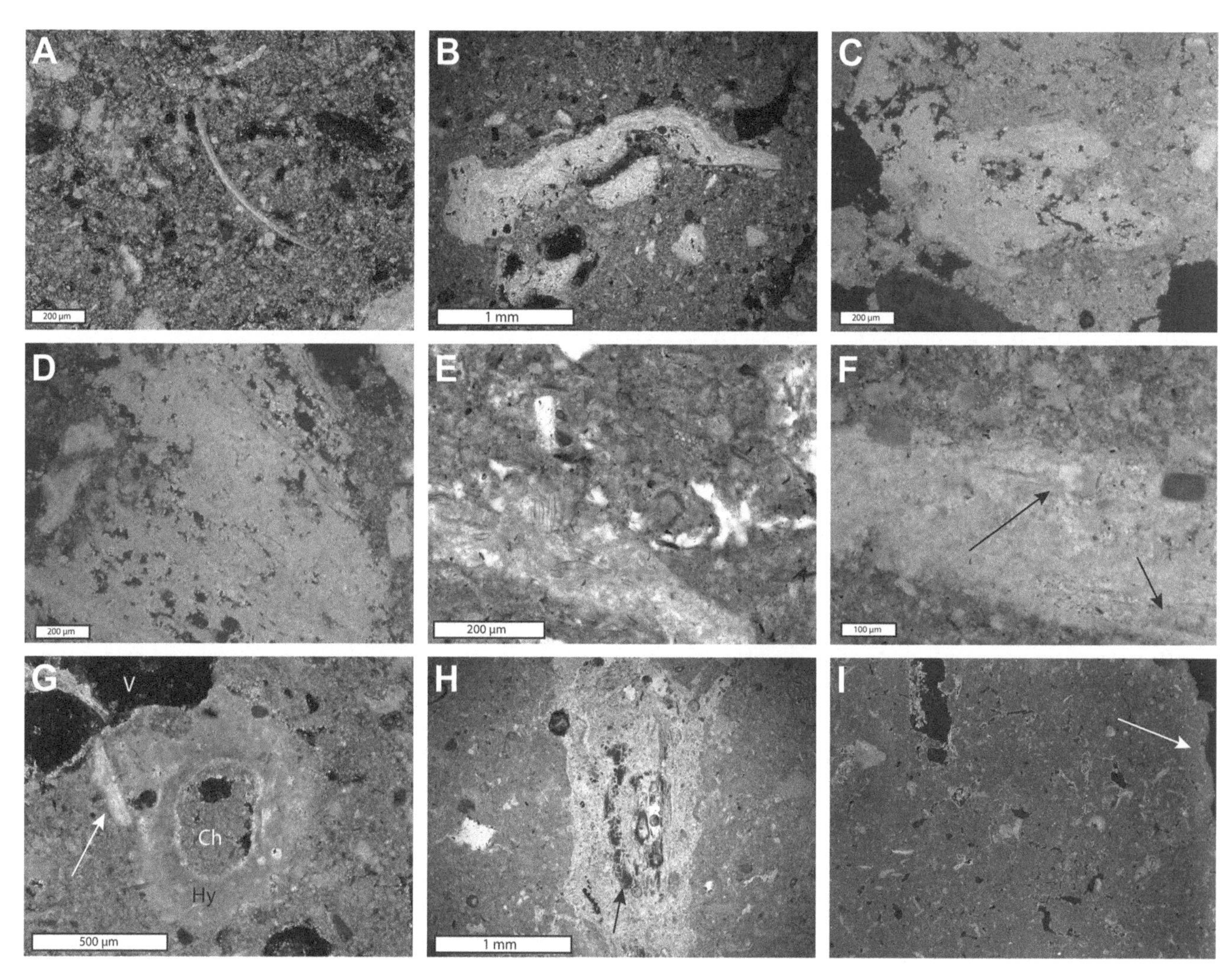

Figure 3.7. Stratum 5. (A) A fragment of shell surrounded by the fine sedimentary matrix, which is rich in carbonate. XPL. (B) A concentration of calcareous intraclasts showing the range in morphology and size. XPL. (C) A calcareous intraclast from the Upper Massive carbonate, for comparison. The lighter color of the surrounding matrix is due to a greater abundance of carbonate. XPL. (D) Another intraclast from Stratum 5 showing an internal vegetal fabric and void system that has been truncated along the intraclast edges. XPL. (E) An intraclast (lower corner) that is rich in intact diatoms, while the surrounding sedimentary matrix is rich in phytoliths. PPL. (F) An intraclast that is internally laminated, with two zones of articulated phytoliths in their original anatomical positions (arrows). Half polars. (G) A channel void (Ch) with a calcareous hypocoating (Hy) and later infilling by sediment identical to the surrounding matrix. A small intraclast is indicated by the arrow. A chamber void (v) formed after the coating of the channel. Calcified fungal hyphae are visible within the chamber void. XPL. (H) A large channel void entirely filled with partially decomposed organic material (arrow) surrounded by fungal hyphae. The morphologies of the organic material indicates that it is part of a plant root. PPL. (I) A sample collected from the edge of the Locus 1 outcrop where Stratum 5 is exceptionally hard and resistant to weathering. Intraclasts are visible as light areas showing roughly horizontal orientation. The arrow indicates the outer crust, which is enriched in Si.

and fabric is nevertheless still indicative of a wetland, with standing water, plants, and faunal activity.

One aspect of the Stratum 4 deposits that should be emphasized is the presence of inclusions of sediment sourced from elsewhere within the wetland system. Towards the base of the sequence, a concentration of clay mixed with phytoliths may be a redeposited clast of sediment sourced from a more clay-rich and plant-rich portion of the wetland. Within the upper samples, vertical voids and burrows infilled with pelletized carbonates indicate either downward movement of material from Stratum 5 or movement of material by insects from nearby calcareous deposits. Pelletized carbonates in voids are a common feature of carbonates formed in palustrine settings (Alonso-Zarza and Wright 2010), although in this stratum, the matrix sediment is siliceous rather than calcareous.

Stratum 5

Two samples are from Stratum 5 (Figure 3.7). Both are very fine in texture with a calcareous matrix. Stratum 5 is characterized by a dramatic increase in carbonate compared to Stratum 4 (Table 2.2; Figure 2.10) with carbonate abundance as high as 59 percent. Micromorphological analyses indicate that carbonate comprises much of the fine fraction, as well as the coarse fraction. The coarse fraction is composed of fragments of shell (Figure 3.7A) and calcareous inclusions (Figure 3.7B–F). These inclusions are irregular in shape, have internal vegetal fabrics, and are composed of microcrystalline carbonate that is purer compared to the surrounding matrix. The inclusions also contain phytoliths and diatoms, sometimes in higher concentrations than the surrounding matrix. The inclusions are mostly unrelated to the existing void network and lack obvious rhizolith morphologies. The size varies from sand to gravel-size, their edges are rounded, and in some cases the internal vegetal fabric is truncated by the fragment edge. Several are bisected or fragmented by channel voids, which extend into the matrix of Stratum 5. The orientations of the inclusions are random but tend toward horizontal and inclined rather than vertical. Some of the phytolith morphologies in these samples are different from those in lower portions of Stratum 4.

Sample FDM-1-13 was collected from the top of Stratum 5 so as to target the outer edge of the outcrop and observe the nature of the crust that appears to protect the Locus 1 island from weathering. The sediment is very similar in composition and fabric to Stratum 5 in sample FDM-1-2; however, the matrix is darker in color and likely contains less calcareous material, an observation that appears to be reflected in the percent of carbonate measured in the loose sediment (Table 2.2). The crust on the edge of the block is visible in this sample as a 2-mm zone of lighter brown color. Under the microscope it appears to have a higher density than the internal sediment. Elemental mapping shows an enrichment in silicon.

Stratum 5 marks a shift from a siliceous biogenic sediment of 4de to a calcareous biogenic sediment (Figure 2.10). The calcareous fine matrix, which is similar to that of a marl, encloses clasts of calcareous material that appear reworked from elsewhere within the wetland system. These calcareous inclusions (termed intraclasts in the carbonate literature) are not rhizoliths, although some do contain phytoliths. Pisoids—coated particles that are similar to intraclasts—have been documented in micromorphological samples from alkaline spring environments (Mallol, Mentzer, and Wrinn 2009). Intraclasts can also form in lacustrine settings in zones exposed to periodic lake level lowering and shoreline processes (Alonso-Zarza and Wright 2010). The inclusions are therefore likely a different facies of paludal carbonate that have become partially lithified, fragmented, and locally reworked, perhaps under higher energy conditions or flowing water. Relative to the underlying Stratum 4, sediment derived from plants and diatoms is still present, although different phytolith morphologies were observed. Finally, the presence of shells in this portion of the sequence makes Stratum 5 most similar to the Upper Massive carbonate sampled in Locus 3.

The very top of Stratum 5 in the Locus 1 sequence exhibits more developed bioturbation features, which suggests that the top of the stratum was exposed to subaerial conditions. The outer crust that protects the locality from weathering is calcareous (Figure 3.7I), but elemental mapping reveals no significant enrichment in calcium compared to the matrix. Instead, the very outer microns of the crust are slightly enriched in silicon. This observation suggests that the crust is actually a lithified zone resulting from recrystallization of the matrix and it may also contain secondary silica.

Vertical Trends in the Locus 1 Sequence

The overall sequence is a transition from alluvial fan deposits with low to moderate pedogenesis, to a pond or lake deposit, to a shallower, more frequently exposed wetland containing plants. The final phase in the sequence includes a shift in the geochemistry of the wetland from an environment that favored silica deposition and preservation to one that favored carbonate deposition.

Strata 2 and 3 exhibit features consistent with an alluvial origin. The presence of diatoms within these, and in variable abundance, may indicate that this area experienced high moisture conditions and perhaps periodic standing water. Localized erosion then incorporated the diatoms and fragments of organic material into the alluvial deposits. Post-depositional features in these two strata are variable in expression. In situ weathering and development of *b*-fabric is more advanced in Stratum 2 compared to Stratum 3, which may indicate that a weak soil had formed on Stratum 2 prior to the channel erosion that scoured out its surface in Locus 1. In contrast, bioturbation by insects is more advanced in Stratum 3 with more abundant chambers infilled with sediment aggregates. Overall porosity is also higher in Stratum 3.

Stratum 4 marks the appearance of standing water in this location. Variability in this unit indicates a dynamic environment. The basal diatomite preserves original depositional fabrics, while the overlying diatomaceous earth is a mixed deposit of diatoms, phytoliths, and organic material. Primary depositional fabrics are generally absent, but localized lenses of pure diatoms and aligned organic material are present, suggesting fluctuations in the depositional conditions and degree of bioturbation. Plants were more abundant during the formation of this part of the sequence as evidenced by phytoliths, organic material, and plant root channels. The microscopic analyses indicate that the change in color within Stratum 4 observed in the field is due to a gradual shift in the relative abundance of diatoms, phytoliths, and included organic material.

The micromorphological thin sections that were produced for this study did not capture the contact between Strata 4 and 5, and these two strata were only recognized as separate entities following the initial observation of primary carbonates in the thin section from sample FDM-1-13 along with determination of calcium carbonate content (Table 2.2, Figure 2.10). At the time of sample collection, the carbonate at the top of the sequence was thought to be a post-depositional alteration of Stratum 4. Sample FDM-1-2 was later processed into a thin section in order to make further observations about the nature of Stratum 5. Stratum 5 is composed of primary carbonate present in a fine, marl-like matrix. Coarse calcareous material within this deposit is not in primary position but is fragmented and has been locally reworked within the wetland system. The composition and internal fabric of this unit is similar to that of portions of the Upper Massive carbonate in Locus 3.

Carbonate elsewhere within the sequence is secondary. In Strata 2 and 3, carbonate is present only in voids as a later phase of void infilling—either as a replacement phase in the weathering of organic material (see below) or as material moved laterally or downward within the entire sequence as a result of insect activity. Within Stratum 4, the measurements of calcium carbonate abundance conducted on loose samples show a very gradual increase in the percent carbonate (see Table 2.2), and the micromorphological analyses indicate that this reflects an increased abundance of infilled bioturbation channels and chambers in the upper part of the unit rather than a gradual shift in the composition of the primary sediment.

The voids within the Locus 1 sequence (and other loci) contain fungal filaments (Figure 3.6C). These filaments are in some places very dense, with tubular morphologies and diameters that range from 0.01 to 0.02 mm. Cross-sections captured in thin section indicate that their centers are hollow, while variable low-order and high-order interference colors indicate both calcium oxalates and calcite, with elemental mapping confirming the presence of calcium. The filaments are associated with degraded (recent) organic matter and insect fecal pellets. These features likely have an origin related to a late phase of plant root penetration into the sequence, with fungal contribution during decomposition, as evidenced by association between the filaments and needle fiber calcite in a few samples. These features are the only remaining indicator of what happened to the Locus 1 sequence after the termination of the wetland system. Locus 1 may have been buried by a deposit similar or laterally equivalent to Stratum 6, with plant root activity extending downwards into the underlying wetland and alluvial deposits. This overlying deposit was then removed, leaving behind the calcareous caprock-like sediment of Stratum 5, which, owing to secondary silicification, was less susceptible to erosion.

CONCLUSIONS

The micromorphological analyses of the major stratigraphic units, as well as the Big Red soil and associated carbonates has contributed to a greater understanding of the local environment at the site from the Pleistocene into the Holocene. Micromorphology aided in determining that none of the carbonates at the site had a purely pedogenic origin, instead there are gradations between groundwater, palustrine carbonates, and possibly lacustrine carbonates. This gradation is often observed horizontally and vertically in wetland environments. None of the carbonates exhibited characteristics typical of a tufa associated with a spring outlet, which supports the field observations that although ancient springs may have provided a source of water, their locations are not visible on the landscape today. Notably, diatoms and microscopic shell fragments were documented in many of the strata and carbonates, including in the alluvial deposits of Strata 2 and 3. These observations, along with the presence of a carbonate containing silicified plants beneath the Big Red soil suggests that localized wetlands may have been a constant feature of the landscape for millennia.

Few observations from the micromorphological samples have direct implications for understanding the human and megafaunal activities at the site. However, the observations of the contact between Strata 3 and 4—where the gomphothere bones and artifacts are located—shows that bioturbation moved sediment from Stratum 3 upward into Stratum 4. At a much later date, vertical root systems introduced rounded aggregates of material from Stratum 5. In addition, the observation of microscopic plant tissues in Stratum 4 provides information about the nature of the organic material that was dated as "bulk organic" radiocarbon samples. Much of the observed organic material is microscopic, intact plant tissues deposited in original position. Unfortunately, no fragments of plant material or charcoal were observed in the micromorphological sample from the very top of Stratum 3, and therefore these analyses can shed no light on the nature of the radiocarbon dating samples from that context.

The complete column of resin-indurated sediment is housed at the University of Arizona in order to provide a permanent archive of the stratigraphic sequence at El Fin del Mundo. The petrographic thin sections are on temporary loan to the Institute of Archaeological Sciences at the University of Tübingen, where they are used as part of the teaching collection to train micromorphology students in the identification and interpretation of wetland deposits.

CHAPTER FOUR

Excavations in Locus 1, Upper Bonebed

Guadalupe Sánchez, Vance T. Holliday, and Ismael Sánchez-Morales

Much of the fieldwork at El Fin del Mundo focused on excavations in Locus 1. Initial site discovery in 2007, including identification of at least one in situ bonebed and recovery of stone artifacts, including a Clovis point, was made at or around the Locus 1 island (Chapter 1). The flat surface of the island upon initial investigation was ~40m long (E-W) and varied in width (N-S) from to 2 to 20m (Figure 1.7). The margins of the island were vertical to sloping depending on accumulation of debris shed from the vertical walls by weathering (Figure 4.1). The weathering and erosion processes were responsible for exposing the bone and lithic artifacts that led to site discovery. A proboscidean tusk was eroding out on the north profile and four stone artifacts were found on the surface around erosional island at Locus 1 (Figure 1.2).

The primary archaeological feature, the Upper Bonebed, along with the Lower Bonebed, were in the eastern 18m of the Locus (Figures 4.1 and 4.2). The area was gridded in 1-m squares. Most were excavated in units of 1 by 2 meters, but a few were 1-m by 1-m or 2-m by 2-m. Each 2-m by 1-m unit was numbered (Figure 4.3) and excavated in 50-cm squares. Excavations generally were in 5-cm levels, but also followed natural stratigraphic breaks. Vertical control was maintained with reference to a primary site datum and local subdatums. The site datum, set at 10 m above the highest point of the excavation area, is 636.69 m above sea level. All elevation measurements taken during excavation refer to vertical datum, in measurements designated meters below datum (mbd). Although subdatums were used during the excavations, all the depths presented here were converted to the main datum. The hardness of the alluvial Stratum 3B is such that it was scraped out with a trowel. All the soil removed from the excavation was screened using both ¼-inch and ½-cm mesh, including water screening in 2008, which allowed us to collect microflakes and microfauna found in the strata.

The exposures around Locus 1 and excavation data showed that the diatomite at the base of Stratum 4 was between 11.75 and 11.91 mbd but dipped down very gently to the east and varied in thickness. Stratum 3B was found between 11.82 and 11.91 to 12.00 mbd. The physical characteristics of Strata 3B and 3A are very similar and the contact between the two tended to be unclear during excavation. The 3B/3A transition zone was diffuse and 10- to 20-cm thick, 12.16 to 12.31 mbd. Stratum 3A was 12.20 to 12.50 mbd with an abrupt contact on Stratum 2.

Bone varying in size from small unidentifiable fragments to large proboscidean elements were found throughout Stratum 3A, in upper 3B, and in the Stratum 4 diatomite. As a result of the excavations three distinct "zones" with bone were recognized:

1. the Upper Bonebed, in the west portion of Locus 1 and the principal reason for the investigations at the site, concentrated in the upper 20cm of Stratum 3B with some bone protruding above 3B and encased in diatomite of Stratum 4 (Figure 4.4);
2. the Lower Bonebed, found throughout Stratum 3A but concentrated on or just above the stratum 2/3A contact, whose extent was not determined; and
3. a highly bioturbated area in the eastern portion of the excavations (Figures 2.2 and 4.3; Table 4.1.).

The Upper Bonebed was in two distinct concentrations and represented the Clovis archaeological feature (Figure 4.4A and 4.4B). The Lower Bonebed was generally not a distinct bed of bones analogous to the upper bed and

Figure 4.1. Excavations underway in Locus 1, December 2007. Upper: View northeast across the Locus 1 island showing the initial stages of excavation in the Upper Bonebed. The heavily eroded nature of the island margins is well illustrated. The southwest corner of the island is in the lower left. The jumbled character of the northwest tip of the island extending down into Stratum 3 also shows well. Lower: View of the south wall of Locus 1 with excavations under way in the Upper Bonebed, January 2008. The full stratigraphic section in Figure 2.3 is at right where the pair of individuals are working. Photographs by Vance Holliday.

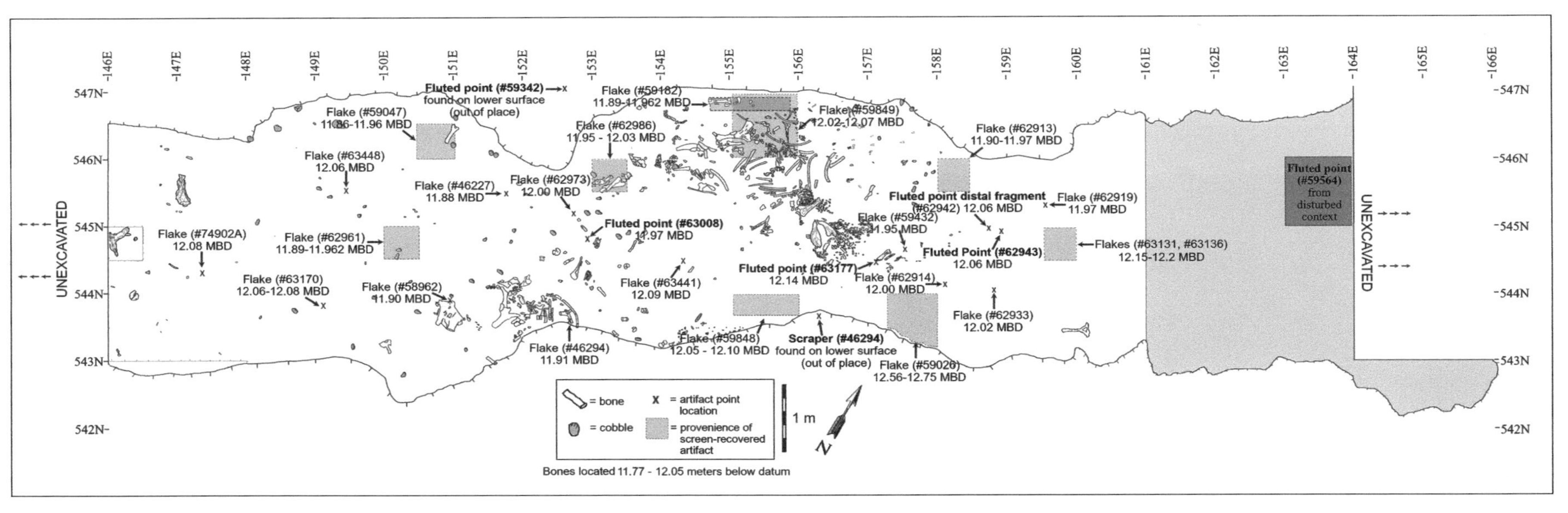

Figure 4.2. The Upper Bonebed in Locus 1 illustrating the spatial distribution of lithic artifacts relative to the two bone concentrations. Modified by Ismael Sánchez-Morales from original map by Edmund Gaines.

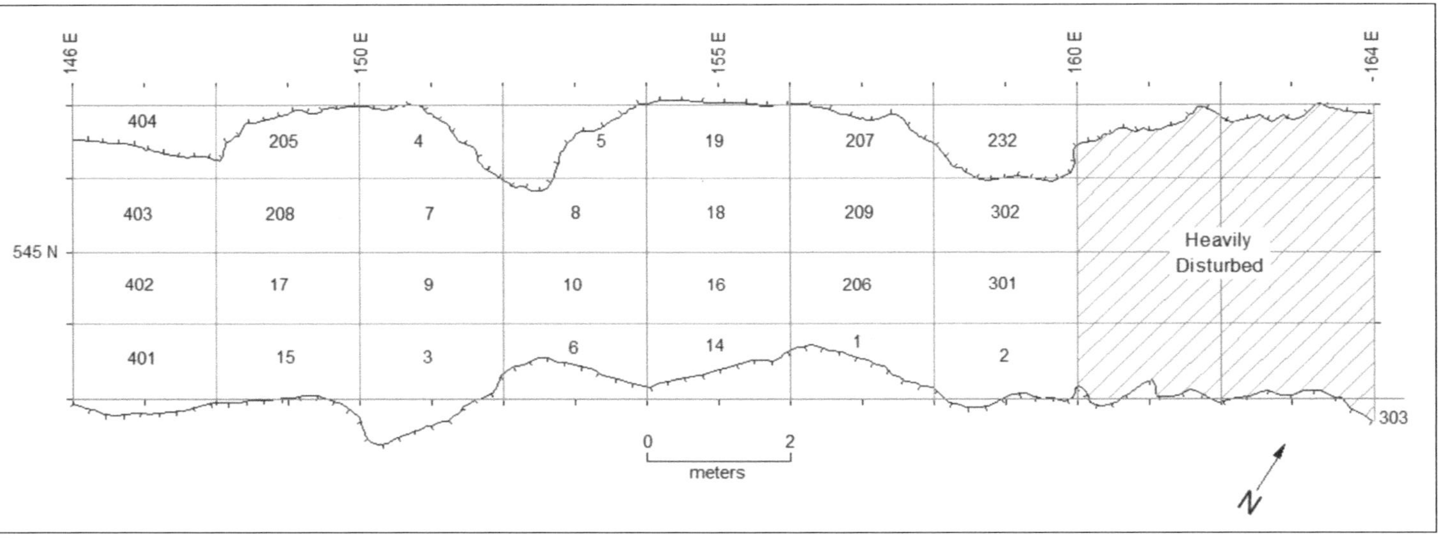

Figure 4.3. Map of the Locus 1 excavations illustrating the grid, the excavation units, and the area of heavily disturbed deposits at the east end. Sampling section 08-1 (for sediments, micromorphology, pollen, phytoliths, diatoms, and ostracodes) was along the north side at about the 150E grid intersection and 10-4 (diatoms in bedded diatomite) was at about the 160E intersection. Prepared by Jim Abbott.

Figure 4.4. Bone concentrations: (Top) Concentration 1 with pelvis in foreground (right and left portions articulated), one rib behind the concentration, one tibia fragment and one rib fragment in the top-left corner of the photograph. (Bottom) Concentration 2 showing the left half of a pelvis at upper left and a tibia fragment in center foreground. Most of the long narrow bones are rib fragments. In the center foreground is a tibia fragment. Photographs by project members.

Table 4.1. Excavated Units, Stratigraphy, Artifacts, and Bones from Locus 1 (all depths in meters below datum)

Unit	*East*	*North*	*Stratum 4 top of Diatomite*	*First bones found, Upper Bonebed, Stratum 3B*	*Bottom of Upper Bone Be, Stratum 3 B*	*Strata 3A/3B transition*	*Lower Bonebed Strata 3A/2 Contact*	*Stratum 2*	*Artifacts and Upper Bonebed (BC1 and BC2)*
1	156–158	543–544	11.945	unknown	unknown		12.7	12.9	#58926: Light yellow chert flake, 4 cm long found in the screen, depth unknown.
2	158–160	543–549	11.94	unknown	unknown		12.74	12.92	
3	150–152	542.5–544	11.75	11.8					#58962: Quartz crystal flake, less than 1 cm, in situ at 11.9 mbd. BC1.
204 (unit 3 in 2007)	150–152	542–544				12.025	12.36	12.58	
4 (unit 235 east in 2008)	150–152	546–547	11.8	11.82					#59047 Orange Chert/Chalcedony flake less than 1 cm. In screen between 11.86–11.96 mbd.
234 (units 4&5 in 2007)	151–153	546–547			12.03		12.43	12.55	
235 (unit 4 in 2007)	150–151	546–547				12.12?	12.53	12.59	
5	152–154	545–546	11.83	11.88			12.671 (2018)	12.671 (2018)	
6	152–154	543–544	11.86	11.86					#46294: Purple chert flake, less than 1/2 cm at 11.91 mbd; in situ BC1 west side of unit.
6	152–154	543–544	2007	2007	12.041				BC1
7	150–152	545–546	11.81	11.85			2018	2018	#46227: Quartz crystal flake; less than 1/2 cm long, in situ, at 11.88 mbd.
304B (unit 7 in 2007)	150–151	545–546			12.042		2018	2018	#62961: Yellow translucent biface thinning flake, 1/2 long, in screen at 11.89–11.96 mbd.
308B (unit 7 in 2007)	151–152	545–546			12.036		2018	2018	
7	150–152	545–546				12.19	2018	2018	
7	150–152	545–546					12.632	12.632	#33111 Dark purple thinning flake with carbonate concretion, <1 cm, in situ at 12.29 mbd.
8	152–154	545–546	11.86	11.87					
8	152–154	545–546			12.041	2015			Two flakes: #62973 reddish chert/chalcedony flake, 1/2cm, in situ at 12.00 mbd. #62986 light brown chert/chalcedony flake, less 1/2cm, in screen at about 11.95–12.032 mbd.
8	152–154	545–546				12.185			
8	152–154	545–546					12.515	12.647	
9 (units 304A & 308A in 2010)	150–152	544–545	11.81	11.81					BC1– Right scapula and other bones.
304A (unit 9 in 2007)	150–151	544–545			12	12.052			
308A (unit 9 in 2007)	151–152	544–545				12.032			
9	150–152	544–545				12.24			

(continued)

Table 4.1. (continued)

Unit	*East*	*North*	*Stratum 4 top of Diatomite*	*First bones found, Upper Bonebed, Stratum 3B*	*Bottom of Upper Bone Be, Stratum 3 B*	*Strata 3A/3B transition*	*Lower Bonebed Strata 3A/2 Contact*	*Stratum 2*	*Artifacts and Upper Bonebed (BC1 and BC2)*
9	150–152	544–545				12.255	12.505	12.745	Two flakes: translucent quartz crystal flakes, less than 1 cm long; #33101 at 12.44 mbd and #33191 in screen between 12.258–12.495 mbd.
10	152–154	544–545	11.83	11.83					BC1: left tibia, a fragment of rib, a left calcaneum and a fragment of left fibula.
10	152–154	544–545			12.041				#63008: Complete Clovis point, purple and white chalcedony/chert, measures: 46.13mm long, 18.4mm wide and 5.9mm thickness, at 11.962–11.968 mbd.
10	152–154	544–545				12.235			
10	152–154	544–545							Two artifacts: #33126 brown translucid chert/chalcedony unifacial retouched flake with a Carbonate concretion; 5 cm long; in situ at about 12.225 mbd.
14	154–156	543–544	11.86	11.87					bones could not be associated to CB1.
14	154–156	543–544			12.07				#59848: Purple and white chert/ chalcedony angular debitage, 1 cm, in screen at 12.05–12.1 mbd.
14	154–156	543–544			12.19		12.46		
15	148–150	543–544	11.78	11.83					
15	148–150	543–544				12.14		12.53	#63170: Basalt flake 3 cm long, in situ at 12.06–12.08 mbd, west of BC1.
16	154–156	544–545		11.91					BC1.
16	154–156	544–545				12.12		12.49	#63441: Quartz crystal flake, 1cm long, in situ at 12.09 mbd.
17	148–150	544–545	11.8	11.87	2012	2012			BC2.
17	148–150	544–545			12.07	12.1			
17	148–150	544–545					12.4		
18	154–156	545–546	11.9	11.9					BC2.
18	154–156	545–546				12.13	12.38		
19	154–156	546–547	11.89	11.89					BC2.
19	154–156	546–547							
204 (unit 3 in 2007)	150–152	542–544					12.36	12.58	
205	148–150	546–548	11.65	11.65					
206	156–158	544–545	11.82	11.82					#59432: Quartz crystal flake, 1.5 cm long, in situ at 11.95 mbd. BC2: mandible, tooth, tusk, right stylohyoid.
206	***156–158***	***544–545***			***12.15***	***12.15***	***12.41***		#63177: Complete rhyolite Clovis point, long: 95.2 mm, wide 30.3mm, thickness 8.4mm; in situ at 12.14 mbd. BC2: cranial fragments.
207	156–158	546–547	11.85	11.91	12.04				#59783: Rhyolite flake, 1.2 cm long, in situ at 12.02 mbd. BC2.
208	156–158	545–546	11.83	11.87					

Table 4.1. (continued)

Unit	*East*	*North*	*Stratum 4 top of Diatomite*	*First bones found, Upper Bonebed, Stratum 3B*	*Bottom of Upper Bone Be, Stratum 3 B*	*Strata 3A/3B transition*	*Lower Bonebed Strata 3A/2 Contact*	*Stratum 2*	*Artifacts and Upper Bonebed (BC1 and BC2)*
208	156–158	545–546			12.06				# 63448: Rhyolite flake, 1 cm long, in situ at 12.06 mbd.
208	156–158	545–546				12.205	12.355–12.405	12.58	
209	148–150	545–546	11.84	11.93					BC2: mandible, scapula, ribs and skull fragments.
209	148–150	545–546				12.16	12.31		
226	128–130	534–536	XXX	XXX	XXX				#59569: Maroon chert/chalcedony complete Clovis point; 5.2mm long, 17.94mm wide and 6,38mm thick; Very disturbed unit; East end.
227	162–164	544–546	XXX	XXX	XXX		12.84	12.94	Very disturbed unit; East end of the excavation area.
227A	162–164	546–547.5	XXX	XXX	XXX		12.62	13.09	Very disturbed unit; East end of the excavation area.
227B	162–164	542.1–544	XXX	XXX	XXX		12.84	13.03	Very disturbed unit; East end of the excavation area.
231	161–162	543.5–546.7	XXX	XXX	XXX		12.59	12.9	Very disturbed unit; East end of the excavation area.
232	158–159	546–547			11.99				Very disturbed unit; East end of the excavation area.
233	160–161	543–547			12.05		12.42		Very disturbed unit; East end of the excavation area.
301A	158–159	544–545	11.87	11.95					Three artifacts: # 62943 reddish brown and gray chert, complete Clovis point; 41.24mm long, 19.52mm wide and 5.99mm thick, in situ at 12.055 mbd. # 62914: Green quartzite flake, <1cm at 11.9 mbd, found in situ. #62933: Yellow and purple chert/chalcedony biface thinning flake, <1cm long, in situ 12.02 mbd. East of BC2.
301A	158–159	544–545				12.22	12.38		
301B	159–160	544–545	11.9	11.97					
301B	159–160	544–545				12.2	12.46		Two flakes: #63131: Petrified wood flake, 1/2 cm long, found in screen at 12.1 mbd. #63136: Yellow and brown chert, <1cm long, found in screen at 12.15–12.2 mbd.
302A	158–159	545–546	11.82	11.91		12.12			#62942: Purple and white chert/chalcedony distal fragment of Clovis point; 49.25mm long, 24.33mm wide and 7.47mm thick; found in situ at 12.043 mbd; #62013: Green quartzite flake, 1 cm long, in screen at 11.9–11.97 mbd
302B		545–546	11.85	11.85		12.12			#62919: Yellow chert/chalcedony flake, 3 cm long; found in situ at 11.957–11.97 mbd.
303	164–166	542–543	XXX	XXX	XXX			13	
cala A	151.5–152	544–546				12.16	12.47	12.6	

Key: XXX indicates mixed deposits at east end on Locus 1 excavations.

mbd = meters below datum; BC1 and BC2 = Bone Concentration 1 and Bone Concentration 2.

Figure 4.5. View east with the crew at work on the Upper Bonebed. Bone Concentration 1 is on the right; Bone Concentration 2 at left. The lens of white diatomite is visible low in the section at left and right (Left to right: Ismael Sánchez-Morales, Érika Mendoza, Alejandro Jiménez, Puin Morales, Alberto Cruz, Rob Miller [standing], and Julio Vicente). Photograph by Guadalupe Sánchez.

appears to be paleontological (Chapter 6). The term "Lower Bonebed" is more of an informal term used to differentiate the older, lower bone from the more obvious, discrete upper layer of bone. The Lower Bonebed and associated bone concentrations are fully described and discussed in Chapter 6.

THE UPPER BONEBED

This feature was found in upper Stratum 3B and in the diatomite of Stratum 4. It consisted of two concentrations of semi-articulated bone, almost exclusively *Cuvieronius* sp. (gomphothere) (more fully described and discussed in Chapter 6) with associated artifacts (Table 4.1; Figures 4.2, 4.4, 4.5). Both bone concentrations were exposed along the eastern end of the north and south margins of Locus 1. They were therefore truncated by the erosion process that created the Locus 1 island.

Bone Concentration 1 (BC1) was the smaller of the two concentrations in terms of both the area it covered, and number of specimens recovered. It was exposed along the south profile and found in parts of 2-m by 1-m units 6, 10, and 16 (Figures 4.2, 4.4, and 4.5) and produced 49 identifiable specimens (see Chapter 6 for a detailed description). It contained tooth enamel and tusk fragments, a scapula, ribs, an innominate and fragments, sacral and caudal vertebrae, a tibia, two fibulas, carpals and tarsals, a metatarsal, and phalanges. Portions of the pelvis protruded above Stratum 3B and were encased by diatomite. A specimen of burned bone (#63045) was recovered from the north margin of the pile. BC1 represented one individual gomphothere; a female aged between approximately 8 and 19 African elephant years (AEY).

Bone Concentration 2 (BC2) was exposed along the north profile of Locus 1 (Figures 4.2, 4.4, and 4.5). It was composed of several distinct groupings of semi-articulated bone including a complete mandible with fragments of the cranium and tooth enamel at the southeast end of the concentration. Much of upper BC2 was encased in diatomite, which also tended to be thicker here. All of the bone appears to be from one individual (Chapter 6). It was significantly larger than BC1 in terms of area covered owing to the scattered groupings of bone, including several large specimens. As such, it covers parts of excavation units 17, 18, 19, 206, 207 and 209 (Table 4.1, Figures 4.2, 4.4). BC2 was also larger in terms of bone recovered, consisting of 88 identifiable specimens including a mandible, enamel fragments, cranial fragments, stylohyoids, a scapula, ribs, an innominate and fragments, lumbar and caudal vertebrae, a radius, a tibia, carpals and tarsals, and phalanges. Portions of the mandible, skull fragments, and clavicle protruded above Stratum 3B and were encased by diatomite. Burned bone was found along the southwest margin of the bone pile. The BC2 individual was a male that was two to eight AEY.

Other bone fragments and lithic artifacts were scattered across Locus 1 in upper Stratum 3B and in the diatomite. Few of these bone specimens were identifiable. More lithic artifacts were found a few meters distant from the two bone concentrations than within them (Figure 4.2).

Sánchez and others (2014) reported recovery of a small (~1cm in diameter), rounded bone with a "V" apparently incised (#59892). Subsequent research suggests that it is a dermal ossicle from Harlan's ground sloth (*Paramylodon harlani*) (Chapter 6).

The two bone concentrations display similar degrees of weathering (Chapter 6). BC1 has a greater percentage of specimens in weathering stage 3 relative to stage 4, whereas BC2 contains a lower percentage of specimens in weathering stage 3 and more in stage 4 (see Chapter 6 for a description of weathering stages). Both concentrations have approximately 10 percent of bones in weathering stage 5. No evidence of human modification was found on bone from either concentration during the zooarchaeological analysis (Chapter 6).

MIXED CONTEXT, EAST END OF LOCUS 1

The east end of Locus 1, east of roughly the 160E grid line, was heavily disturbed by animal burrowing (Figures 2.2, 4.2, and 4.3); this area includes excavation units 227, 227a, 227b, 231, and 233. The relatively soft diatomite at the base of Stratum 4 and the somewhat looser sediments in Stratum 3 apparently attracted burrowing animals. The result was fracturing and collapse of Stratum 4 diatomaceous earth onto the eroded upper contact of the remains of Stratum 3 and, locally, Stratum 2 (Figure 2.2). That process, in turn, mixed the bone and artifacts from all levels. Over 50 nacreous shell beads dated to the Middle Archaic period (~5,560 cal yrs B.P. see Table 2.3) were recovered from the bioturbated diatomaceous earth of Stratum 4 (Chapter 2). Virtually all of the shell beads were found in a vertical distribution of approximately 70 cm from excavation Level 2 through Level 9 of Excavation Unit 227, pointing to a severe stratigraphic mixing and vertical movement of materials in this area of the landform. The uniformity in material, size, and shape of the beads indicates that they could have come from the same necklace. In the same excavation unit, a complete Clovis point (#59569) was recovered on the screen but from the mixed deposits (Figures 2.2B and 4.2). The taxa of most of the bone recovered from the disturbed deposits have affinities to the intact Lower Bonebed (Chapter 6). The assemblage includes a proboscidean tusk and a molar belonging to the Lower Bonebed in Stratum 3A. Cobbles also were found together with the bones. When the site was first discovered, teeth plates from a mammoth were recovered from a jumbled pile of mixed sediment immediately northeast of the east end of the Locus 1 island. No Stratum 2 deposits were found in the area and elevations show the fragments to be at the elevation of Stratum 3. They are likely from Stratum 3A.

ARTIFACTS FROM THE UPPER BONEBED

From the 2007 field season through the 2012 field season, 34 artifacts (all but one made of stone) were recovered from within Stratum 3B within or at the same level as the Upper Bonebed in the Locus 1 island or found out-of-place but considered to be associated with the feature. Four Clovis projectile points were directly associated with the gomphothere remains. Three Clovis points were found out of place (one near Locus 1 during the initial site discovery, one adjacent to the excavated surface following the first field season, and one on the screen from an area that was heavily mixed). Six artifacts were recovered from disturbed contexts. Three additional tools likely or probably linked to the Upper Bonebed were recovered in eroded contexts around or near the eroded margins of Locus 1. One uniface was found that fell out of the bonebed and two biface tools were found adjacent to the eroding island. Twenty flakes were recovered (13 in place, 10 on the screen). One burned bone was found on the screen.

Clovis Points

In addition to the original surface find south of Locus 1, six Clovis points, four in primary context and two in disturbed contexts, were recovered during the excavations in Locus 1 (Table 4.2). The points are described next.

Table 4.2 Clovis Points from Locus 1 with Provenience

Artifact Number	*Artifact Condition*	*Unit*	*Provenience Coordinates*	*Stratum*	*Depth MBD*	*Raw Material*	*Length mm*	*Width mm*	*Thick mm*
58342	Complete	Disturbed context	152.77E, 547.59N	Surface		Quartz crystal	49.16	23.18	4.8
59659	Complete	227 disturbed context	162–164E, 544–546N	Disturbed	12.67–12.77	Brown chert	52.5	17.94	6.38
62942	Distal fragment	302–A	158.74–158.79E, 544.995–545.02N	3B	12.043–12.067	Purple chalcedony	49.25	24.33	7.47
62943	Complete; reworked	301–A	158.89–158.93E, 544.95–544.97N	3B	12.055–12.063	Brownish chert	41.24	19.52	5.99
63008	Complete	10	152.95–152.989E, 544.827–544.877N	3B	11.962–11.968	Pink chert/ chalcedony	46.13	18.4	5.9
63177	Complete	206	157.07–157.17E, 544.41–544.46N	3B	12.14	Rhyolite	95.28	30.35	8.4

Clovis Point #46023

This specimen was the first Clovis point found on the site during the initial survey of Locus 1 in 2007. It was located on the surface about 28 m south of the Locus 1 island (Figure 1.7). The area included a weathered lump of Stratum 4. A test pit (Unit 11) was excavated at the find site but no buried context nor any archaeological or paleontological remains were found. This complete point is made of light gray (2.5Y 7/1) chert with grayish brown (2.5Y 5/1) bands (Figure 1.4). The body displays a lanceolate morphology with a concave base and rounded corners. The basal thinning was accomplished through the detachment of a single channel flake on one face and three shorter channel flakes removed on the other side. The flaking pattern is horizontal, and the body displays fine edge retouch. The lateral margins near the base and the basal edge are heavily ground.

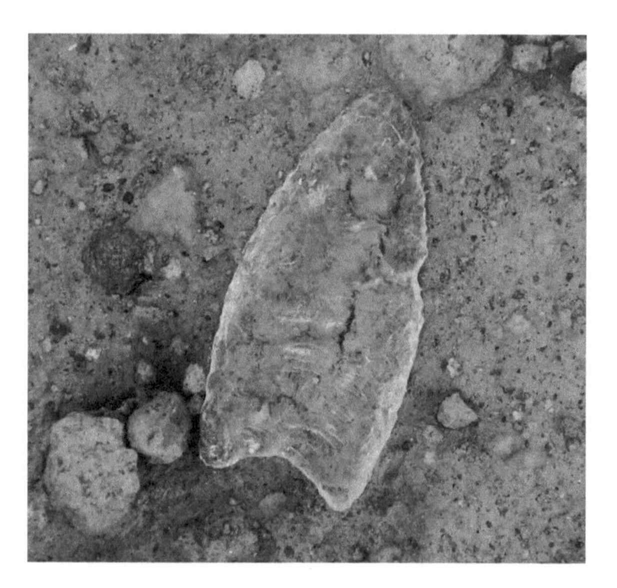

Figure 4.6. Clovis point #58342 in situ. Photograph by *Proyecto* member.

Clovis Point #58342

This complete Clovis point (Figure 4.6) was manufactured from a translucent quartz crystal; a hard but very fine-grained raw material with conchoidal fracture. The shape is lanceolate with concave base. One face exhibits basal thinning by two parallel channel flakes that reach a length of 17.95mm and 12.79mm, the longest reaching almost half of the total length to the tip. On the other side it was thinned by means of four shorter and narrower channel flakes reaching lengths of 13.56mm, 8.76mm, 6.53mm, and 6.68mm. It has a marked basal polishing; the polishing extended along both lateral edges up to 19.1mm, from the base to a little less than half the length of the body. It exhibits an extensive, continuous retouching that covers the entire piece. It is possible to see some overwhelmed flake scars, as well as a finer retouching in the margins, which are very sharp and in which you can still feel the edge to the touch. It has a notch in the upper sector of the lateral edge; probably this damage occurred when hafted.

The artifact was found on the surface along the northern edge of the Locus 1 island at the level where excavation stopped in the 2007 season in an area that was used as a passage to get off the work area, prone to deterioration by erosive agents (Figure 1.7). For this reason and due to the fragility of the artifact and its exceptional state of conservation, we infer that it must have been exposed from a buried context associated with the upper bed of bones between the 2007 and 2008 seasons. This is a reasonable interpretation, but the artifact is considered as coming from a disturbed context.

Clovis Point #59569

This complete Clovis point was manufactured from brownish chert (brown 7.5YR 5/2). One of its faces exhibits basal thinning by means of a 10.28-mm long channel flake located in the center of the base. The other face was thinned in the same way by means of an 11.23-mm long channel flake in the central part of the base. It displays polishing of the base and extends 19.71 mm along the two lateral edges, about a third of the total length. Retouching occurs along all margins of the artifact, with finer retouching observed on the edges.

This projectile point was recovered from unit 227 at a depth of 12.67–12.77 mbd. The stratigraphy of the east end of the island was heavily disturbed by burrowing and root growth that mixed Strata 4 and 3, as described previously, but the artifact was likely associated with the Upper Bonebed .

Clovis Point #62942

This thin distal fragment of a Clovis point (Figure 4.7) was made of purple chalcedony with white veins (white 10YR 8/1 speckled with dark reddish gray 10R 4/1 and reddish gray 6/1 and banded by red 2.5YR 4/4). One face displays a channel-flake scar that ends in a hinge fracture. Overshot flaking covers one complete face of the point. The artifact was found in unit 302A, at 158.74–158.79E/544.995–545.02N, and a depth of 12.043–12.067 mbd. It was located a meter to the southeast of BC1 and near the complete Clovis point (#62943) in the adjacent unit.

Clovis Point # 62943

This small Clovis point (Figure 4.7) is probably a distal fragment of a larger point reworked to again obtain the basal shape characteristic of a Clovis point. The shoulders probably resulted from edge retouch while hafted. It was manufactured on a brownish chalcedony/chert (light gray 2.5Y 7/1 with dark yellowish brown 10YR 4/4, speckled with dusky red 10R 3/2). On one side it exhibits basal thinning by means of a single channel flake measuring 16.9mm long, while on the other side it was thinned by two channel flakes 15.76 and 10.32mm long. The artifact exhibits polishing along the proximal third of both margins with extensive, continuous retouching across the entire artifact as well as a finer retouching in the margins. It was in situ in unit 301A at 158.89–158.93E/544.95–544.97N, and a depth of 12.0055–12.063 mbd, three meters east of C2.

Clovis Point #63008

This Clovis point (Figure 4.8) is also small, made on chalcedony or chert (white 10R 8/1 mottled with weak red 10R 4/2 and light olive brown 2.5Y 5/4). One side exhibits basal thinning by means of a single channel flake 11.56mm long. The other face was thinned by two channel flakes, one measuring 25.21mm and extending along more than half of the

Figure 4.7. Clovis points #62942 (left) and # 62943 (right) in situ. Photograph by Greg Hodgins.

Figure 4.8. Clovis point #63008 in situ. Photograph by *Proyecto* member.

total length of the blade, ending in a hinge fracture, and the other flake scar measures 11.26mm. The point was polished on both side edges from the base to approximately the first third of the length of the body. Extensive continuous retouch covers the entire piece with finer retouching along the margins. Like #62943, this Clovis point was used until exhaustion. The edges were reworked to the maximum possible and then the artifact was discarded.

The point was found in Unit 10 at 152.95–152.989E/544.827–544.877N and a depth of 11.962–11.968 mbd. It is the only Clovis point found above 12 mbd. The point was located less than 1 m northwest of BC1; it is the only Clovis point found close to BC1 and west of BC2. All other points were found east of BC2.

Clovis Point #63177

Clovis point #63177 (Figure 4.9) is made of rhyolite (dark brown 7.5YR 3/2 with reddish brown 5YR 5/3). It weighs 28.1 grams, four times more than any of the other points found in situ. On one side it exhibits basal thinning by a single channel flake 19.66mm long. The other side was thinned by a one channel flake 22.07mm long with two other much shorter posterior scars of 7.13 and 5.8mm in length. The artifact exhibits very marked polishing across the base as well on both lateral edges from the base up to 33.07mm long. It exhibits an extensive continuous retouching that covers the entire artifact, with several overshot scars. The retouch along both edges is still very sharp.

This point was recovered from Unit 206, at 157.07–157.17E/544.41–544.46N. It is the deepest point found at 12.14 mbd. It was ~15 cm below the elevation of the gomphothere mandible and other large elements of BC2 and a meter southeast of the mandible. Many small fragments of bone were found throughout the 15 cm of sediment between the mandible and the point. At least 40 fragmented cranial bones (2–5cm long) were recovered at the same elevation as the artifact, suggesting some sort of downward mixing of bone fragments and the point.

Figure 4.9. Clovis point #63177 in situ. Photograph by *Proyecto* member.

Point #63177 is the only one of the in-situ points that is not reworked or damaged. We suspect that this point, being large and thin with very sharp edges, was lost under the cranium during the hunting and butchering of gomphothere 2. Rhyolite boulders occur 400 meters southeast of Locus 1 (in Locus 25; Figure 1.6). The point was probably made from that local rhyolite.

Flakes Found in-situ during Excavation of Locus 1

Thirteen flakes were found in situ during the excavations of Locus 1. The flakes vary in size and are made from diverse raw materials (Table 4.3).

Flake #46227

This artifact (Figure 4.10) is a complete and very small micro-flake of opaque quartz crystal. It was found in Unit 7, at 151.85E 545.4N, and a depth of 11.88 mbd. The flake was found in the west side of the excavation, north of BC1 and west of BC2, in an area with very few bones.

Flake #46294

This artifact (Figure 4.10) is a very thin thinning-flake made of purple chalcedony or chert with white spots. The material is very similar to the Clovis point distal fragment #63008 found on the surface of Locus 10 in 2010. The flake was found in Unit 6, at 152.63E 543.56N and a depth of 11.91 mbd. It was associated with a large rib and other bones of the gomphotheres in BC1.

Flake #58962

This flake (Figure 4.10) is relatively thick and made of a white-opaque quartz crystal. It was collected during the excavation in Unit 6 and at 150.95E 543.87N and a depth of 11.9 MBBD. It was found in association with a gomphothere innominate in BC1.

Flake #59432

Artifact #59432 is made of translucent quartz crystal. It is a long flake that expands toward one end and exhibits a prominent bulb of percussion. It was found in unit 206 at 157.47E 544.66N and a depth of11.95 mbd. The flake was

Table 4.3. Flakes from Locus 1 Recovered in situ

Artifact Number	*Raw Material*	*Unit*	*Provenience Coordinates*	*Depth MBD*	*Length mm*	*Width mm*	*Thickness mm*
46227	Quartz	7	151.85E 545.4N	11.88	4.41	4.83	1.33
46294	Purple chalcedony	6	152.63E 543.56N	11.91	4.85	4.32	0.94
58692	Quartz	3	150.95E 543.87N	11.9	8.22	8.73	3.16
59432	Quartz	206	157.47E 544.66N	11.95	10.52	5.88	1.55
59667	Brown chalcedony	227 disturbed	162.4E 545.75N	12.71	13.48	7.69	2.7
59783	Rhyolite	207	156.38E 546.07N	12.02	11.79	7.88	2.08
62919	Light-yellow chalcedony	302–B	159.53–159.55E 545.32–545.345N	11.957–11.97	32.4	21.7	4.5
62914	Quartzite	301–A	158.07E 544.15N	11.995	5.4	7.7	1.8
62933	Chert	301–A	158.83E 544.07N	12.02	6.5	6.8	1
62973	Jasper	8	152.77E 545.16N	12	1.2	7.5	2.7
63170	Basalt	15	149.1–149.13E 543.81–543.83N	12.06–12.08	26.7	22	6.75
63441	Quartz crystal	16	154.2–154.25E 544.5–544.56N	12.09	8.78	5.6	1.6
63448	Rhyolite	208	149.43E/545.6N	12.06	7.55	5.96	1.6

Figure 4.10 Flakes found in situ and during screening: left to right, #46227, 46294, and 58962 (found in situ); 59047 and 59182 (found during screening in 2007). Photograph by Guadalupe Sánchez.

located east of the mandible of gomphothere 2 (BC2) in a level with many bone fragments.

Flake #59667

This flake is a reddish-brown chalcedony or chert. It is relatively long and expands toward the distal end with a notable bulb of percussion. It could be a biface thinning flake. The artifact was found in unit 227, at 162.4E/545.75N and 12.71 mbd depth in the far east of the excavation area where the stratigraphy was mixed by burrowing and rooting.

Flake #59783

This rhyolite/rhyodacite flake is elongated with a prominent bulb of percussion but is not a biface thinning flake. It was collected in unit 207, at 156.38E 546.07N and about 12.02 mbd. The flake is located at about 1 m north of the mandible

Figure 4.11. A biface thinning flake (# 62919) found in situ in 2010. Photograph by Guadalupe Sánchez.

of gomphothere 2 (BC2) in a level with many bone fragments and two fragmented ribs.

Flake #62919

This flake (Figure 4.11) was found near the three Clovis points that were found east of BC2. It is made of a light-yellow chalcedony or chert not common in local deposits. The dorsal face has flake scars that show that it is a biface thinning flake, with a clear bulb and platform of percussion. It was found in unit 302B, at 159.53–159.55E/545.32–545.345N, and a depth of 11.957–11.97 mbd.

Flake #62914

This artifact is a small bifacial retouch flake (Figure 4.12) made of a greenish quartzite. It was found in unit 301A at 158.07E/544.15N, and a depth of 12.0 mbd. The flake was found 1 m southwest of BC2 in an area with #62933 but without bone fragments.

Flake #62933

A bifacial retouched flake (Figure 4.12), this artifact is made of a yellow and purple chalcedony or chert. It is very thin and concave. It was found in unit 301A, at 158.83E/544.07N, and about 12.02 mbd. The flake is in the same unit as flake #62914 and about 2 m southeast of BC2 in an area with few bones.

Flake #62973

This complete micro-flake (Figure 4.12) is made of a dark red jasper not common in local deposits. It is an elongated bifacial thinning flake, found in Unit 8, at 152.77E/545.16N and 12.00 mbd depth. It was found in the empty space between BC1 and BC2 and only 25 cm northwest of Clovis point #62973 in association with a tooth fragment.

Flake #63170

This complete flake is made of basalt, which is not common in local deposits. The flake has 60 percent cortex. The bulb and platform of percussion are notable. The flake was found in Unit 15 at 149.1–149.13E/543.81–543.83N, and 12.06–12.08 mbd. The flake was found near three fragments

Figure 4.12 Flakes found in situ in 2010 (left to right, 62914, 62933, and 62973). Photograph by Guadalupe Sánchez.

Table 4.4. Flakes from Locus 1 Recovered from the Screen

Artifact Number	*Unit*	*Coordinates*	*Depth mbd*	*Raw Material*	*Length mm*	*Width mm*	*Thickness mm*
59026	1	543.5N–544N, 157.30E–158E	12.10–12.75	Beige chert	45.1	38.3	6.32
59047	4	150.5–151E, 546–546.5N	11.86–11.96	Orange translucent chert	7.51	5.11	1.88
59182	19	154.7–155.86E, 546.7–546.9N	11.89–11.99	Yellow chalcedony or silex	5.74	4.25	2.9
59848	14	155–156E, 543.5–544N	12.05–12.1	Light brown chert	7.6	10.48	4.01
59849	19	155–156E, 546–547N	12.02–12.07	Maroon chalcedony or chert	7.18	11.81	3.7
62913	302A	158E–545.5N	11.9–11.97	Dark green quartzite	4.5	9.6	2.43
62961	304A	150E–544.5N	11.89–11.962	Yellow translucent chert	4.3	4.6	1.6
62986	8	153E–545.5N	11.952–12.032	Brown chert	7.6	4.6	1.9
63131	301B	159.5–160E, 544.5–545N	12.15–12.2	Undetermined	6.8	3.48	1.3
63136	301B	159.5–160E, 544.5–545N	12.15–12.2	Brownish orange chert	3.43	7.35	1.89

of charcoal or charred plant fragments including Sample #63176, which was dated by radiocarbon to 11,880+/-200 ^{14}C yr B.P. (AA-100182B; Table 2.3).

Flake #63441

The artifact is an elongated flake of translucent quartz crystal with a notable bulb and platform of percussion. It was recovered in Unit 16 at 154.2–154.25E/544.5–544.56N and 12.09 mbd. The flake was found in the empty space between the two gomphotheres, about 1 m south BC2 and 2 m east of BC1.

Flake #63448

This very thin retouch flake made on rhyolite has a visible bulb of percussion made on rhyolite. It was found in unit 208 at 149.43E/545.56N at a depth of 12.06 mbd. It was from the transition zone between Strata 3B and 3A in an area where the boundary is unclear. It was also in association with charred plant remains (Sample #63447) dated by radiocarbon to 11,550+/-60 ^{14}C yr B.P. (AA-100181A; Table 2.3).

Flakes Collected in the Screen During Excavation

Ten flakes were found on the screen during the excavation (Table 4.4). All of them are made of chalcedony or chert from a variety of sources. Most of the excavation was carried out in subunits of 50 by 50 cm, so the provenience of the flakes has an accuracy of 50 cm.

Flake #59026

This flake was found during the initial stages of excavation before Strata 3A and 3B were differentiated and when units below the Clovis bonebed were excavated in 50-cm levels. The flake is cream-beige chalcedony/chert produced when making a too larger than a biface. The flake was found between 543.5N–544N/157.30E–158E. We calculate that the depth is between 12.10 and 12.75 mbd. Association of the artifacts with the Upper Bonebed is not clear.

Flake #59047

This very small, thin, irregular flake (Figure 4.12) is made on an orange semi-translucent chalcedony/chert not common in local deposits. The flake is not the result of bifacial thinning. It came from Unit 4, at 150.5–151E/ 546-546.5N in the 10-cm level 11.86–11.96. Unit 4 is about 2 m west of BC2, and not directly associated with either bone concentration.

Flake #59182

This artifact is a long and thick flake fragment made of opaque yellow chalcedony/silex (Figure 4.12). It is not a biface thinning flake. It came from Unit 19 in association with BC2 near the left tibia. It came from a small excavation area of 98 cm by 29 cm, at 154.7–155.86E 546.7–546.9N, in the 10-cm level between 11.89 and 11.99 mbd.

Flake #59848

This artifact is an indeterminate angular fragment made on a light brown chalcedony/chert with flake scars on one face. It could be a flake fragment without bulb of platform of percussion. The artifact was found in Unit 14, 3 m south of BC2. It came from an area of 90 cm x 30 cm, at 155-156E/543.5-544N, and between 12.05 and 12.10 mbd. It was not directly associated with the gomphotheres.

Flake #59849

This is a thick triangular flake made on a maroon chalcedony/chert. It came from Unit 19 in a 1-m by 1-m area at 155–156E/546–547N in association with BC2 in the vicinity of the innominate, ribs, and vertebrae. It was from a depth of 12.02–12.07 mbd.

Flake #62913

This fragment of a biface thinning flake is made of a dark green quartzite. It is from Unit 302A in a 25-cm by 50-cm area, at 158–158.25E/545.5–546N and in a 7-cm level, 11.9–11.97 mbd. The artifact was not associated with BC2 but was located 50 cm north of Clovis point fragment #62842.

Flake #62961

This bifacial thinning flake is made of a translucent yellow chert and has a square shape. It came from Unit 304A, 1 m northwest of the scapula in BC1. It came from an area of 50 by 50 cm, at 544.5–545N/150–250.6E and a 6-cm level between 11.89 and 11.962 mbd.

Flake #62986

This complete elongate bifacial thinning flake is made of a brownish chalcedony or chert. The flake was found in Unit 8 in association with a group of weathered, unidentifiable bones west of BC2. The flake was recovered from a 50-cm by 50-cm square at 153–153.5E/545.5–546N, in an 8-cm level from 11.952 to 12.032 mbd.

Flake #63131

This very small elongate flake is probably a tool retouch flake. It is too small for raw material identification. It was recovered from Unit 301B, 3 m east of the mandible in BC2, in a 50-cm by 50-cm square at 159.5–160E/544.5–545N, in a 5-cm level between 12.15 and 12.2 mbd.

Flake #63136

This small, complete retouch flake expands toward its distal end with a well-defined bulb and platform of percussion. It is made from a brownish orange chalcedony or chert not common in local deposits. It was found along with flake #63131 in unit 301B, 3 m east of the mandible in BC2. The excavation area was a 50-cm by 50-cm square at 159.5–160E/544.5–545N, in a 5-cm level between 12.15 and 12.2 mbd.

Miscellaneous Artifacts

Burned Bone #63045

A small fragment of burned bone was found in Unit 10 from a 50-cm by 50-cm area at 544–544.5N/152/152.5E and a depth of 11.956 to 12.036 mbd. The burned bone fragment came from an area near the phalanges of BC1.

Unifacial Core Tool #45980

This large unifacial side scraper is made of pale yellow (2.5Y 7/3) chalcedony or chert (Figure 4.13). The texture is waxy, suggesting possible heat treating to facilitate the

Figure 4.13. Two views of the unifacial core tool made of chert (#45980) found on the south side of the island where it had fallen out of the Upper Bonebed, leaving a negative impression among the bones (Figures 1.2 and 1.7). Photographs by Guadalupe Sánchez.

Figure 4.14. Two views of the distal fragment of a biface made of rhyolite (#46021), found on the surface on the north side of the Locus 1 island (Figure 1.7). Photographs by Guadalupe Sánchez.

knapping of the piece. It was the first tool found during the first visit to the site. It had fallen out of place from the Upper Bonebed exposed in the profile at 156.20E/543.4N (Figure 1.2).

Biface #46021

This artifact is the distal fragment of a biface (Figure 4.14), found on the lower surface southeast of the Locus 1 island (Figure 1.7). It is made of pinkish gray local rhyolite (5YR 6/2) at 160E/550N.

Biface Fragment #46022

This midsection of a biface is made of clear epidote quartz (Figure 4.15), locally available in the arroyo that crosses through the site. It was found on the lower surface below the north edge of the Locus 1 island at 150E/552N.

Figure 4.15. A biface midsection made of clear epidote quartz (#46022). It was found on the surface on the north side of the Locus 1 island (Figure 1.7). Photograph by Guadalupe Sánchez.

DISCUSSION

The Upper Bonebed of Locus 1 at El Fin del Mundo represents an association in primary context of extinct megafaunal bone elements and diagnostic Clovis stone artifacts. Lines of evidence that point to the archaeological nature of this context include the following:

1. The discovery of the two bone concentrations, each representing the semi-articulated skeleton of a gomphothere, in direct association with four Clovis points and multiple flakes, most of them of raw materials exogenous to the matrix of the deposit (i.e., chert).
2. The relatively undisturbed condition of the context in the main block of excavation of Locus 1, sealed by the white diatomite indicating fast burial of the skeletons and artifacts. The chances of a fortuitous association of two gomphothere carcasses, four Clovis points (possibly 7 if those found in disturbed context are included), and more than 20 flakes surrounding the bone concentrations, buried over a short period of time are very slim.

3. Virtually all of the Clovis points found in Locus 1, including the three points found out of context, are complete except for the distal fragment found in primary context. The complete nature of the points associated with the bonebed and the pristine condition of at least two of them with no evidence of resharpening or repair (#46023 and #63177) suggest that the hunters could not retrieve these points from the carcasses. The snapped distal fragment from Locus 1 is consistent with contrasting patterns observed between the Upper Bonebed and the upland loci. Projectile points might have snapped during hunts, and the retrieved shafts with the bases still attached to them must have been transported back to the campsites, where the broken points were thrown away and replaced. Virtually all of Clovis points from the upland loci are either snapped basal fragments or highly reduced and damaged points, suggesting that unusable implements were discarded at the campsite. This pattern indicates a clear contrast between the artifacts found in Locus 1 and the uplands, reflecting different behaviors at each area: hunting in Locus 1 and discard and replacement of damaged implements in the campsite (Chapter 5)

4. The general context of the Upper Bonebed is remarkably similar to other Clovis and later Paleoindian megafauna kill sites reported in the U.S. Southwest and neighboring areas. Hunting and/or butchering of Pleistocene megafauna in a wetland, stream, or lake environment in association with Paleoindian artifacts have been reported from the San Pedro River Valley sites in Arizona including Naco, Lehner, and Murray Springs (Haynes and Huckell 2007; Chapter 5), Water Canyon, New Mexico (Holliday, Dello-Russo, and Mentzer 2020), the Clovis type site (aka Blackwater Draw Locality 1), New Mexico (Hester, Lundelius, and Fryxell 1972), and Lubbock Lake, Texas (Johnson 1987). In addition, Murray Springs contains a Clovis campsite associated with the kill localities (Chapter 5) in more and better-drained elevated terrain nearby, mirroring the association of Locus 1 and the upland campsite Loci in El Fin del Mundo. The configuration of El Fin del Mundo and the elements contained in the Upper Bonebed of Locus 1 are consistent with Paleoindian patterns of megafauna exploitation in the larger geographic area.

Locus 1 likely represents a single hunting event of two gomphotheres, although the possibility of two separate events over a short period of time cannot be ruled out (Chapter 6). Seven Clovis points were recovered from the archaeological feature, three from secondary contexts (#58342 quartz crystal point north from the excavation, #59569 brown chert east end of the excavation, and #46023 chalcedony point south of Locus 1). The other four Clovis points were found within the archaeological feature. Of these, two are highly reduced and probably repaired, one is a distal fragment, and only one seems to be a pristine specimen. That point (#63177) is the largest from the site and was found deeper than the others, surrounded by small skull fragments of BC2. Two broken bifaces (#46022 and #46021) and a unifacial tool (#46294) were found on the surface immediately adjacent to the margins of Locus 1. They may have been associated with the Clovis feature.

Thirteen flakes were recovered in situ during the excavation, along with 10 in the screen. Except for the quartz crystal flake (#58962) found in association with the pelvis of BC1 and a purple chert flake (#46294) found in association with the ribs of BC1, the flakes were found around the two concentrations of bone. These flakes suggest that the hunters at Fin del Mundo revived the cutting edges of their stone tools on site. This task was carried out in clean areas adjacent to the animals; flakes, biface thinning flakes, and Clovis points were found in these openings.

Most of the lithic artifacts found during the excavation were at a depth of 11.86 to 12.14 mbd. The Clovis point of rhyolite (#63177) beneath the skull fragments in BC2 was the deepest at 12.14 mbd. In the 2018 season, four artifacts were found between 12.25 and 12.29 mbd including a unifacial tool and three flakes. The unifacial tool (#33126) was made on semi-translucent chert and is 5cm long, with most of its surface encased in carbonate concretion. Two flakes of quartz crystal were recovered. flake #33101 was recovered at 12.55 and #33191 was found on the screen. Flake #33111 is a purple chert flake found in situ encased in carbonate concretion. These artifacts may be part of an older archaeological context, but most likely are part of the Upper Bonebed. They were found in the general area of the rest of the artifacts. They may have moved down when 3B was still wet and muddy, perhaps during butchering, or later when 3B was subjected to cracking.

CHAPTER FIVE

The Clovis Lithic Assemblage from El Fin del Mundo and Early Paleoindian Land Use in North-Central Sonora and the Upper San Pedro River Valley

Ismael Sánchez-Morales

El Fin del Mundo offers a unique contribution to our understanding of Clovis lithic technology and patterns of landscape use. This is particularly true for the transnational cultural region of Northwest Mexico and the southwestern United States, given the proximity of the site to the extensively studied Clovis localities of the San Pedro River Valley in southeastern Arizona, which are around 250 km northeast of El Fin del Mundo (Figure 1.1). Furthermore, unlike most of the Sonoran Clovis record, El Fin del Mundo is relatively pristine and has not been collected by amateur archaeologists. The Clovis stone tool technology of the southwestern United States has been extensively studied (e.g., Bradley, Collins, and Hemmings 2010; Hester, Lundelius, and Fryxell 1972; Huckell 2007), in contrast to the relatively few systematic archaeological investigations focused on the Paleoindian occupations in Northwest Mexico. Recent studies have, however, yielded new information on abundant Paleoindian sites and artifacts in the region (Gaines, Sánchez, and Holliday 2009; Sánchez and Carpenter 2012; Sánchez and others 2014; Sánchez 2016). Among these recent investigations, research at the sites of El Fin del Mundo and El Bajío is particularly important for understanding early Paleoindian stone tool technology and associated behaviors in Mexico, because these sites have produced the largest Clovis lithic assemblages south of the modern U.S.-Mexican border. In the United States, the San Pedro River Valley contains the highest density of buried Clovis sites in North America. The lithic assemblage from Murray Springs in the San Pedro Valley is the largest single sample of Clovis stone tool production known in the western United States and is associated with multiple megafauna kills and a campsite (Huckell 2007). The configuration of this site and its Clovis lithic assemblage are similar to those from El Fin del Mundo, making this an ideal case for comparison of Clovis stone tool technology and landscape use strategies.

Generally speaking, landscape use refers to the "integrated suite of behaviors in which foragers adjust the way they organize resource acquisition and processing, manufacture and use of technology, and group size in space and time" (Barton and Riel-Salvatore 2014:341). This term is often used in archaeological research of foraging societies to refer to:

1. settlement patterns and intensity of site occupation, including the location and types of places that people chose to settle in and the diversity and duration of activities carried out at those particular locations;
2. the array of natural resources exploited and the strategies of exploitation (e.g., game and stone tool raw materials); and
3. mobility, or the relative frequency, extent and purpose of group movements across the landscape.

The material culture of hunter-gatherer societies, including stone tool technology, is profoundly influenced by aspects of forager land-use such as mobility and the selective exploitation of tool stone sources (Binford 1979, 1980; Torrence 1983; Bleed 1986; Kelly 1988; Kelly and Todd 1988; Nelson 1991; Kuhn 1994). Therefore, the study of lithic technology is one way by which we can aim to understand landscape use among Clovis foragers.

Previous research in Sonora (Gaines, Sánchez, and Holliday 2009; Sánchez 2016) suggests that Clovis land-use in the region was characterized by mobility patterns restricted to local landscapes and more intensive use of readily available resources relative to Clovis groups farther north. These interpretations were based on the predominance of locally available lithic raw materials such as basalt, rhyolite, obsidian, milky quartz, and quartz crystal in the known collections of Clovis points from Sonora. It was suggested that this apparent selection of resources contrasts to the Clovis assemblages from the United States, which tend to be dominated by exotic, very high-quality cryptocrystalline tool stones (i.e., chert, chalcedony, jasper, and agate). Gaines, Sánchez, and Holliday (2009) argue that rather than transporting raw materials from long-distance sources or possibly trading for high-quality exogenous tool stones, which apparently took place among Clovis foragers to the north, Clovis groups in Sonora learned where to find lower quality but locally available raw materials, and intensively exploited and relied on these more readily accessible sources. However, other features of Clovis stone tool technology in Sonora that may have been equally influenced by landscape use have not been sufficiently investigated.

This chapter describes in detail the diagnostic Clovis stone tools from El Fin del Mundo, drawing from previously published work (Sánchez-Morales 2018; Sánchez-Morales, Sánchez, and Holliday 2022) and presents a comparative analysis of this assemblage and those from the Clovis sites of El Bajío, Sonora, and the San Pedro River Valley, Arizona. The results of this analysis are used to offer interpretations about Clovis land-use in this region, including site use, lithic raw material exploitation patterns, and group mobility.

THE SITES

The geographic and geological contexts of El Fin del Mundo are discussed in Chapters 1, 2, and 3 and the archaeological context of the Upper (Clovis) Bonebed is presented in Chapter 4. In the following section the background on the geographic context and previous research at the site of El Bajío and the sites of the San Pedro River Valley are briefly summarized.

El Bajío (SON K:1:3)

El Bajío is located in a broad (>30-km wide), north-south trending valley drained by the Río Zanjon, which flows south and joins the Río Sonora in Hermosillo. The Río Zanjon Valley is a typical structural basin of the region. The site is on the east side of the valley within a small, slightly inclined, irregularly shaped *bajío* or low-lying area formed by the piedmont of the granitic Sierra San Jerónimo to the northeast, and a series of basaltic hills that trend northwest-southeast at the extreme southern edge of the site (Sánchez 2016). El Fin del Mundo is approximately 110 km due west of El Bajío in a complex structural and volcanic series of hills and mountains west of the Río Zanjon Valley (Figure 1.1). Several streams arise from the Sierra San Jerónimo and cross the *bajío* to the southwest, forming alluvial fan deposits, possibly creating conditions under which water may have accumulated during the late Pleistocene and early Holocene. Today water sources do not appear to exist within this arid zone (Sánchez 2016).

El Bajío is one of the most extensive Clovis sites in western North America (Sánchez 2016). The site covers 4 km^2 with a moderate density of lithic artifacts and consists of 22 loci defined by concentrations of archaeological materials on the surface. The localities are distributed along the piedmont and on the hilltops. The most significant locus is a primary source of a fine grained, high-quality raw material of greenish black color first identified as basalt by Robles Ortiz and Manzo Taylor (1972) and later identified by Sánchez (2001, 2016) as semi-vitreous basalt and vitrified basalt. This source of raw material outcrops on Cerro La Vuelta (Locus 20) located in the southwestern portion of the site. Clovis and Archaic groups exploited this vitrified basalt source, as indicated by the thousands of artifacts and debris produced from the quarrying of this outcrop (Sánchez 2016).

The site has been intermittently studied by amateur and professional archaeologists since 1971, resulting in the collection of Paleoindian and Archaic artifacts from the surface and excavation units. Background on this previous research is reviewed by Sánchez (2010, 2016). The investigations carried out by Sánchez in 2003 resulted in the excavation of trenches and test units at El Bajío (Sánchez 2016). In addition, 10 hearths and 2 knapping stations were excavated, and 110 hearths exposed on the surface were documented (Sánchez 2016). Samples for radiocarbon dating were collected from different trenches and features but no buried contexts of Paleoindian age were found (Sánchez 2016). Lithic materials on the surface that were indicative of a significant Clovis occupation were collected.

Although no water sources or spring deposits were documented during the fieldwork, the presence of black clay (undated paleo-wetland) deposits suggest that water accumulated in the low-lying sink near the central group of loci, fed by runoff from the piedmont and runoff water from the small range that includes Cerro La Vuelta. The lithic assemblage from the site indicates that this area was used repeatedly over a long period of time, from the terminal Pleistocene to the late Holocene (Sánchez 2016).

Clovis Archaeology of the Upper San Pedro River Valley

The upper San Pedro River valley is a broad, relatively flat, alluvium-filled basin flanked by mountains within the Basin and Range physiographic province in southeastern Arizona (Haynes and Huckell 2007). The San Pedro River rises to the south in Sonora, north and east of the city of Cananea, and flows north to join the Gila River. The upper valley extends northward across the international border to the area around Charleston, Arizona (Haury, Sayles, and Wasley 1959). Historic arroyo cutting exposed several late Pleistocene-early Holocene archaeological sites along tributary drainages of the San Pedro. The upper San Pedro Valley contains the highest concentration of excavated, in situ Clovis sites in North America (Haynes 2007). These sites are distributed over approximately 30 km south to north in the river valley, and approximately 250 km northeast of El Fin del Mundo. Pleistocene vertebrate remains, including mammoth and/or bison, associated with Clovis artifacts have been reported at the sites of Naco, Lehner, Leikem, Murray Springs, Escapule, and Navarrete (Figure 1.1). The geological contexts of these localities indicate wet conditions in marsh-like environments. The excavation methods, archaeological and zooarchaeological findings, and geologic contexts from these sites have been extensively published and can be referenced in Haury (1953); Antevs (1953); Haury, Sayles, and Wasley (1959); Mehringer and Haynes (1965); Hemmings and Haynes (1969); Mead, Haynes, and Huckell (1979); Haynes (1982, 1992, 2008); and Haynes and Huckell (2007).

FIELD AND DATA COLLECTION METHODS

Lithic materials from El Fin del Mundo that were not recovered during the excavation of Locus 1 were collected during systematic surveys in the different loci of the site. Most archaeological artifacts were recovered from surface contexts that contain a palimpsest of Paleoindian and Archaic occupations. For this reason, only diagnostic materials that could be associated with temporal-cultural horizons based on typology and technology were collected. Most often, these diagnostic artifacts are projectile points. However, artifacts displaying technological traits that are characteristic of Clovis lithic technology and which are not observed in Archaic components from other sites in the region were collected as well. Provenience information including UTM coordinates was recorded.

The Clovis lithic artifacts were classified into three categories: bifaces, unifaces, and unretouched blades and blade manufacture by-products (Table 5.1). The micro-flakes recovered during the excavation of Locus 1 are not included in this study (see Chapter 4). The classification of each artifact was based on both technological features such as retouch type and blank production technology and typology or assumed function (e.g., Clovis point, end scraper), after Bradley, Collins, and Hemmings (2010), Huckell (2007), and Collins (1999). Databases containing all the technological, morphological, qualitative, metric, and provenience information of each artifact were developed following the format used by Sánchez (2010, 2016) and are included in Appendix C. Metric data were collected using a digital caliper and weights were measured with a digital scale. All measurements were recorded in millimeters (mm), grams (g), or degrees (°). In the case of fragmentary artifacts, the maximum length, width, and thickness of the fragment were recorded, indicating with an asterisk (*) that the measurement corresponds to an incomplete artifact. When possible, maximum length and maximum width/basal width estimations of bifacial implements were recorded and included in the database within parentheses next to the actual measurements. The identification of raw materials was based on macroscopic observations of texture, flaking quality, inclusions observable with a portable 10x magnifying glass, Munsell color, and comparisons of these features with those of selected specimens from the assemblage that represent different raw materials identified using a binocular microscope.

Metric and qualitative data for the lithic collection from El Bajío were provided by Guadalupe Sánchez (personal communication 2015). Metric data for the materials from Murray Springs, Naco, Lehner, Leikem, Navarrete, and Escapule were collected by the author from the collections at the Arizona State Museum. Although published data on these artifacts are available (Haury 1953; Haury, Sayles, and Wasley 1959; Hemmings and Haynes 1969; Haynes 2007; Huckell 2007), this step was necessary to make sure that the measurements of the artifacts from the Sonoran sites and the San Pedro River Valley sites were taken using the same methods. Some of these artifacts were not available for study. Consequently, some measurements and raw material color descriptions are missing for some unifacial tools and blades from Murray Springs. When available, published data on these materials were included. The Clovis points

Table 5.1. Frequencies of Lithic Categories and Provenance Counts of Clovis Stone Tools from El Fin del Mundo

Lithic Category	*Locus 1*	*Locus 2*	*Locus 5*	*Locus 8*	*Locus 9*	*Locus 10*	*Locus 14*	*Locus 17*	*Locus 19*	*Locus 21*	*Locus 22*	*Locus 23*	*Locus 24*	*Locus 25*	*Total Number of Artifacts*
Bifaces															
Secondary bifaces	2		5			15		2			2				**26**
Clovis point preforms		1	4			2		2		1					**10**
Clovis points	7		7	1	1	2									**18**
Unifaces															
End scrapers			32	1											**33**
Side scrapers	1														**1**
Blades															
Wedge-shaped blade core			1												**1**
Conical core			1												**1**
Core tablet flakes			4												**4**
Crested blades			2			2			1		1				**6**
Cortical blades			1			1					1	1		1	**5**
Noncortical blades			24	1		7	1				1	1	1	11	**47**
Total Number of Artifacts	10	1	81	3	1	29	1	4	1	1	5	2	1	12	152
Percentages	6.6	0.7	53.3	2.0	0.7	19.1	0.7	2.6	0.7	0.7	3.3	1.3	0.7	7.9	100.0

from Naco were on display at the Arizona State Museum during the period this research was conducted, thus measurements were taken from casts curated in the School of Anthropology at the University of Arizona.

THE CLOVIS LITHIC ASSEMBLAGES

In the following pages, the Clovis lithic assemblage from El Fin del Mundo is described. Brief descriptions of the Clovis components from El Bajío and the San Pedro River Valley sites are also presented. Definitions of the technological and typological categories used in these descriptions can be found in Collins (1999), Huckell (2007), Bradley, Collins, and Hemmings (2010), and Sánchez (2010).

El Fin del Mundo

The Clovis lithic assemblage from El Fin del Mundo consists of 152 artifacts as of 2021 classified into bifaces ($n = 54$), unifaces ($n = 34$), and unretouched blades (cortical and noncortical) and blade manufacture byproducts ($n = 64$) (Table 5.1). Except for five Clovis points recovered during the excavation of Locus 1, the totality of the assemblage was collected from the surface of different loci. The densest concentration of Clovis artifacts (53.3% of the total assemblage) was identified in Locus 5, approximately 600 m south from Locus 1. Locus 10, located immediately to the east of Locus 5, produced the second densest concentration of Clovis artifacts (19.1% of the total assemblage). The rest of the assemblage was recovered from dispersed scatters of lithic materials in the upland loci to the south and northwest of Locus 1.

Bifaces

The assemblage includes 26 secondary bifaces, 10 Clovis point preforms or fluted secondary bifaces, and 18 Clovis points. Most bifaces (n= 35), including nine Clovis points, were recovered from the surface of the contiguous Loci 5 and 10. Seven Clovis points were found in Locus 1, five during excavation and two on the surface. In addition, two fragments of secondary bifaces were recovered from the surface in Locus 1, near the slopes of the landform that contained the Clovis gomphothere kill feature. The rest of the bifaces were recovered from the upland loci. Rhyolite is

Table 5.2. Raw Materials by Lithic Category from El Fin del Mundo

	Raw Material Type											
Lithic Category	*Chert*	*Dacite*	*Jasper*	*Obsidian*	*Quartz*	*Quartz Crystal*	*Quartzite*	*Rhyolite*	*Vitrified Basalt*	*Not Identified igneous*	*Not Identified Sedimentary*	*Total Number of Lithics*
Bifaces												
Secondary bifaces						1		23	1	1		26
Clovis point preforms	1		1		1		1	5		1		10
Clovis points	11					2	1	4				18
Unifaces												
End scrapers	27		4	1							1	33
Side scrapers	1											1
Blades												
Wedge-shaped blade core	1											1
Conical core									1			1
Core tablet flakes							2		2			4
Crested blades			1				1	4				6
Cortical blades							2	1			2	5
Noncortical blades	12	2	1				8	17	3		4	47
Total Number of Lithics	53	2	7	1	1	3	15	54	7	2	7	152
Percentages	34.9	1.3	4.6	0.7	0.7	2.0	9.9	35.5	4.6	1.3	4.6	100.0

the most common raw material among bifaces, accounting for 59.3 percent of these artifacts. Chert represents 25.9 percent, and the remaining ~15 percent includes other poorly represented raw materials (Table 5.2).

Secondary Bifaces. Secondary bifaces were classified as such based on techno-morphological features that suggest a more advanced stage of the manufacture process in comparison to the primary bifaces from the Murray Springs site (Huckell 2007:191--192). These features include a smaller size, more closely spaced flake scars that were often removed in serial fashion, fewer irregularities in shape and thickness, a more elongated shape, and differentiated distal and proximal ends. The secondary bifaces are all fragmentary and exhibit some features characteristic of Clovis biface reduction including overshot and overface biface-thinning flake scars and large lanceolate morphology. Thirteen are proximal fragments, seven are distal, five are medial (Figure 5.1), and one is an indeterminate fragment. The distal and medial fragments exhibit straight or slightly convex parallel lateral margins (Figure 5.1a–g). The proximal fragments display slightly convex basal morphologies, except for one with a straight base and one that exhibits a slightly concave base (Figure 5.1h–r). Most of these fragments broke transversely to their long axes during the manufacture process in angles that vary from 90° to 75°. One of the medial fragments found in Locus 5 (Artifact #59096) was manufactured on vitrified basalt that may have come from Cerro La Vuelta in El Bajío (Figure 5.1s). This raw material identification is tentative, however, as it is based only on macroscopic observations.

There is an inherent degree of uncertainty to surface assemblages, and some of these biface fragments determined to be Clovis may be part of later components. However, four of them exhibit overshot flaking (Figure 5.1a, q, r, and t), and the rest were included in the Clovis assemblage due to the presence of overface flake scars (Smallwood 2010, 2012), as well as sizes and morphologies comparable to those of some of the finished Clovis points and Clovis point preforms from the site, and to Clovis secondary bifaces from El Bajío. Furthermore, no similar bifaces have been observed in other sites nearby that have produced exclusively or predominantly Archaic lithic assemblages such as La Playa (Ochoa D'Aynés 2004), La

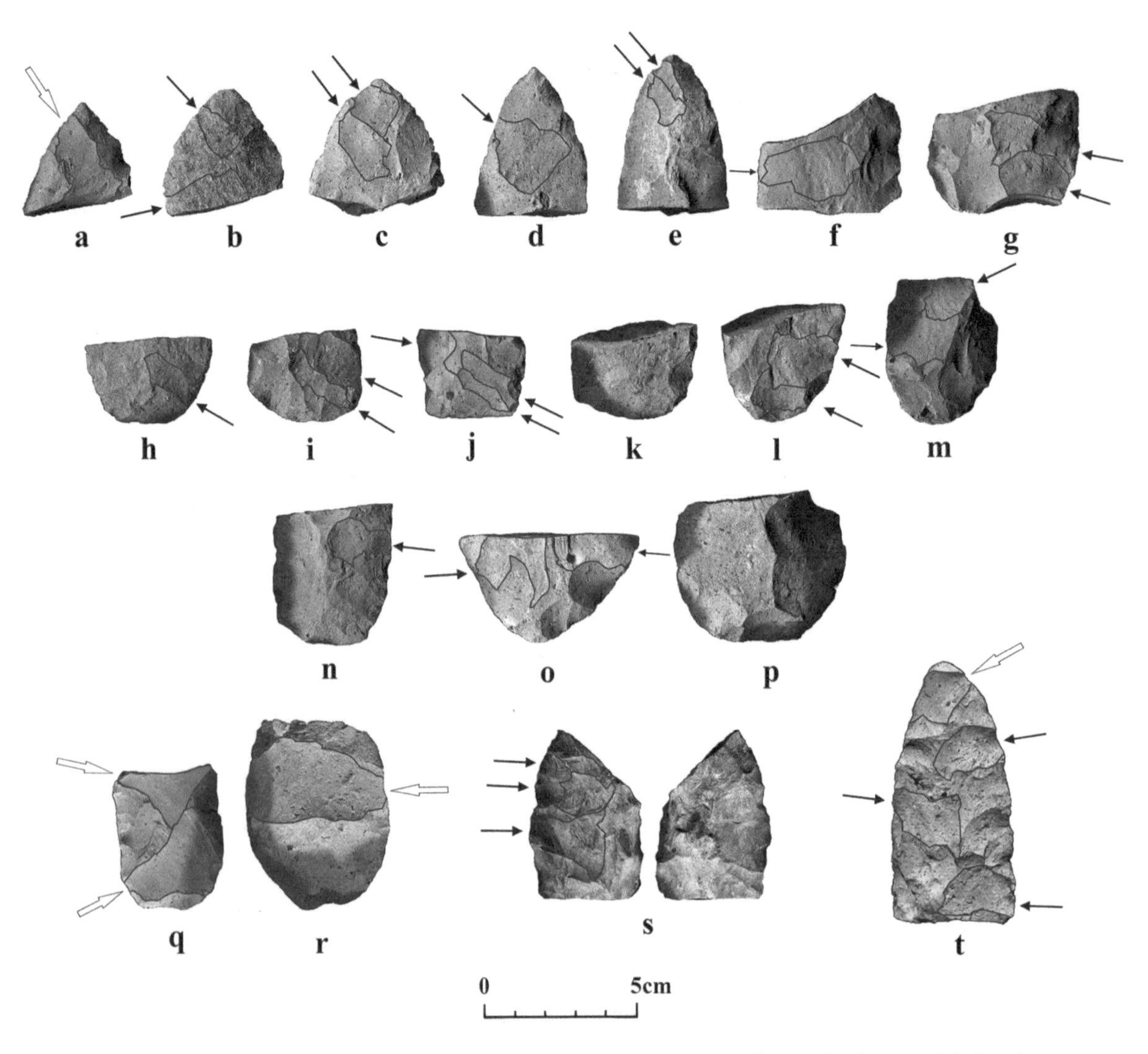

Figure 5.1. Selected examples of secondary bifaces from El Fin del Mundo: (a–e and t) distal fragments; (f–g) medial fragments; (h–r) basal fragments; (s) secondary biface fragment possibly made on vitrified basalt; and (t) distal fragment of large biface found on the surface of Locus 1. White arrows indicate the direction of an overshot flake scar; gray outlines indicate the location and black arrows indicate the direction of an overface flake scar. Modified from Sánchez-Morales 2018:Figure 2.

Mona and La Hilareña (Martínez Ramírez 2012), suggesting that these artifacts are less likely to be associated with the Archaic component of El Fin del Mundo. In addition, the differences in raw material selection observed between the Clovis points and the Archaic points (Figure 5.2) from the site, as well as their overall dimensions provide support for our classifications. For instance, in a sample of 121 Archaic points from the upland loci, the best represented raw materials are milky quartz, accounting for 57.9 percent; rhyolite, representing 14 percent; and basalt and chert, both accounting for 11.6 percent (Sánchez-Morales 2012). Meanwhile, among Clovis points these same rocks account for 1.8 percent, 21.1 percent, 0.0 percent, and 57.9 percent respectively.

Clovis Point Preforms/Fluted Secondary Bifaces. Ten bifaces have been classified as Clovis point preforms or fluted bifaces based on the presence of basal fluting on at least one face, dimensions comparable to those of finished Clovis points, and, in some specimens, more regular edge morphology than secondary bifaces (Sánchez-Morales 2018). However, they lack basal/lateral grinding, fine

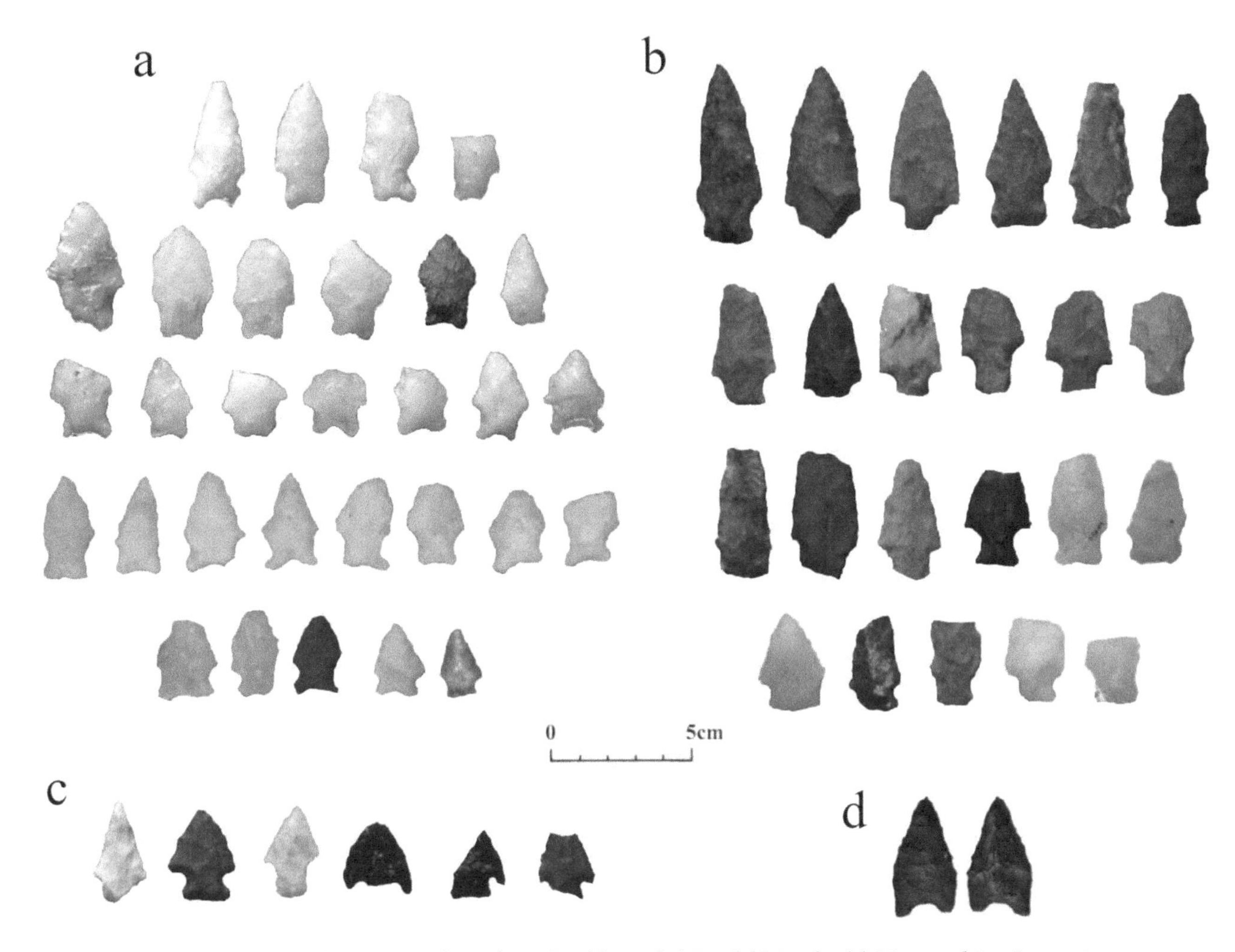

Figure 5.2. Selected Archaic points from the upland loci of El Fin del Mundo: (a) Pinto and San Jose points (Middle Archaic); (b) San Pedro points (Early Agricultural Period); (c) Cienega points (Early agricultural Period); and (d) possible Dalton/Golondrina point with beveled body (Late Paleoindian/Early Archaic). Photographs by Ismael Sánchez-Morales.

marginal retouch, and, in most cases, the basal concavity characteristic of finished Clovis points. Nine of these fluted bifaces are basal fragments and only one is a complete specimen broken into two fragments (Figure 5.3). Most of them (n= 6) were found in Loci 5 and 10.

Four fluted bifaces made on rhyolite (Figure 5.3a, c, d, and f) exhibit basal thinning through a single channel flake on only one side. Only three of these specimens preserve the full length of the channel flake scar, and one of them (Figure 5.3d) also exhibits overshot flaking. Another rhyolite biface is a possible basal fragment of a Clovis point preform (Figure 5.3h), which was reworked on its distal end to produce a working edge. It broke diagonally, possibly around the mid-length of the body, and the resulting margin was retouched bifacially to create an asymmetrical convex edge. This specimen is not fluted but it displays a concave base and slightly convex lateral edges. It is narrower and thinner than the other four rhyolite, fluted bifaces, and it is similar in size and morphology to some of the Clovis points in the assemblage. The artifact also exhibits lateral grinding, suggesting that it was used hafted or prepared for hafting. This biface is not considered to be a reworked Clovis point because of the absence of basal fluting, and because the basal and lateral retouch is not as fine and regular as that observed on the finished Clovis points from the site and the outlines of the basal and lateral edges are relatively irregular.

Fluted bifaces made on jasper (Figure 5.3b), quartz (Figure 5.3e), and an unidentified igneous rock, probably dacite, exhibit a single flute on one face. The basal fragment

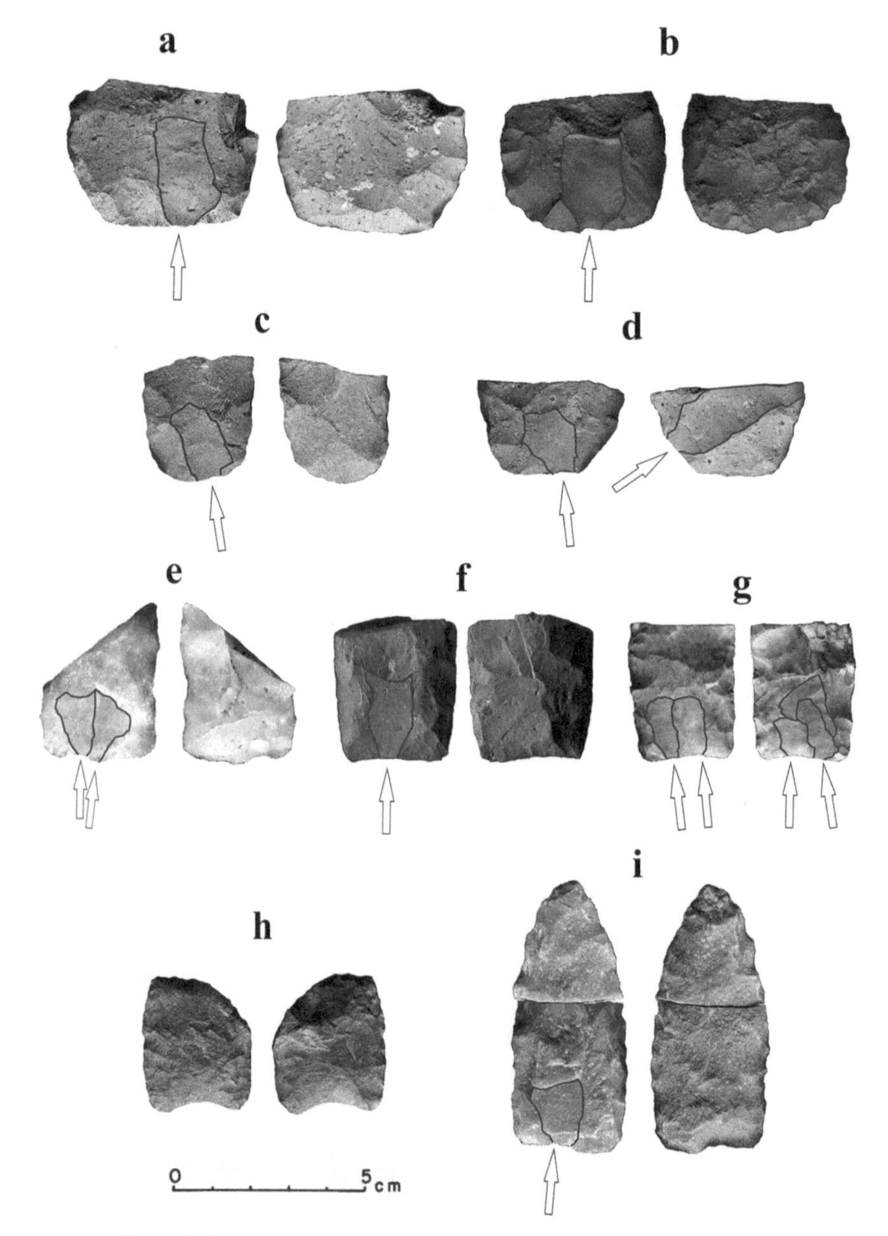

Figure 5.3. Selected Clovis point preforms/fluted bifaces from El Fin del Mundo: (a-f) basal fragments of secondary bifaces that exhibit single fluting on only one face; (g) basal fragment of Clovis point preform that exhibits multiple fluting on both faces; (h) basal fragment of a possible unfluted Clovis point preform reworked on its distal end to form a convex cutting edge; (i) Clovis point preform broken into two fragments displaying single fluting on only one face. Arrows indicate the direction of a channel or overshot flake negative. From Sánchez-Morales, Sánchez, and Holliday 2022:Figure 4.

of a preform made on chert (Figure 5.3g) exhibits multiple fluting on both faces, showing at least two channel flake scars on each side. It exhibits a slightly concave base and straight lateral margins that are not ground. It broke transversely, possibly at one third of the total length of the biface. The only complete Clovis point preform was made on quartzite (Figure 5.3i) and is broken into two fragments. It displays a lanceolate morphology with an asymmetrical, slightly concave base and shows a short channel flake scar on one face. The body has a thicker section towards its center on the same face as the flute. Apparent thinning of this thicker area of the body was intended through the detachment of flakes from the closest lateral margin but the attempts were unsuccessful, terminating in step fractures.

Clovis Points. Eighteen whole and fragmentary Clovis points have been found at El Fin del Mundo. The collection includes seven complete points, two nearly complete specimens with a snapped basal corner, one specimen with a catastrophic impact fracture, six proximal fragments, and two distal fragments (Figure 5.4). Seven points were recovered in Locus 1 (Figure 5.4a–g). Four of these were in direct association with the Upper Bonebed; three are complete and one is a distal fragment that exhibits a transverse snap break (Figure 5.4d). The other three Clovis points from Locus 1 are complete specimens. One was found during the excavation of a bioturbated area on the east side of the excavation area (Figures 2.2, 4.2, 5.4f) and the other two were found on the surface in proximity to the landform that contains the bonebed (Figure 5.4e, g).

All the Clovis points from Locus 1 exhibit fine retouch and symmetrical or nearly symmetrical bodies. Two of these points are large (~9cm long), finely made specimens (Figure 5.4a,e) that could be described as "classical Clovis," resembling in size and shape points from the San Pedro Valley (Huckell 2007) and the Blackwater Draw site (Hester, Lundelius, and Fryxell 1972). One of these large points (#63177) was found in association with the gomphothere bones, and another (# 46023) was found on the surface a few meters to the south of the landform. The distal fragment from Locus 1 probably belongs to a large point of similar dimensions, as indicated by its comparable width and thickness (Table C.3). The other four points are smaller, complete specimens. The two smaller points associated with the bonebed are probably reworked from distal fragments of larger points, as suggested by the width and thickness of their distal halves, which are comparable to those of the two complete large points, their low length-to-thickness (7.1:1 and 7.7:1) and low length-to-width ratios (2.1:1 and 2.5:1), and a high degree of basal tapering (Figure 5.4b, c).

The point found during the excavation of a bioturbated area is a complete, small specimen (~5 cm long) that displays a fine and symmetrical outline with straight lateral margins and does not exhibit evidence of extensive resharpening, suggesting that it was made purposely smaller (Figure 5.4f). The remaining Clovis point from Locus 1 is made on clear quartz crystal and displays a short and wide body (Figure 5.4g). It was found on the surface of the north slope of the landform of Locus 1 during the inspection of the excavation units at the beginning of the 2008 field season. It was approximately 1 m from the north edge of the area excavated in 2007. This was an area very likely subjected to erosion. For this reason and due to the exceptionally good preservation of the point despite its fragility, it likely eroded out from a buried context in the Upper Bonebed between the 2007 and 2008 field seasons.

Eleven Clovis points were collected from the surface of the upland loci (Figure 5.4h–r). These include one distal and six basal fragments, three heavily refurbished points, and one point with a catastrophic impact fracture. Most of these artifacts were found in Loci 5 and 10 (n= 9), and the other two points were recovered in Loci 8 and 9, which represent low density scatters of lithic materials on the west and north edges of Loci 5 and 10, respectively.

The six proximal fragments from the upland loci exhibit transverse snap fractures: four of them nearly at 90° from the lateral margins, and the other two in a more diagonal fashion at about 75° (Figure 5.4p–q). Two specimens (Figure 5.4m, n) seem to have broken at approximately one third of their total length when compared to the largest complete points from Locus 1, which have similar basal widths (Table C.3). Another three proximal fragments with similar dimensions to the other specimens seem to have broken even closer to the base (Figure 5.4o–r). The distal fragment exhibits a diagonal snap fracture and belongs to a large, finely retouched point (Figure 5.4h). The dimensions of this fragment suggest that the complete point was similar in size to the two largest complete specimens from Locus 1. Given that the diagnostic elements of the base are not preserved in this specimen, its classification as Clovis was based on its dimensions, flaking pattern, fine craftsmanship, and morphology comparable to the large specimens from Locus 1. In addition, it was found in the central area of Locus 5, where the densest concentration of Clovis materials was collected.

It has been reported in previous descriptions that two points from the upland Loci (Figure 5.2d and Figure 5.4k) that were classified as Clovis display heavily refurbished distal portions that resulted in short and thick beveled bodies with irregular lateral margins (Sánchez-Morales 2018; Sánchez-Morales, Sánchez, and Holliday 2022). This alternate-beveled retouch is characteristic of transitional late Paleoindian/Early archaic types such as Dalton/Golondrina, best known from central and southern Texas and northeastern Mexico (Jennings, Smallwood, and Greer 2016). This could potentially indicate a previously undocumented early Holocene component at El Fin del Mundo. Archaeological sites with Late Paleoindian-Early Archaic components in Sonora are unknown and diagnostic artifacts from this period are extremely rare, but four isolated surface finds of beveled points with no basal fluting were reported by Gaines, Sánchez, and Holliday (2009) from three different sites south of Hermosillo. This may well be the case for one of the specimens from El Fin del Mundo

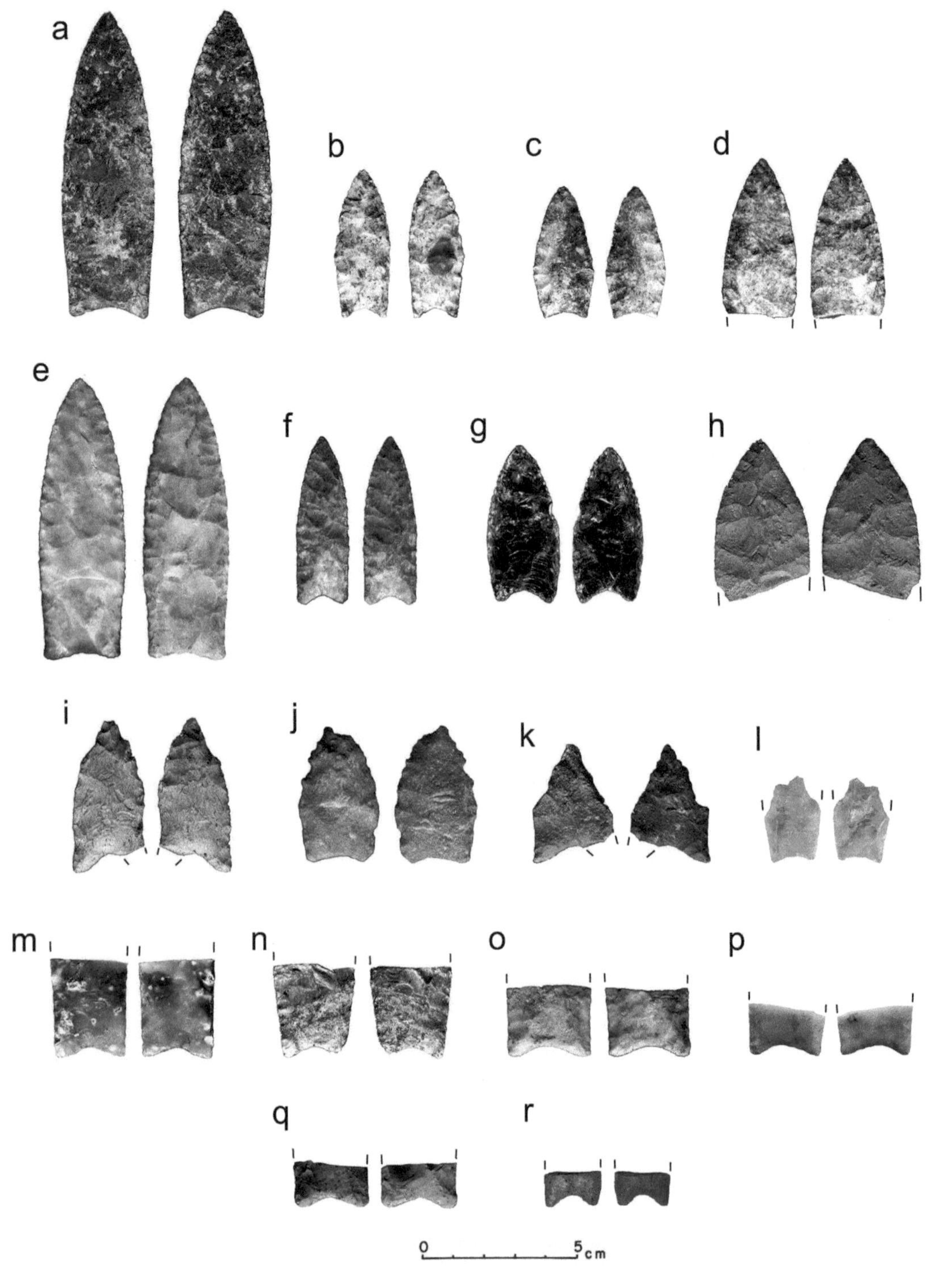

Figure 5.4. Clovis points from El Fin del Mundo: (a–d) found in direct association with the bonebed in Locus 1 (a, #63177; b, #63008; c, #62943; d, #62942); (e–g) found in disturbed contexts in Locus 1 (e, #46023 and g, #58342, found on the surface; f, #59569, found during the excavation of a highly bioturbated area of Locus 1); (h–r) found on the surface of the upland campsite loci. The collection includes four complete, unrefurbished points (a, e, f, and g), five repaired or highly refurbished points (b, c, , j, and- k), six basal fragments (m–r), two distal fragments (d and h), and one point displaying catastrophic impact damage (l). Photographs by Ismael Sánchez-Morales.

(Figure 5.2d), which is made on an igneous rock, probably dacite, displays poor workmanship compared to most Clovis points, and is not truly fluted but rather basally thinned through pressure flaking. However, the other specimen (Figure 5.4k) is still considered to be Clovis. This point is made on high-quality chert. It shows very fine craftmanship, is fluted on both surfaces, and the basal corner that is present flares slightly out towards the end, a feature also observed on six other Clovis points from the site (Figure 5.4a, e, i, m, o, q). The artifact is therefore considered to represent a rare case of refurbishing of a Clovis point, which resulted in a highly reduced, beveled body.

Two other heavily refurbished points are somewhat less reduced on their distal ends, and they are not beveled. One of these points (Figure 5.4i) also displays poor craftmanship, although it exhibits multiple fluting on one face and one single channel scar on the other. The last Clovis point from Locus 5 is a small, proximal fragment, comparable in dimensions to the smaller points from Locus 1. This point exhibits a complex fracture that probably resulted from a catastrophic impact, destroying the distal end and most of the medial section of the point (Figure 5.4l).

Chert was the raw material most often utilized in the manufacture of Clovis points at El Fin del Mundo (Table 5.2). Eleven points were made on chert, two of them on the same variety (Figure 5.4e and o). Rhyolite was utilized in the manufactured of four Clovis points, including one of the large complete points from Locus 1, one distal fragment from Locus 5, one basal fragment from Locus 10, and the heavily reduced point with a snapped basal corner from Locus 9. Two other points were made on clear quartz crystal, including a complete specimen found on the surface of Locus 1 and the small specimen with an impact fracture from Locus 5. Lastly, one point was made on quartzite.

Unifaces

The unifacial implements from El Fin del Mundo include 33 end scrapers and the chert side scraper (#45980) found on the surface of Locus 1 (Figure 5.5). The latter lacks any diagnostic features and its inclusion in the Clovis component of the site is based purely on its possible association with the Clovis context in Locus 1. This artifact is described in Chapter 4(Figure 4.13); here the focus is on the collection of end scrapers. Assignment of these artifacts to the Clovis component is based on the presence of features observed on end scrapers reported from several Clovis sites in different regions such as El Aigame and El Bajío in Sonora (Sánchez 2016), Murray Springs in Arizona (Huckell 2007), Gault in Texas (Waters, Pevny, and Carlson 2011), and Paleo Crossing in Ohio (Eren and Redmond 2011). These features include the recurrent, though not exclusive, utilization of blades as blanks; the small size; and the presence of gravers or spurs on one or both distal corners (Morrow 1997:71). Raw material was also considered given that high quality tool stones such as cherts are common among the Clovis points and blades from the site but rare in the Archaic projectile point component (Sánchez-Morales 2012).

Virtually all end scrapers were found in Locus 5 (n= 32), except for one recovered in Locus 8. Twenty-two were made on prismatic blades (Figure 5.5l–ff) and 11 on biface thinning flakes with distally expanding lateral margins (Figure 5.5a-k). In almost all the specimens the bulb of percussion was removed through invasive flaking from one or both lateral margins (Sánchez-Morales 2018). All of them exhibit fine, abrupt or semi-abrupt marginal retouch on the distal end and a working edge angle that varies between 40° and 80°. Seventeen specimens also display continuous marginal retouch on one or both lateral margins. When retouch is present on the lateral margins it is restricted to a section and not the full length of the edge, most often on the distal half. In addition, 19 specimens display a spur or short projection on one of their distal corners. Features suggestive of hafting are present in four scrapers, which exhibit shallow notches on both lateral edges (Figure 5.5m, q, s, and dd) and on four others with a notch on only one lateral edge (Figure 5.5b, n–o) on the proximal half.

Chert is the dominant raw material among the unifaces (82.4%; Table 5.2) with several chert varieties present in the scraper collection. In fact, each end scraper is made on a different variety of chert. The remaining end scrapers were made on jasper, obsidian, and an unidentified raw material of sedimentary origin. The obsidian end scraper (Figure 5.5y) is the only diagnostic Clovis artifact made on this raw material recovered at the site.

Unretouched Blades and By-products of Blade Manufacture

Blade production at El Fin del Mundo is represented by 52 unretouched blades classified into cortical (n = 5) and noncortical (n = 47). In addition 12 byproducts of blade manufacture have been recovered including 4 blade-core platform rejuvenation tablet flakes, 6 crested blades, 1 wedge-shaped blade core, and 1 conical blade core (Figures 5.6 and 5.7). These materials were assigned to the Clovis assemblage of the site based on the presence of technological features as described by Collins (1999:35–72), which include the production of "true" blades from conical and wedge-shaped cores, the preparation and detachment of

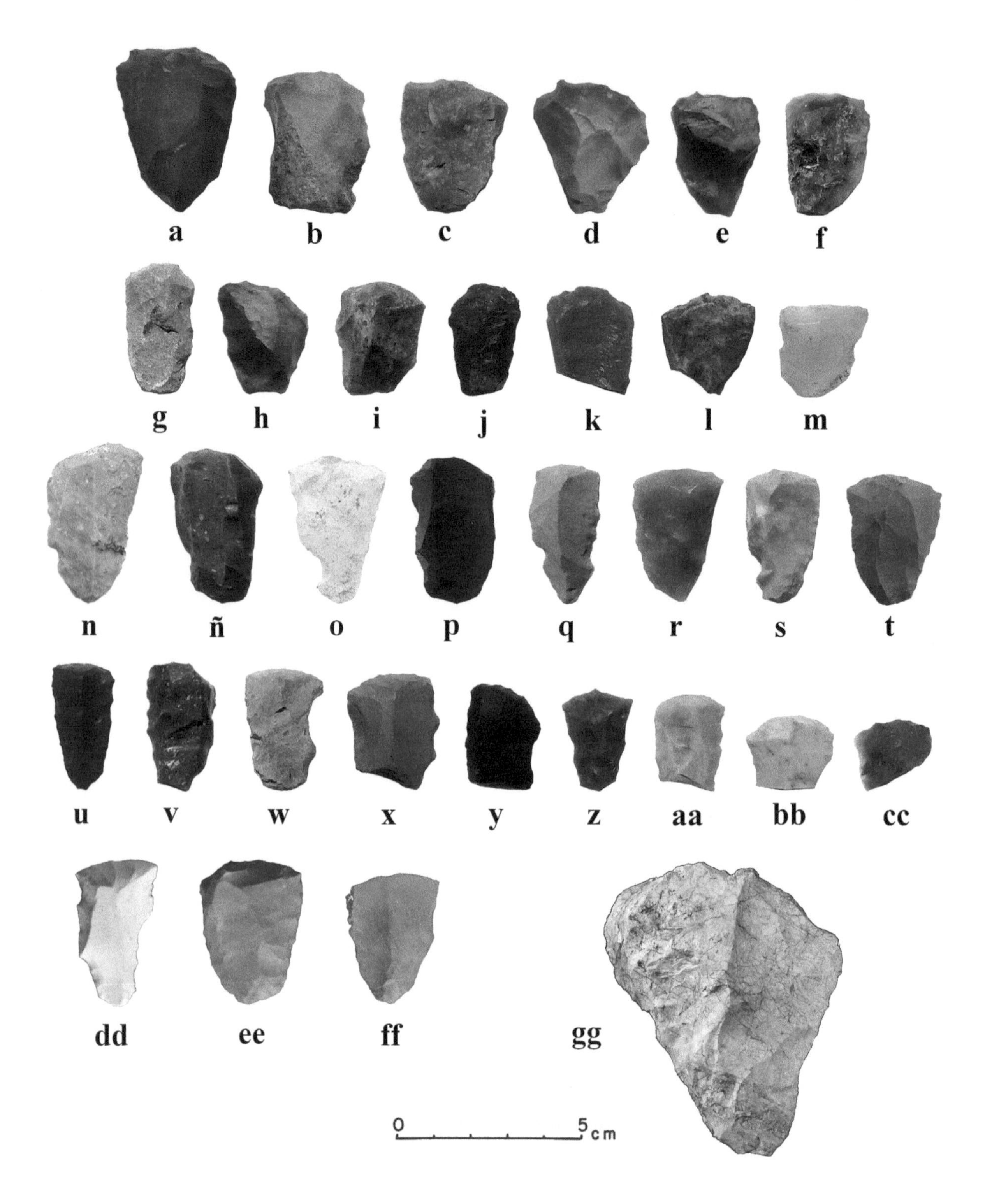

Figure 5.5. Clovis unifaces from El Fin del Mundo. End scrapers: (a–j) made on flakes; (k–ff) made on blades. Scrapers k, l, aa–cc are incomplete broken specimens; (gg) side scraper from the surface of Locus 1. Modified from Sánchez-Morales 2018:Figure 4.

Figure 5.6. Blade production byproducts from El Fin del Mundo: (a) blade core platform rejuvenation tablet flakes, the first and third from the left possibly made on vitrified basalt; (b) wedge shaped blade core; and (c) exhausted conical blade core possibly made on vitrified basalt. Arrows indicate striking directions. Modified from Sánchez-Morales, Sánchez, and Holliday 2022:Figure 7.

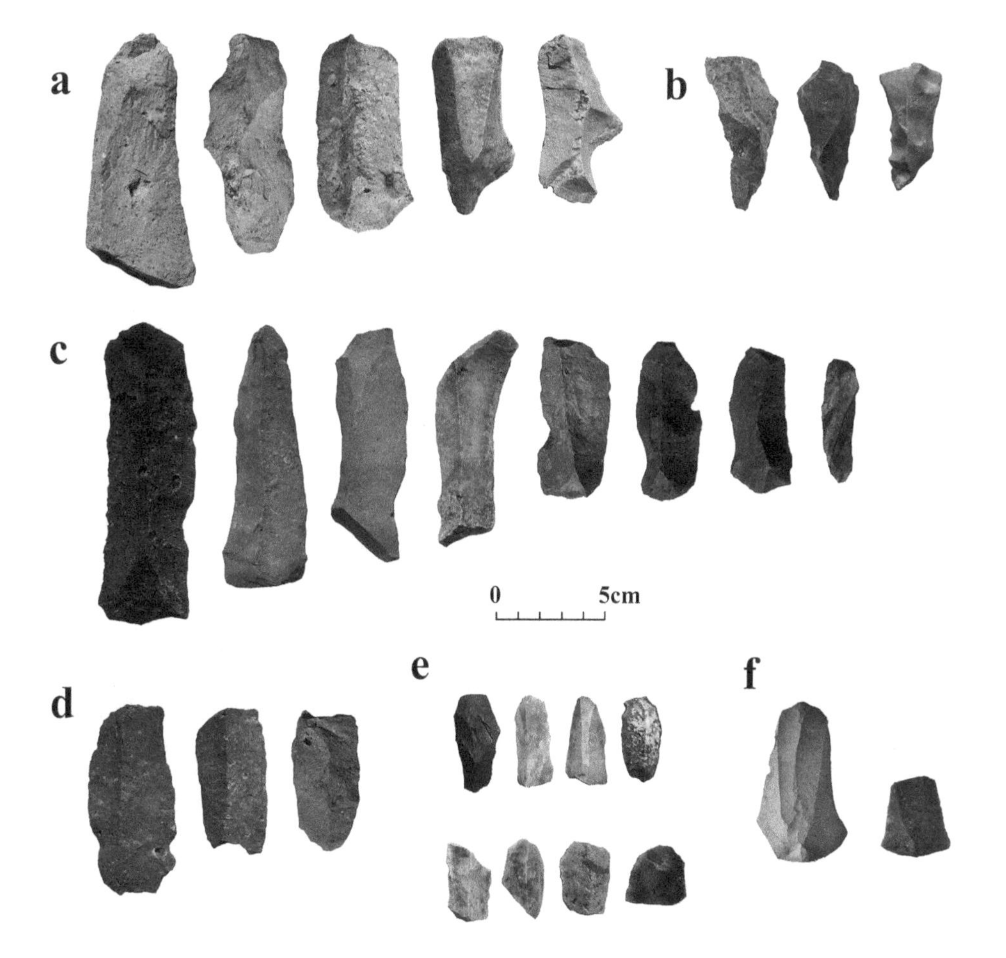

Figure 5.7. Selected blades from El Fin del Mundo and El Bajío: (a) complete and fragmentary rhyolite and quartzite cortical blades; (b) crested blades; (c) rhyolite and quartzite noncortical blades; (d) fragmentary rhyolite noncortical blades; (e) blades made on different varieties of chert (upper row are complete, lower row are proximal and distal fragments); and (f) complete overpassed blade from El Bajío made on vitrified basalt (left) and distal fragment of overpassed blade from El Fin del Mundo made on quartzite (right). Artifacts are oriented with the striking platform upwards. Modified from Sánchez-Morales 2018:Figure 5.

crested blades to guide following detachments, and the rejuvenation of blade core platforms through the detachment of tablet flakes. In addition, just as in the case of secondary bifaces, the production of prismatic blades has not been observed in lithic assemblages from sites in the region with Archaic or later components, but it is present at El Bajío (Sánchez 2016) and Murray Springs (Huckell 2007), suggesting that this technology is more likely associated with the Clovis occupation at El Fin del Mundo. Over half of these artifacts (51.6%) were collected from Locus 5, representing diverse stages of the blade production sequence.

Rhyolite was the raw material most frequently utilized for blade production at El Fin del Mundo (34.4% of the total blade mode). Chert (20.3%) and quartzite (20.3%) are also well represented among blades and blade manufacture byproducts. Six artifacts (9.4%) including the only conical blade core, two tablet flakes, and three blades were made on vitrified basalt that may be from El Bajío. As mentioned before, this identification is only tentative as it is based on macroscopic observations. The remaining blades were produced on other poorly represented raw materials (Table 5.2).

Wedge-Shaped Core. One wedge-shaped blade core made on yellow chert was collected from Locus 5 (Figure 5.6b). Blades were obtained from only one face of the core, which exhibits four blade negatives. The core is small (41.7 mm maximum dimension) and the angle between the platform and the last blade negatives on the flaked face is greater than 90°, suggesting that it was deemed an exhausted core and discarded.

Conical Blade Core. One exhausted conical core was found on the surface of the central area of Locus 5 (Figure 5.6c). It was made on vitrified basalt that may have come from El Bajío as suggested by the texture, flaking quality, and the color of the patina that covers the whole surface of the artifact. The platform is more or less flat and roughly circular in plan view. The dimensions and morphology of the platform are comparable to those of the blade core rejuvenation tablet flakes also found in Locus 5. Blades were obtained in a unidirectional pattern and four blade negatives are preserved. The core is short in lateral view and may have been discarded due to exhaustion, as suggested by its small dimensions (51.96 mm maximum dimension) and the negatives of failed detachments that ended in step fractures.

Blade Core Rejuvenation Tablet Flakes. Four tablet flakes were identified in the collection, all of them from Locus 5 (Figure 5.6a). Two were detached from quartzite cores, both of the same yellowish-brown variety. The other two were detached from cores possibly made on vitrified basalt, and one of them exhibits heavy patination. All the tablet flakes display plain platforms, prominent bulbs of percussion, and one convex multifaceted face that retains the negatives of blades and covers approximately half of the perimeter of the tablet.

Crested Blades. Six complete crested blades were collected at the site (Figure 5.7b), four made on rhyolite, one on jasper, and one on quartzite. Most of them (n=5) exhibit plain platforms, triangular cross sections, and bifacial flaking, typical of the preparation of the crest on the core to produce the first blade and set up the flaking face for subsequent blade detachments.

Cortical Blades. Five cortical blades, two made on quartzite, one on rhyolite, and two on unidentified raw materials of sedimentary origin, were found in the upland Loci (Figure 5.7a). These blades exhibit cortex on 25 to 50 percent of their dorsal face. On three of them the original cortical surface of the core was used as the striking platform; on the other two the platform is plain.

Noncortical Blades. This category includes 47 blades, of which 26 are complete, 4 are semi-complete, and 17 are fragmentary (Figure 5.7c, d, e). Most of them were found in Loci 5 (n= 24) and 25 (n=11). All of these specimens display at least two parallel unidirectional facets or negatives of previous blade detachments and triangular or trapezoidal cross sections; most have a plain platform. Twenty-three noncortical blades display chipping and macroscopic striation on at least one of their lateral margins, suggesting that they were used, although some of these marks could have resulted from post-depositional damage such as trampling. One distal fragment of a large prismatic blade made on quartzite overpassed the core and completely removed the core's distal end (Figure 5.7f). Similar examples of overpassed Clovis prismatic blades have been identified at El Bajío (Sánchez 2016). Most noncortical blades were made on rhyolite (n=17), with all of the blades from Locus 25 made on this material. Other well-represented raw materials are chert (n=17) and quartzite (n=12). In addition, three were possibly made on vitrified basalt. The other seven noncortical blades were made on other uncommon raw materials (Table 5.2).

Given that all cortical and noncortical blades are surface finds, it remains possible that some of them may not have derived from specialized Clovis methods of blade production and are rather blade-like flakes that resulted from other Clovis or later methods of tool-blank manufacture. More detailed studies are needed to better support their association with the Clovis component of El Fin del Mundo. Our future research incorporates comparative statistical analyses of metrics of complete artifacts with published data from other Clovis blade assemblages such as the Keven Davis cache in Texas (Collins 1999) and the Sinclair site in Tennessee (Tune, Jennings, and Deter-Wolf 2022).

Raw Materials

Overall, the Clovis lithic assemblage from El Fin del Mundo shows variety in raw materials but it is largely dominated by rhyolite (35.5% of total assemblage) and chert (34.9 %) (Table 5.2). Rhyolite is the most common raw material among blades and bifaces, representing over half of the bifacial implements (Table 5.2). It is naturally abundant at the site and can be found in a primary source in the outcrop of Locus 25 and as cobbles in the nearby arroyos and riverbeds. Chert is most common among end scrapers, representing over 80 percent of them, and is also the second most represented raw material among bifaces and blades (Table 5.2). Primary sources of chert are unknown in the region and no secondary sources of this material have been identified during the systematic survey of the site area. Thus, we consider chert to be an exogenous tool stone. Geological maps indicate that the closest concentrations of limestone are present approximately 40 km to the southeast (González-León y Moreno-Hurtado 2021), in the region between El Fin del Mundo and El Bajío. Potentially undiscovered chert sources could be present in this area (Sánchez-Morales, Sánchez, and Holliday 2022). Quartzite is the third most common raw material in the assemblage and is the most commonly found in the blade mode. Quartzite is locally available in the form of cobbles that occur in the nearby arroyos and riverbeds, particularly in Loci 10, 17, and 22.

Other raw materials present in the assemblage, although poorly represented, include jasper, dacite, quartz crystal, quartz, and obsidian (Table 5.2). All of these rocks, except for obsidian, which is an exogenous raw material of unknown provenance, are locally available in the stream beds that dissect the site. Small outcrops of igneous rocks such as basalt have also been observed in Locus 22. In addition, quartz crystal and white quartz can be found along a quartz vein that outcrops approximately 5 km southwest of the site and continues for approximately 2 km reaching a nearby hill known as El Cerro del Cuarzo (Sánchez-Morales 2018). Even though concentrations of lithic artifacts and debris have been observed at this outcrop, no diagnostic Clovis stone tools have been found here to link the Clovis occupation of El Fin del Mundo to this locus. Finally, vitrified basalt from El Bajío may be present at El Fin del Mundo, raising the possibility of links between the two sites.

El Bajío

El Bajío has produced the largest Clovis stone tool assemblage yet recovered in Mexico. The analysis of this collection conducted by Sánchez (2016) includes 334 lithics that were divided into blade industry, unifacial industry, bifacial industry, expedient miscellaneous cores, and hammers. Most of these artifacts (94.3% of the complete assemblage) were recovered from surface scatters of lithic materials that also included an Archaic component (Sánchez 2016; Sánchez-Morales 2012). In her analysis, Sánchez (2016) included several lithic types that are not diagnostic of Clovis technology and are common in Archaic and Ceramic traditions as well as in other Paleoindian assemblages. These lithic types include unifacial implements such as side, composite, and circular scrapers; denticulates; gravers; notches;, bifacial drills; and the lithic classes of expedient miscellaneous cores and hammers. The association of these materials (n= 80) with diagnostic Clovis tools on the surface and the heavy patina that most of them exhibit prompted Sánchez (2016) to include them in the Clovis assemblage from El Bajío. However, given that these are not diagnostic artifacts of Clovis technology and their association with the rest of the Clovis assemblage from the site is uncertain, they have been excluded from this study in order to be consistent with the assemblage from El Fin del Mundo, which also excludes nondiagnostic lithic tool types from surface contexts.

The assemblage of diagnostic Clovis stone tools from El Bajío used in this analysis includes 254 artifacts (Table 5.3). Blades and byproducts of blade manufacture are the most numerous, consisting of 122 artifacts (48% of total sample) ranging from blade cores to numerous blades and blade production byproducts. The category of bifaces includes 102 artifacts (40.2%) representing varying stages of the manufacture process. Unifaces include 30 end scrapers (11.8%) made on flakes and blades. Sánchez (2016) provides an extensive description of the lithic assemblage and the results of her analysis; a short summary is included here. The metric and qualitative information of these artifacts can be found in Sánchez 2016.

Table 5.3. Frequencies of Lithic Categories and Provenance Counts of Clovis Stone Tools from El Bajío

Lithic Categories	*Montané collection*	*Cerro Rojo*	*Locus 1*	*Locus 2*	*Locus 4*	*Locus 5*	*Locus 6*	*Locus 7*	*Locus 8*	*Locus 10*	*Locus 12*	*Locus 15*	*Isolated*	*Total Number of Lithics*
Bifaces														
Primary bifaces	6	1			5	1		1					2	16
Secondary bifaces	18	2		1	6			1		2	4		2	36
Square-based bifaces	14		1		3				1		9		5	33
Clovis point preforms	2				6	1		1	1	2		1	2	16
Clovis points													1	1
Unifaces														
End scrapers	14	9					1	1	1	2			2	30
Blades														
Wedge-shaped blade core	5					1	2			1				9
Conical core					1		1	1						3
Core tablet flakes	4					1								5
Crested blades	5						1	3						9
Cortical blades	9					1	1				1			12
Non-cortical blades	30				6	7	4	10		9	1		4	71
Blade core error recovery flakes	2					1		1					2	6
Platform maintenance flakes	4					1		2						7
Total Number of Lithics	**113**	**12**	**1**	**1**	**27**	**14**	**10**	**21**	**3**	**16**	**15**	**1**	**20**	**254**
Percentages	**44.5**	**4.7**	**0.4**	**0.4**	**10.6**	**5.5**	**3.9**	**8.3**	**1.2**	**6.3**	**5.9**	**0.4**	**7.9**	**100.0**

Bifaces

The bifaces (n = 102) were divided into five categories (Table 5.3; Sánchez 2016). Secondary bifaces are the most numerous (35.3% of all bifaces) (Figure 5.8b). Square-based bifaces, a separate category of secondary bifaces included by Sánchez due to their recurrent shared base morphology, are the second largest group (32.4%) (Figure 5.8c). Sánchez (2016: 113–117) suggests that square-based bifaces are most likely associated with the Clovis component of the site, as indicated by the presence of overshot flaking and basal thinning through the removal of one or several channel-like flakes on most of the specimens. Primary bifaces are also present in the assemblage (15.7%) (Figure 5.8a).

Sixteen Clovis point preforms (15.7% of all bifaces), 13 of which are basal fragments, display concave bases, lateral margins that expand from the base, and basal fluting (Figure 5.8e; Sánchez 2016). Only one complete Clovis point is present in the biface sample studied by Sánchez (2016). This point is housed at the museum of the town of Carbó, Sonora, and its exact provenience within the site is unknown. It displays single, short flutes on both faces, and basal and lateral grinding (Figure 5.8f). In her analysis Sánchez (2016) also included one fluted basal fragment (# 37628) recovered during the 2003 field season in the Clovis point category. During the reanalysis of the specimen carried out for this study, however, it was concluded that it is a fragment of an unfinished point as indicated by the irregular outline and the lack of grinding of the basal and lateral margins and was included in the Clovis point preform category.

Sánchez (2010:293) comments that at least a dozen more Clovis points have been collected from El Bajío by amateur archaeologists and were not available for her analysis. Photographs and drawings of some of these artifacts are included in an article by amateur archaeologists Manuel Robles and Francisco Manzo (1972). All the Clovis points from El Bajío were made on the local vitrified basalt

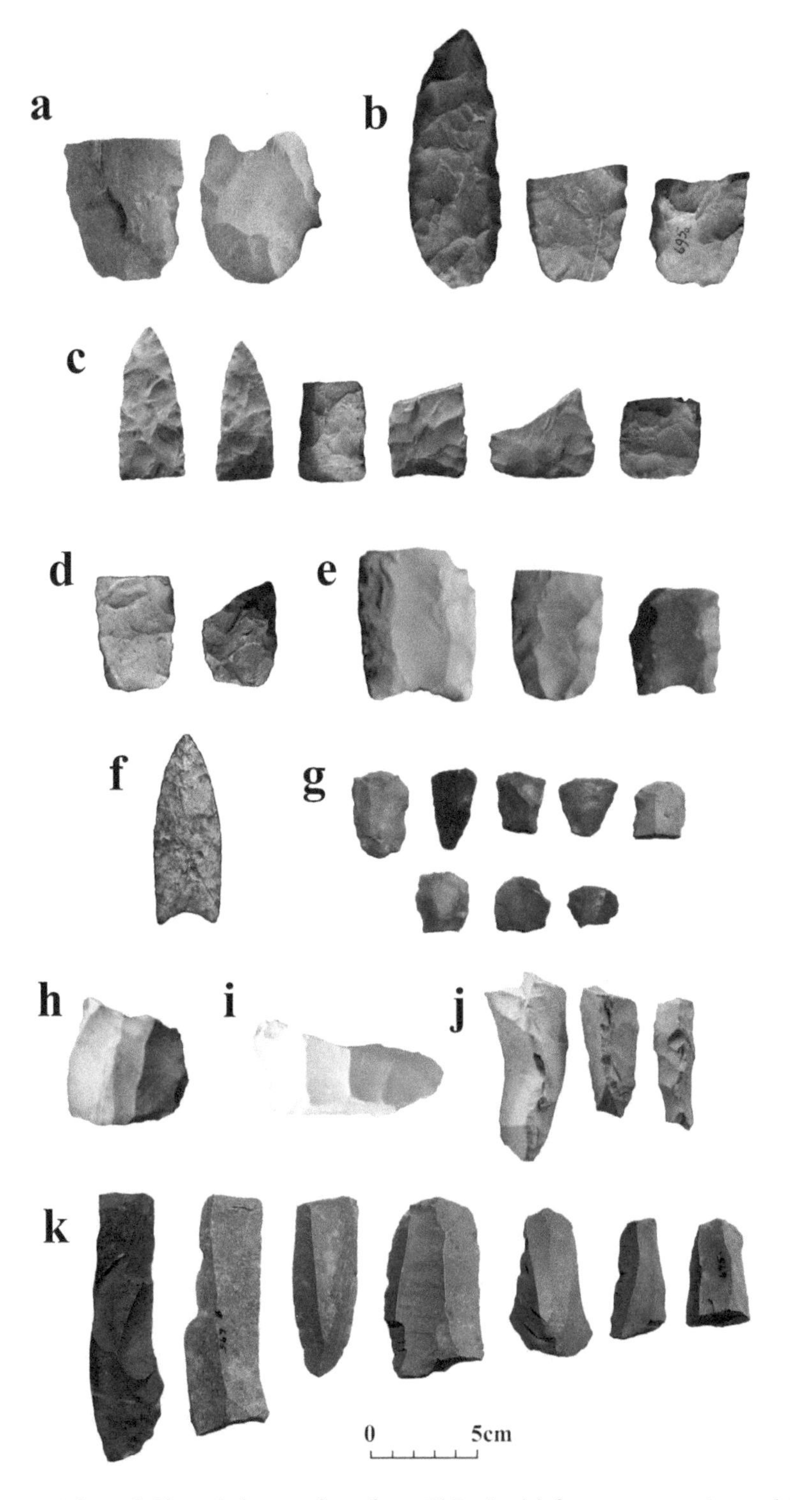

Figure 5.8. Selected Clovis lithic artifacts from El Bajío: (a) fragmentary primary bifaces; (b) complete and basal fragments of secondary bifaces; (c) complete and fragmentary square-based bifaces; (d) basal fragments of secondary bifaces with single fluting on only one face; (e) basal fragments of Clovis point preforms; (f) Clovis point housed at the museum of Carbó; (g) end scrapers made on different varieties of chert; (h) wedge-shaped blade core; (i) blade-core rejuvenation tablet flake; (j) crested blades; and (k) complete and fragmentary noncortical blades. All artifacts, except for end scrapers, were made on vitrified basalt.

Table 5.4. Raw Materials by Lithic Category from El Bajío

Lithic Category	Raw Material			*Total Number of Lithics*
	Vitrified Basalt	*Chert*	*Rhyolite*	
Bifaces				
Primary bifaces	16			16
Secondary bifaces	35	1		36
Square-based bifaces	33			33
Clovis point preforms	16			16
Clovis points	1			1
Unifaces				
End scrapers	19	11		30
Blades				
Wedge-shaped blade core	9			9
Conical core	3			3
Core tablet flakes	5			5
Crested blades	9			9
Cortical blades	9	1	2	12
Noncortical blades	65	6		71
Blade core error recovery flakes	6			6
Platform maintenance flakes	7			7
Total Number of Lithics	**233**	**19**	**2**	**254**
Percentages	**91.7**	**7.5**	**0.8**	**100.0**

except for one basal fragment made on obsidian, which was collected by professional archaeologist Julio Montané (1996) in the 1980s.

Sánchez (2016) concludes that the production of projectile points was an important part of the biface industry at El Bajío but that other bifacial tools, likely intended for use as knives, gravers, and for other tasks employed in campsite activities, were also produced at the site. In addition, Locus 12 represents a knapping station where square-based bifaces were recovered along with debris related to the manufacture of these tools (Sánchez 2016:59–60). Virtually all the bifaces from El Bajío were manufactured using the local vitrified basalt (99% of all bifaces) (Sánchez 2016).

Unifaces

Unifacial tools are represented by 30 end scrapers. Seventeen were manufactured on flakes and 13 were made on prismatic blades (Figure 5.8g; Sánchez 2016). Fourteen of these tools display a spur on one or both distal corners. Fifteen specimens exhibit notches in the lateral margins near the base that may be hafting-related. Twelve display marginal retouch on one or both lateral margins. When present, lateral retouch is irregular compared to the shaping of the distal end. Most end scrapers were made on vitrified basalt (63.3%), but exogenous chert is also well represented (36.7%).

Unretouched Blades and Byproducts of Blade Manufacture

Sánchez (2016) identified 122 artifacts representing all stages of Clovis blade production (Figure 5.8h–k). These artifacts were classified into nine categories including conical and wedge-shaped cores, crested blades, cortical and noncortical blades, and byproducts of blade production such as tablet flakes (Table 5.3). Vitrified basalt is the most represented raw material among blades and blade manufacture byproducts (92.7%). Chert (5.7%) and rhyolite (1.6%) are also present among blades.

Raw Materials

The lithic assemblage from El Bajío shows very low diversity of raw materials and is highly dominated by the locally abundant vitrified basalt, which accounts for roughly 92 percent of the collection (Table 5.4). The rest of the

Table 5.5. Frequencies of Lithic Categories and Provenance Counts of Clovis Stone Tools from San Pedro Valley Sites

Lithic Category	*Naco*	*Lehner*	*Leikem*	*Navarrete*	*Escapule*	*Murray Springs*	*Total Number of Lithics*
Bifaces							
Secondary bifaces						7	7
Clovis points	9	13	1	1	2	19	45
Unifaces							
End scrapers						4	4
Other unifacial tools		7				19	26
Blades							
Noncortical blades						13	13
Expedient Tools		1				3	4
Total Number of Lithics	**9**	**21**	**1**	**1**	**2**	**65**	**99**
Percentages	**9.1**	**21.2**	**1.0**	**1.0**	**2.0**	**65.7**	**100.0**

assemblage is represented by exogenous chert varieties (7.5%) of unknown provenance, which are most common among end scrapers. Locally available rhyolite is also present, but it accounts for less than 1 percent of the assemblage.

Clovis Lithic Assemblages from the San Pedro River Valley

In this section a short summary of the Clovis lithic assemblages from the San Pedro River Valley sites is presented. The organization of this summary is based on Huckell's (2007) analysis of the Clovis lithic technology from the Upper San Pedro River Valley conducted from 1974 to 1976, expanded in 1981, and revised in 1986. This analysis included the lithic materials from the Murray Springs, Naco, Lehner, Leikem, Escapule, and Navarrete sites. The summary is presented by lithic categories combining all sites, except for Clovis points that are described by site. The metric and qualitative data of the Clovis stone tools from each site are included in Appendix C.

The combined Clovis lithic assemblage from the San Pedro River Valley includes 99 artifacts (excluding associated debitage and debris). Clovis points are the dominant tool type in the assemblage and over half of the Clovis stone tools from the San Pedro River Valley sites are included in the category of bifaces (52.6% of total assemblage) (Table 5.5), contrasting with the assemblages from El Fin del Mundo and El Bajío where blades and byproducts of blade production are the most numerous. The category of unifaces is the second best represented (30.3%), followed by blades (13.1%), and expedient tools (4%). Most of the artifacts in the assemblage were recovered from Murray Springs (65.7% of total assemblage), as this is the most extensive site excavated in the region and the only one that contains an extensive campsite.

Bifaces

Clovis Points. The collection of bifacial tools from the San Pedro River Valley includes 45 Clovis points: 31 that are complete or nearly complete, 2 points with snapped bases, 1 distal fragment, 3 tip fragments, 2 mid-sections, and 6 proximal fragments (Figures 5.9 and 5.10). The sample displays a high morphological and metric variation even in assemblages recovered from a single animal carcass (Haury 1953; Huckell 2007).

Naco. Nine points were recovered at the Naco site. Eight were found in association with the mammoth remains and one was collected on the surface in the arroyo upstream (Haury 1953). The morphology of most of the points is lanceolate with lateral margins that curve from the point towards the base, with the maximum width located at the approximate mid-point of the body (Figure 5.9r-y). Points A-10900, A-11912, and A-11913 (Figure 59s, t, and w respectively) display lateral margins that recurve slightly near the base. Seven points exhibit basal thinning through one single channel flake on both sides. Clovis points A-10901 and A-10902 (Figure 59u and x) display multiple fluting (three channel flakes) on one face and one single flute and some smaller flake scars on the other face. Grinding of the basal edge and the lateral margins near the base is present in all the points, extending upwards for about one-third to one-half of the total length of the point.

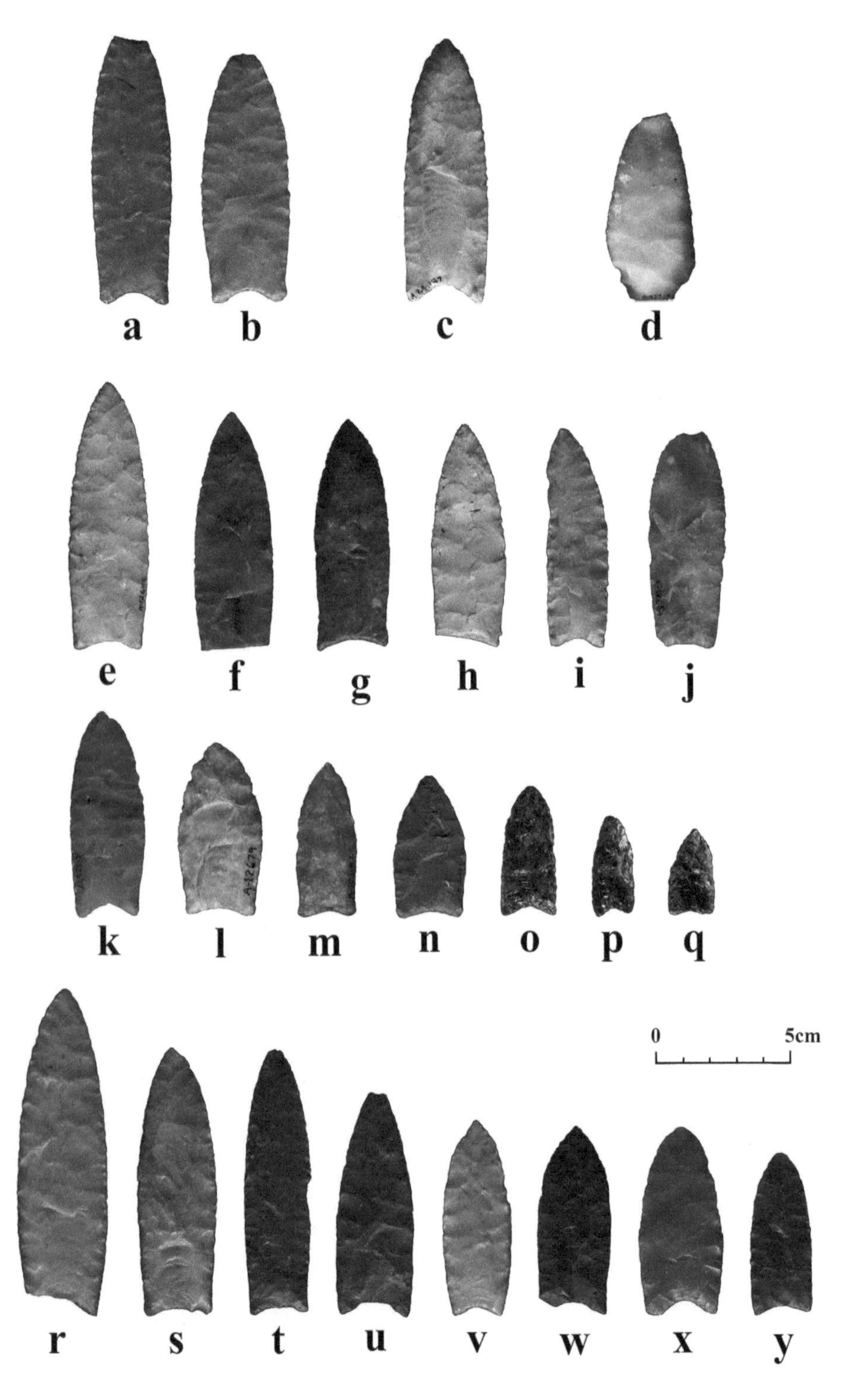

Figure 5.9. Clovis points from the San Pedro River Valley: (a and b) Escapule; (c) Leikem; (d) medial fragment from the Navarrete site; (e–q) Lehner; and (r–y) Naco. The Clovis points from Naco are casts of the originals. Cast of the Clovis point No. A-11914 from Naco was not available for study. Photographs by Ismael Sánchez-Morales.

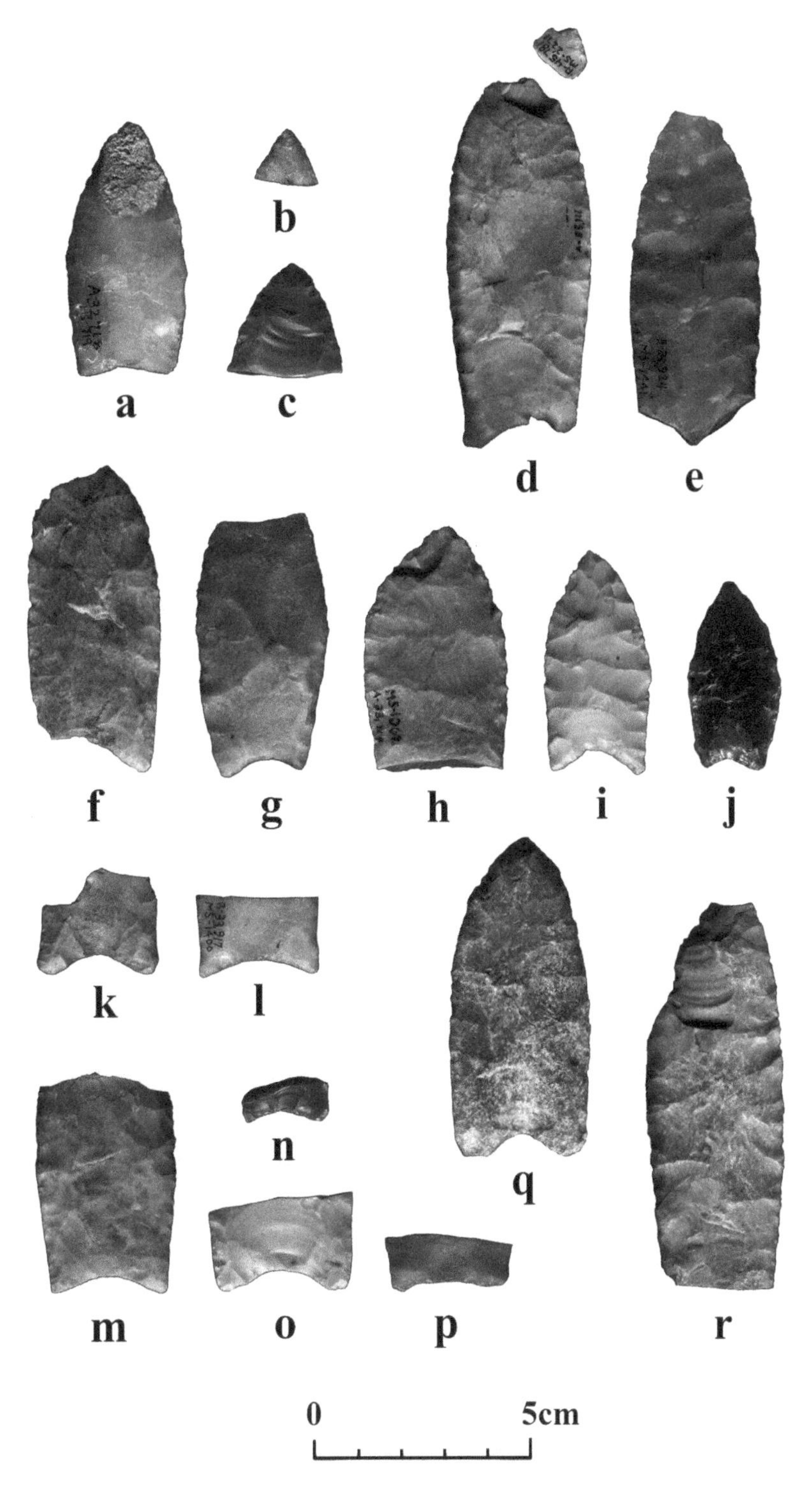

Figure 5.10. Clovis points from Murray Springs: (a–c) Area 3 (bison/mammoth kill); (d–j) Area 4 (bison kill); (k–l) Area 5; (m–p) Area 7 (campsite); and (q–r) Area 6 (campsite). Clovis point A-33111 from Area 4 (specimen d) exhibits impact damage that corresponds to one impact flake found in Area 7. The impact flake is shown above d. Clovis point A-32991 from Area 6 (specimen r) exhibits an impact flute that corresponds to a channel flake found in Area 4. Clovis point A-33925 from Area 4 was not available for study. Photographs by Ismael Sánchez-Morales.

The size range between the largest and smallest Clovis points from Naco is significant with specimens measuring from approximately 58 mm to 117 mm in length. The small size of Clovis point A-10904 (Figure 5.9y) cannot be attributed to reworking, repair, or heavy refurbishing as in the case of some small specimens from El Fin del Mundo. This point from Naco does not exhibit evidence of heavy resharpening of the blade and the general outline of the margins is symmetrical. Also, the relative proportions of its length, thickness, and basal width are consistent with the larger specimens, suggesting that this point probably was crafted as a small artifact.

Two other Clovis points in the Naco collection, A-10900 and A-10902 (Figure 5.9w and x), are relatively small, measuring 67.8 mm and 67.7 mm long respectively. These specimens show refurbishing of the distal portion that resulted in a shortened, wide body, and in the case of Clovis point A-10902 the distal lateral margins are irregular and asymmetrical. Clovis point A-10899 (Figure 5.9v) also exhibits irregular and asymmetrical distal lateral margins that resulted from the resharpening of the body. Point A-11912 (Figure 5.9t) shows a major burin break originating at the base and traveling up one margin toward the tip (Huckell 2007). The burin scar was removed by bifacial pressure flaking, and a new base was created by recentering the concavity and re-fluting the point (Huckell 2007). The same burin breakage and repair pattern is observed on Clovis point A-12675 (Figure 5.9i) from Lehner.

Lehner. The excavations at Lehner yielded 13 Clovis points. All the points display a lanceolate morphology with straight to convex lateral margins, rounded corners and concave bases (Figure 5.9e–q). However, Clovis point A-12676 (Figure 5.9g) exhibits a recurvate outline of the lateral edges near the base with acute slightly expanding corners (Figure 5.9g). Haury, Sayles, and Wasley (1959:14) suggest that this basal narrowing resulted from the grinding of the lateral edges near the base, and it is barely discernable on some of the other points in the collection. Fine marginal retouch is present and is most obvious in Clovis point A-12685 (Figure 5.9e).

The variability in size of the Clovis points from Lehner is high and similar to that observed in the collections from El Fin del Mundo, Naco, and Murray Springs. The points from Lehner vary in length from about 31 mm to 96 mm. The smallest specimens are made on quartz crystal and display poor craftsmanship compared to the other Clovis points from the site (Figure 5.9o – q). In addition, these small points are not truly fluted. Basal thinning was accomplished through the removal of several short and narrow flakes from the base towards the distal end (i.e., end-thinning). The rest of the points exhibit a single flute on both faces, sometimes flanked by shorter flake scars. The basal edge and the lateral margins exhibit grinding that extends about one-third of the total length of the points.

The Clovis point collection from Lehner includes examples of damaged and refurbished points. Clovis points A-12679 and A-12686 (Figure 5.9l and m) display heavy refurbishing of the distal end that resulted in a short wide body. Points A-12674 and A-12680 (Figure 5.9f and h) exhibit snapped bases likely broken where they were hafted. The basal fragments, which could have remained attached to the haft, were probably discarded later at a camp location. Something similar must have happened at El Fin del Mundo, where several basal fragments have been found in the upland loci and one distal fragment was recovered from the bonebed. Clovis point A-12684 (Figure 5.9j) exhibits an impact fracture that produced a flute extending from the tip baseward (Haury, Sayles, and Wasley 1959). Point A-12675 (Figure 5.9i) displays a burin break fracture originating at the base along one of the lateral margins. As in Clovis point A-11912 (Figure 5.9t) from Naco, the burin scar was removed by bifacial pressure flaking, and the base was recentered and fluted again resulting in an asymmetrical with a markedly narrow proximal portion. These points suggest that retrieval and reuse of projectile points was practiced (Haury, Sayles, and Wasley 1959).

Leikem, Navarrete, and Escapule. One Clovis point was found at the Leikem site, a reworked midsection was recovered from the Navarrete site, and two nearly complete Clovis points were recovered at the Escapule site. Because this assemblage of Clovis points is small they are grouped and briefly described here. All the points exhibit lanceolate morphology with straight to convex lateral margins. The points from Escapule (Figure 5.9a and b) exhibit slightly recurving lateral margins near the corners as in some examples from El Fin del Mundo, Naco, Lehner, and Murray Springs. All the points exhibit parallel marginal flaking, but fine edge retouch is lacking. Grinding of the basal edge and the proximal lateral margins is present in all the specimens. Basal thinning via a single channel flake on both faces was observed on the point from Leikem (Figure 5.9c) and one of the points from Escapule (Figure 5.9a). Point A-31232 (Figure 5.9b) from Escapule exhibits multiple fluting on both sides. The two points from Escapule exhibit tip fractures; point A-31231 (Figure 5.9a) also shows an impact flute.

The biface fragment found at the Navarrete site (Figure 5.9d) has been considered as a reworked Clovis point fragment by Haynes (2007:3). But the fragment does not retain any diagnostic features, and its width and thickness are greater than those of the largest Clovis points from the studied collections. Its inclusion in the Clovis lithic components of the San Pedro Valley is based on its stratigraphic and spatial association with the mammoth remains but may represent a different type of bifacial implement.

Murray Springs. The sample of complete and fragmentary bifacial tools from Murray Springs includes 19 Clovis points and 7 non-projectile point bifaces (Figure 5.10). Thirteen complete and damaged Clovis points were associated with the kills. Six points were found in the campsite areas, most of them basal fragments. In addition, the mid-section of a Clovis point (A-32991; Figure 5.10r) from Area 6 was linked to the bison kill in Area 4 by a refitted impact flake, and a Clovis point (A-33111; Figure 5.10d) from Area 4 was linked to the campsite in Area 7 by another refitted impact flake.

The collection includes specimens that exhibit extensive damage and breakage, far more than the points from Naco or Lehner (Huckell 2007), but comparable to the Clovis points from El Fin del Mundo. The complete points (A-32718, A-32992, A-33110, and A-33116; Figure 5.10a, q, i, and j respectively) range in length from approximately 42 mm to 72 mm. These specimens were refurbished or repaired resulting in a significant reduction of their original size. The nearly complete specimens (A-33109, A-33111, A-33115, A-33924; Figures 5.10d, f, and e) display tip fractures and one or both (A-33924; Figure 5.10e) snapped corners. Proximal fragments A-33917, A-33922, A-47139, A-47140 and A-47283 (Figures 5.10k, l, n–p, respectively) represent basal snap breaks (Huckell 2007). Two of these fragments (A-33917 and A-47283; Figures 5.10l and o) broke in a transversal fashion at about 90° from the lateral margins likely on their hafted portion, displaying the same breakage pattern observed on the basal fragments from El Fin del Mundo. The distal fragment A-33114 (Figure 5.10h) exhibits lateral damage in the form of crushing near the tip; the basal portion snapped off in a transverse way. Point A-32991 (Figure 5.10r) displays a snapped base, a snapped, crushed, and impact-fluted tip, with lateral burin breakage near the tip. Specimen A-33111 (Figure 5.10d) also exhibits an impact flute at the tip and snapped basal corners. Another example of an impact flute is observed on tip fragment A-32716 (Figure 5.10c). Points A-33115 and A-33924 (Figure 5.10e and f) display tip fractures and snapped corners. All of these points were damaged beyond repair (Huckell 2007:200).

In addition, bifacial implements at Murray Springs include three complete and four fragmentary bifaces that represent unfinished tools, likely due to the difficulty of thinning excessively thick areas (Huckell 2007). The complete specimens, one of them broken into two pieces in a transversal fashion at about two thirds of length from the base, are secondary bifaces with a lanceolate morphology and slightly convex or straight bases (A-33918, A-33923, A-47170; Figure 5.11a). The distal fragment belongs to a secondary biface that broke in a diagonal way, possibly similar in morphology and size to the largest complete biface (Figure 5.11a). The other fragmentary bifaces possibly belong to large primary bifaces broken during early stages of the reduction process (A-33113, A-46365, A-47163/A-47199; Figure 5.11a). Lastly, Huckell (2007) identified 42 debitage clusters that represent episodes of biface retouching.

Unifaces

The unifacial tools include 23 artifacts from Murray Springs and 7 from Lehner. Huckell (2007) classified the assemblage into laterally retouched flakes, composite tools, gravers, and end scrapers (for detailed descriptions of these implements see Haury, Sayles and Wasley 1959 and Huckell 2007). Huckell (2007) included unifacial implements in his "flake and flake tool mode," which also includes utilized unmodified flakes. In addition, in this summary two artifacts from Lehner that were classified as keeled scrapers were included in the category of unifaces.

Laterally retouched flakes are the most abundant tool type among unifacial implements (71% of unifaces). This category includes 11 artifacts from Murray Springs made on thin flakes, probably products of biface reduction, that display one slightly retouched lateral margin through scalar continuous flaking (Figure 5.11c). It also includes sidescrapers, two from Murray Springs and five from Lehner, made on larger flakes that display invasive retouch on one lateral margin (Figure 5.11e). Two other artifacts from Lehner were included in the category of unifaces. The two keeled scrapers from Lehner (Haury, Sayles, and Wasley 1959) are fragmentary pointed tools with converging retouched margins. According to Huckell (2007) they may have been manufactured from large blades (Figure 5.11f).

There are only four end scrapers from Murray Springs (Figure 5.11b), three complete and one broken that is missing the proximal end. This contrasts with the abundant end scrapers from El Fin del Mundo and El Bajío. These artifacts are very similar in size and morphology to the scrapers from the Sonoran sites in that they exhibit abrupt marginal retouch on their convex distal margins and working edge

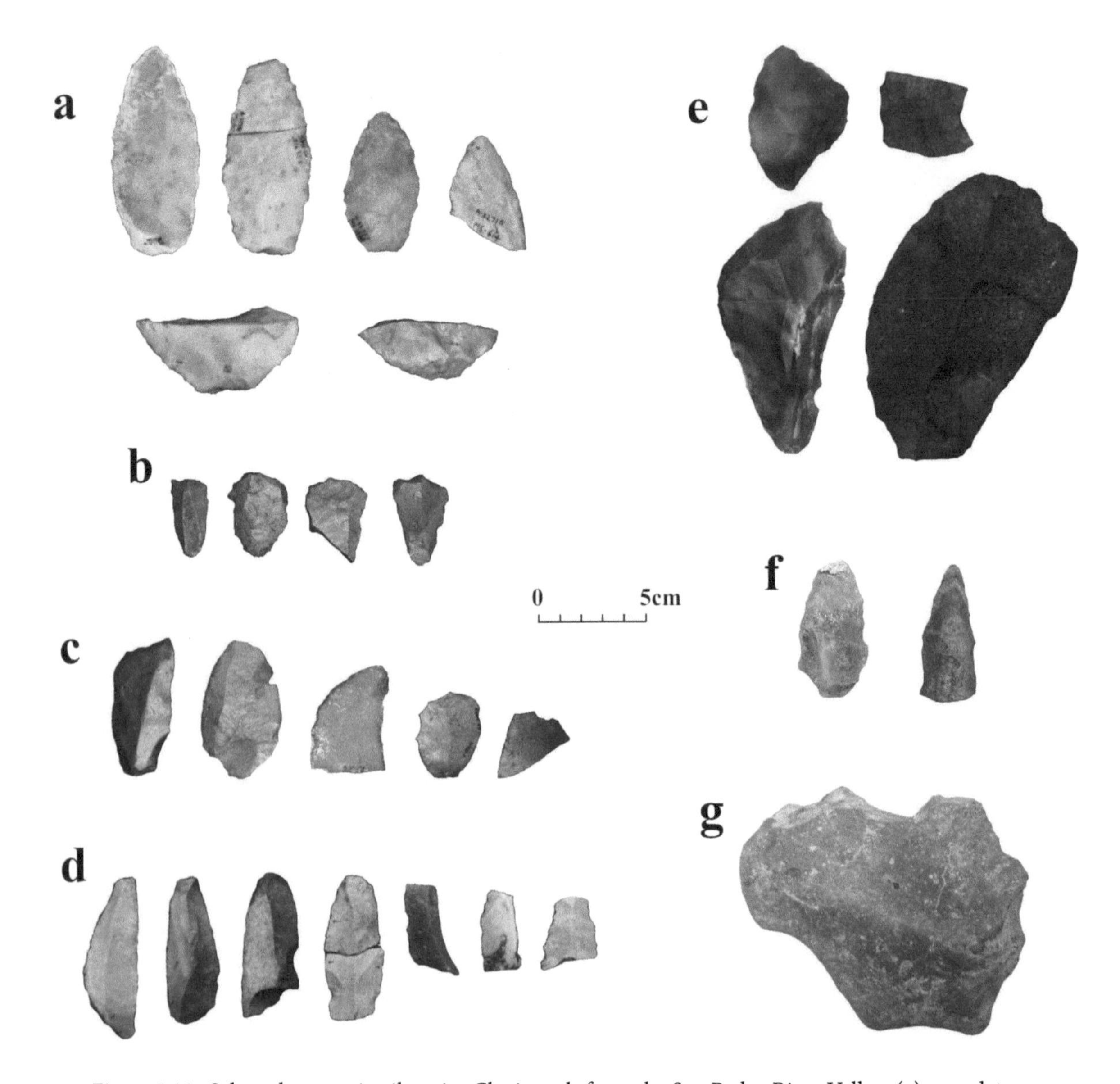

Figure 5.11. Selected nonprojectile point Clovis tools from the San Pedro River Valley: (a) complete and fragmentary primary (lower row) and secondary (upper row) bifaces from Murray Springs; (b) end scrapers from Murray Springs; (c) selected laterally retouched flakes from Murray Springs; (d) selected complete and fragmentary noncortical blades from Murray Springs; (e) selected side scrapers from Lehner; (f) keeled scrapers from Lehner; and (g) cobble chopper from Lehner. Photographs by Ismael Sánchez-Morales.

angles between 60° and 90°. There is a spur on one of the distal corners in two specimens (A-33928 and A-33943). The lateral edges of all four end scrapers show irregular marginal retouch, just like in some examples from El Fin del Mundo and El Bajío. Huckell (2007) suggests that the lateral edges of these scrapers were retouched to dull the margins and adjust the shapes to fit handles. The four end scrapers from Murray Springs were made on prismatic blades.

Blades

Thirteen complete and fragmentary prismatic blades are part of the lithic assemblage from Murray Springs. Most of these blades show a single ridge and triangular cross-sections, and three complete specimens display a marked curvature in longitudinal section (Figure 5.11d). Five blades show nicking and microspalling along their lateral margins, suggesting utilization (Huckell 2007).

Table 5.6. Raw Materials by Site from the San Pedro River Valley

Site	Raw Material													Total
	Chalcedony	*St. David Chalcedony*	*Chert*	*Felsite*	*Jasper*	*Metasandstone*	*Obsidian*	*Petrified wood*	*Quartz crystal*	*Quartzite*	*Rhyolite*	*Silicified limestone*	*Silicified siltstone*	
Naco			7	2										9
Lehner	7		5		1				3	1	1	3		21
Leikem			1											1
Navarrete		1												1
Escapule	1		1											2
Murray Springs	7	13	32			2	2	1				1	7	65
Total	15	14	46	2	1	2	2	1	3	1	1	4	7	99
Percentages	15.2	14.1	46.5	2.0	1.0	2.0	2.0	1.0	3.0	1.0	1.0	4.0	7.1	100.0

Eleven of the 13 blades were found in the campsite areas, whereas the other two were found in Areas 3 and 4.

Expedient Tools

Another Clovis stone tool mode included by Huckell (2007) comprises tools that were produced expediently from cobbles at hand and were either modified by a few percussion blows or used without modifications. Three expedient tools were recovered from Murray Springs and one from Lehner (Figure 5.11g). Two of the modified cobbles from Murray Springs and the one from Lehner can be classified as choppers. The remaining one from Murray Springs shows a rounded and battered end, suggesting it was used as a hammer. These tools may have been manufactured to aid in butchering tasks (Huckell 2007:210). The association of these expedient implements with their Clovis assemblages is unquestionable as they were found during the excavation of the buried contexts, unlike the surface finds from El fin del Mundo and El Bajío.

Raw Materials

The Clovis stone tool assemblage from the Upper San Pedro River Valley includes a variety of raw materials but is largely dominated by chert (46.5%) and chalcedony (29.3%) (Table 5.6). The identification of specific raw material sources has proved difficult (Huckell 2007: 187), although exposures of chert deposits are scattered through the Huachuca and Whetstone Mountains on the west side of the upper San Pedro Valley and in the Tombstone Hills, Mule Mountains, and Naco Hills on the east side (Figure 5.12; Huckell 2007:187). These sources contain some cherts that are lithologically similar to those found at the sites but pieces larger than 10 cm free of flaws are rarely found (Huckell 2007:187). This situation has prompted Huckell (2007) to suggest that most of the chert varieties present in the assemblages are of nonlocal origin.

More evidence for the use of exogenous raw materials is provided by the petrified wood utilized on the Clovis point tip fragment A-32716 (Figure 5.10c) from Murray Springs. The nearest known sources are north of the Mogollon Rim, approximately 225 km north of the site (Huckell 2007). In addition, the obsidian artifacts found in Murray Springs have been traced through X-ray fluorescence to a source in east-central Arizona (Huckell 2007), more than 200 km northwest from the site (Figure 5.12). Finally, a distinctive pinkish gray, occasionally banded variety of chert is present at Murray Springs, Naco, and Lehner. The source of this material is unknown. Two Clovis points from Naco and two from Lehner were manufactured on this chert, as well as several flake clusters, points, other tools, and debitage from Murray Springs (Huckell 2007).

The most abundant chalcedony variety in the assemblage is known as St. David chalcedony and is considered to be semi-local. It can be obtained from pocket-like outcrops located west of the town of Saint David, between the San Pedro River and State Route 90, approximately 35 km northwest of Murray Springs (Figure 5.12; Huckell 2007). Another local raw material present in the assemblage, although rarely used, is a greenish gray to black silicified limestone that crops out in the Lewis Hills, on the east side of the San Pedro River approximately 4 km northeast of the site (Figure 5.12; Huckell 2007:188). Unifaces A-12690,

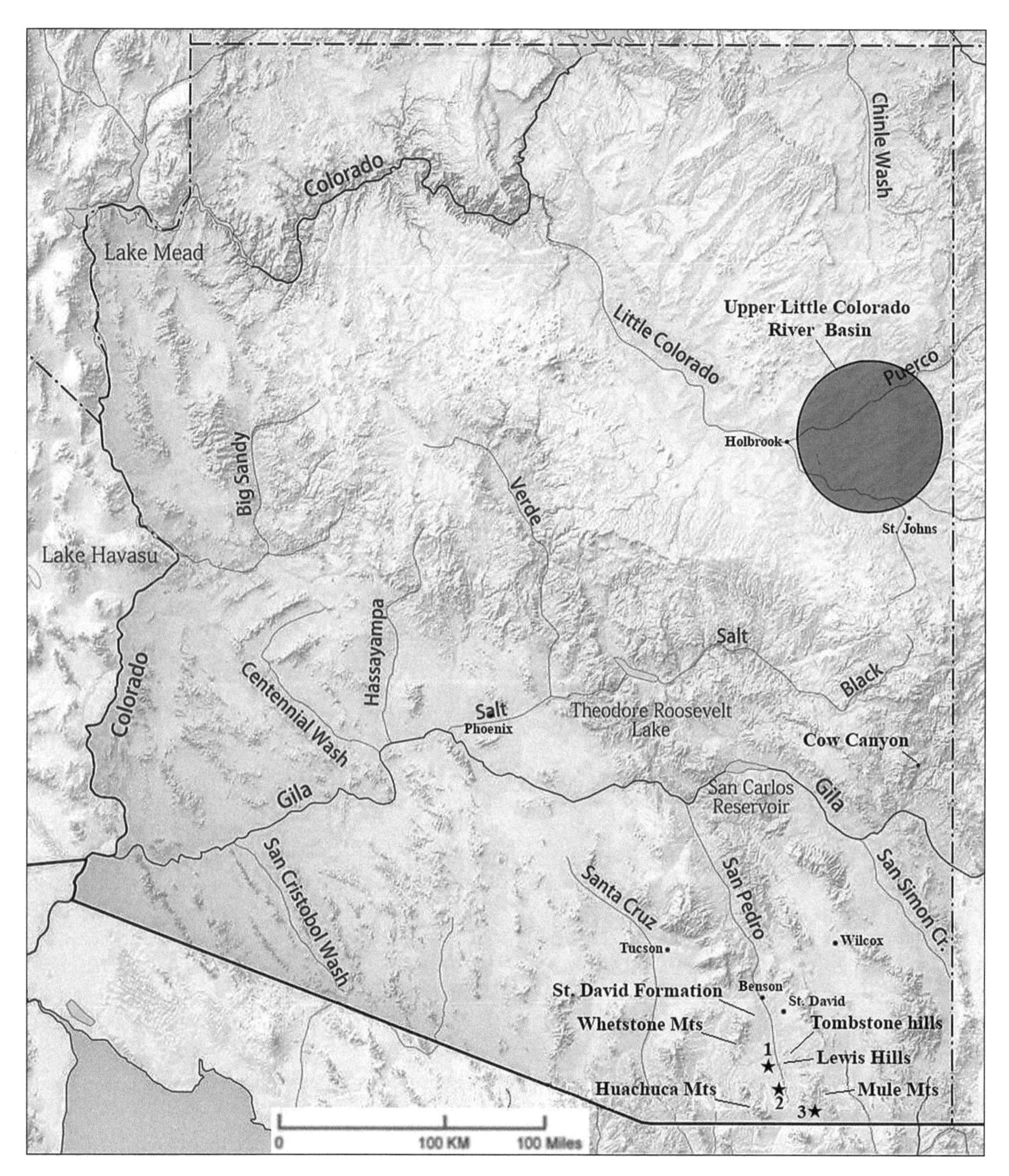

Figure 5.12. Map of Arizona showing locations of lithic raw material sources exploited by Clovis flintknappers in southeastern Arizona as indicated by the stone tool assemblages from the San Pedro River Valley (Huckell 2007): petrified wood from the upper Little Colorado River Basin; obsidian from Cow Canyon; chalcedony from St. David Formation; silicified limestone from the Lewis Hills; quartz crystal from the Huachuca Mountains; and possible sources of some chert varieties located in the Whetstone, Huachuca, and Mule mountains and in the Tombstone and Naco hills. The sources of most chert varieties present in the Clovis lithic assemblages from the San Pedro River Valley are unknown. Key: 1,=Murray Springs and Escapule; 2=Lehner; and 3=Naco, Leikem, and Navarrete sites.

A-12692, and A-12693 (Figure 5.11e) from Lehner were made on this material. As for the quartz crystal used in the manufacture of three Clovis points at Lehner, a possible source is located at the Huachuca Mountains where deposits of sizable crystals are known (Haury, Sayles, and Wasley 1959:15).

Local and nonlocal procurement of raw materials was carried out by the Clovis foragers of the San Pedro Valley. Careful selection of materials of sufficient quality to produce large bifaces and flakes was fundamental for Clovis stone tool production as indicated by the predominance of high-quality exogenous cherts in the assemblage (Huckell 2007). Locally available raw materials of lesser quality, such as metasandstone and quartzite, were also used to produce expedient tools, as observed in the assemblages from Murray Springs and Lehner (Huckell 2007).

DISCUSSION

The three stone tool categories identified in the Clovis lithic assemblage from El Fin del Mundo—bifaces, unifaces, and blades—represent major stone tool production strategies observed at other Clovis sites such as El Bajío in Sonora (Sánchez 2016), Murray Springs in Arizona (Huckell 2007), Blackwater Draw in New Mexico (Hester, Lundelius, and Fryxell 1972), Gault in Texas (Bradley, Collins, and Hemmings 2010; Waters, Pevney, and Carlson 2011), Paleo Crossing in Ohio (Eren and Redmon 2011), and Carson-Conn-Short in Tennessee (Broster and others 1996).

The numerous complete and fragmentary bifaces showing evidence of different stages of the reduction process from blank preparation to finished, highly reduced and discarded Clovis points conform to the idea that biface production seems to be the mainstay of Clovis lithic technology (Huckell 2007). The lack of primary bifaces and biface reduction debitage in the assemblage from El Fin del Mundo is due to a biased field collection strategy that was focused only on artifacts with diagnostic technological and typological features. However, five complete and fragmentary primary bifaces, mostly made on rhyolite, were collected from Locus 5 during the first field season at the site and could potentially be part of the Clovis assemblage (Sánchez-Morales, Sánchez, and Holliday 2022). The production of blades also appears to have been a major part of Clovis lithic technology in Sonora, as indicated by the abundant products and byproducts of blade manufacture recovered from the upland loci at El Fin del Mundo and from El Bajío. In southeastern Arizona, Murray Springs has yielded prismatic blades similar to those observed in the Sonoran sites. However, blade technology seems to be much less ubiquitous in the San Pedro Valley.

The large numbers of end scrapers in the upland loci at El Fin del Mundo and in El Bajío indicate that Clovis foragers heavily relied on these implements. These tools have also been found in other sites in north-central Sonora with Clovis components such as El Aigame, where four Clovis points and 40 Clovis end scrapers have been reported from a private collection (Gaines, Sánchez, and Holliday 2009; Sánchez 2016). End scrapers were also recovered from the campsite areas at Murray Springs indicating that they were widespread implements in the regional Clovis toolkit. The presence of these small, sometimes spurred end scrapers made on blades in a buried Clovis context at Murray Springs strengthens the association of the specimens found in El Fin del Mundo, El Bajío, and El Aigame with their Clovis components. Other unifacially retouched artifacts have been recovered at Murray Springs and Lehner. The non-end scraper uniface types analyzed by Sánchez (2016) from El Bajío lack a definitive Clovis association as they were collected on the surface. But the presence of different laterally retouched unifacial implements from the excavated contexts at Murray Springs and Lehner indicate that Clovis knappers produced varied unifacial tools on flakes and blades other than end scrapers.

Associations between Loci at El Fin del Mundo

The occurrence of Clovis points in Loci 1, 5, 8, 9, and 10 is evidence of occupation by Clovis foragers on the alluvial and wetland environment that existed at the site during the terminal Pleistocene and on the surrounding uplands (Figure 5.13). The presence of bifaces and blades displaying Clovis technological features in virtually all the loci in the uplands further support use of the areas surrounding Locus 1 by Clovis groups. Contemporaneity of these occupations cannot be unequivocally determined because no artifact refits were found between Locus 1 and any of the camp loci. However, the utilization of the same unusual chert variety for one of the complete Clovis points from Locus 1 (# 46023) and one of the basal fragments from Locus 5 (# 59603) suggests that groups of foragers utilized the same sources of exogenous raw materials for tool making.

Rhyolite is abundant in the Clovis assemblage from El Fin del Mundo, but it is also present in the Archaic component (Sánchez-Morales 2012). Its occurrence cannot be used as evidence of linkages among loci. The outcrop in Locus 25 is the only area of the site where rhyolite blocks large enough to produce large blanks for bifaces can be obtained. One of the Clovis points and the secondary

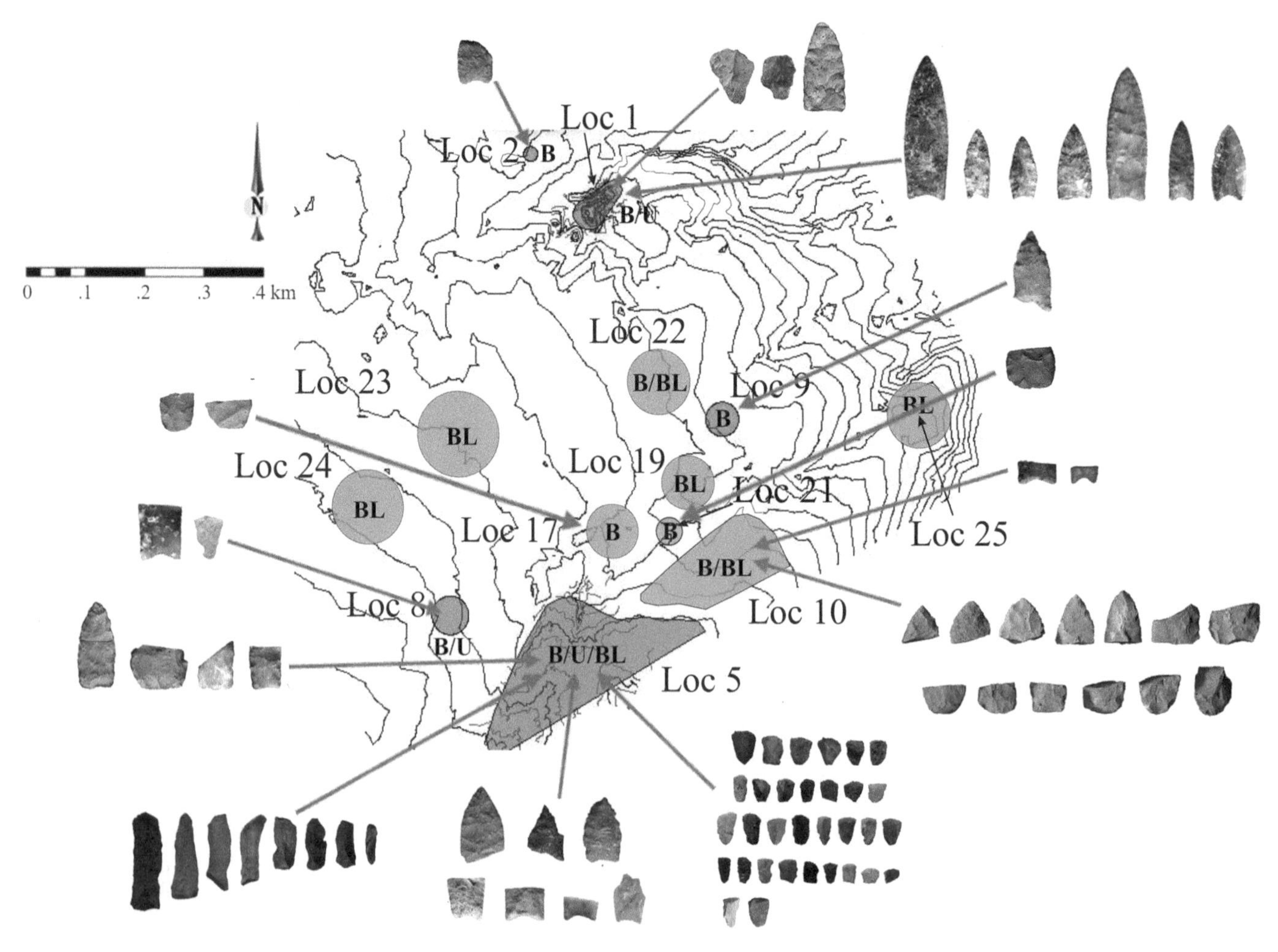

Figure 5.13. Topographic map of El Fin del Mundo indicating the distribution and approximate area of the loci where Clovis tools have been found, showing a selection of artifacts and their provenience (artifacts not to scale). Letters indicate the lithic categories represented at each Locus: B = bifaces; U = unifaces; BL = blades and/or byproducts of blade production.

biface fragment from Locus 1 (# 63177 and #46021 respectively), one of the points from Locus 5 (# 63983), the Clovis point from Locus 9 (# 59727), and other multiple bifaces (22 secondary bifaces and 5 Clovis point preforms) were manufactured using this local rhyolite. This suggests a possible linkage between Locus 25, the rhyolite outcrop, and Locus 1 and the surrounding uplands.

Site Use

The association in Locus 1 of the gomphothere remains with Clovis points suggests that the Pleistocene wetlands in the lowland areas of El Fin del Mundo may have been at least occasionally used as a megafauna hunting area (Chapter 3). Processing of the gomphothere carcasses is indicated by the presence of the microflakes associated with the bonebed, and perhaps also by the relatively large, laterally retouched unifacial tool (i.e., side scraper) found on the surface near the profile of the landform and suspected to have come from the bonebed. Unfortunately, this artifact cannot be associated unequivocally with the Clovis kill feature.

Locus 5 contains the highest density of Clovis stone tools in the site and is the only area where all three lithic categories (bifaces, unifaces, blades) are represented. Locus 10 is an extension of Locus 5 to the northeast and contains the second densest concentration of Clovis artifacts, comprised of bifaces and blades. Taken together, Loci 5 and 10 represent the core area of the Clovis occupation or episodes of occupation in the uplands (Figure 5.13). The lithic component in this area includes fragmentary, damaged, and heavily reworked Clovis points; broken Clovis point preforms; fragments of secondary bifaces; complete and fragmentary blades; blade production byproducts; and end scrapers. Implements that are not usable anymore due to breakage or exhaustion are often replaced and thrown away at campsites and lithic workshops (Walthall and Holley 1997:157). The fragmentary and heavily reworked Clovis points are examples of discarded items due to unrepairable damage or exhaustion. In addition, none of the secondary bifaces collected from these loci are complete and the Clovis point preforms are all fragmentary. The unfinished bifaces in various stages of the manufacturing process, the discarded implements due to use-wear or breakage, and tool features such as tool breakage due to production failure are suggestive of biface production at these loci. Blade production is also indicated by blades and the byproducts of blade production including exhausted wedge-shaped and conical cores, blade core rejuvenation tablet flakes, crested blades, and numerous blades and blade fragments. This lithic assemblage suggests the presence of lithic workshops or knapping stations. The possibility of identifying discrete activity areas in these loci is hindered by the surface nature of the assemblage, the lack of buried contexts in the uplands, and the likely movement and mixing of archaeological materials through several episodes of burial and exposure over millennia.

More evidence of Clovis campsites in the uplands is provided by the numerous end scrapers recovered in these loci. Ethnographic and use-wear studies indicate that end scrapers were typically hafted tools employed in hide-working (Shott 1995; Walthall and Holley 1997) and their presence is suggestive of nonhunting activities related to domestic tasks. Virtually all of the end scrapers from El Fin del Mundo were scattered in the central portion of Locus 5, suggesting that domestic activities were performed in this place. The presence of these artifacts, the variability of tool types, the fact that several seem to have been discarded due to manufacture breakage or exhaustion, and the high density of materials suggest that Loci 5 and 10 represent a residential campsite, or campsites, close to the wetland environment in Locus 1 where water and animal resources could be found close to the rhyolite source in Locus 25 (Sánchez-Morales, Sánchez, and Holliday 2022).

The rhyolite outcrop in Locus 25 contains large blocks of this raw material and thousands of pieces of lithic production debris on the surface. This rhyolite was used to manufacture Clovis bifaces, unifaces, and blades. However, the only artifacts of likely Clovis affiliation found in this locus are 12 blades that were scattered across this area. Given the absence of cores, secondary bifaces, and finished Clovis points, only the early stages of the manufacture process may have been carried out here, such as preparation of cores and blanks for transport. More research is needed in this locus including the study of debitage on the surface in order to explore the possibility of biface reduction on site. A limiting factor for such study, however, is the very likely possibility that the archaeological debris on the surface of this locus represents a palimpsest of multiple exploitation episodes over thousands of years as this raw material is also present in the Archaic component of the site (Sánchez-Morales 2012). As for the other loci in the uplands, the quantity of Clovis materials recovered from them is too small to make any interpretations about the nature of these areas. Due to their location and proximity to Loci 5 and 10 they likely represent peripheral zones of the core area of the Clovis campsites at El Fin del Mundo.

Determining if the Clovis lithic component at El Fin del Mundo is the product of short-term use or the result of a long-term occupation is important for addressing mobility and land use patterns but has proven to be difficult. Given that (1) Clovis groups are generally considered wide-ranging, highly mobile foragers and that (2) site structure is partly a function of the way in which a site is reused over time, Clovis sites are expected to have a structure indicative of short-term use, that is, a number of small separate debris and tool concentrations, rather than continuous undifferentiated scatters resulting from decades of occupations (Binford 1980; Kelly and Todd 1988). Although the expansive area of the site and the numerous artifacts point to multiple occupational episodes, determining if the Clovis lithic component from El Fin del Mundo is the product of short-term, repeated use or resulted from a fewer number of long-term occupations is one of the shortcomings of this analysis (Sánchez-Morales, Sánchez, and Holliday 2022). Given that no undisturbed, buried contexts have been found in the uplands and that the lithic materials may have gone through several episodes of redeposition, the undifferentiated scattering of artifacts on the surface and their mixture with materials from later periods could lead to biased interpretations related to the duration of the Clovis occupation of the site.

Similarities and Differences in Site Use between El Fin del Mundo and El Bajío

The most significant feature shared by El Fin del Mundo and El Bajío is the presence of lithic raw material sources within the sites that were intensively exploited. The high frequency of broken and unfinished bifaces, projectile point preforms, and artifacts associated with blade production indicate that stone tool manufacture and retooling were carried out in both sites near the sources of rhyolite at El Fin del Mundo and vitrified basalt at El Bajío. The flintknapping station of Locus 12 in El Bajío where square-based bifaces were produced supports the idea that the complete reduction process of these tools was carried out on site. In addition, the high frequencies of end scrapers, implements related to domestic tasks, suggest the presence of camps and possibly retooling of these artifacts near the raw material sources at both sites.

Unlike at El Fin del Mundo, there is no evidence of big game hunting at El Bajío. The large number of Clovis points collected from the site by amateur archaeologists might indicate that hunting of large animals was common in the area. However, these points are likely artifacts discarded during retooling given (1) the source of good quality raw material in the site, (2) the abundant Clovis point preforms recovered from the surface, and (3) the damaged, broken, and highly reduced Clovis points from the site illustrated by Robles and Manzo (1972: 205, figures 5 and 6). As suggested by Sánchez (2016), the vitrified basalt source was likely the main reason Clovis groups visited the area, so the manufacture of new implements to replace damaged and worn-out tools was probably the main activity performed at El Bajío.

The occurrence of blades, a conical blade core, two blade core rejuvenation tablet flakes and a biface fragment at El Fin del Mundo that were made on vitrified basalt that may have come from the El Bajío source would indicate links between the two sites, which are 110 km apart. This would indicate extensive knowledge by Clovis groups of the north-central Sonoran landscape and the available tool stone sources in the region. In addition, based on macroscopic observations, Sánchez (2010:310) identified five different varieties of chert at both sites that were used in the manufacture of unifaces, including end scrapers. This strengthens the possibility of movements between the two sites with stops to obtain good quality stone tool raw materials between the largest high quality raw material source in Sonora at El Bajío and the strategic landscape where water, game, and lithic raw material resources were available at El Fin del Mundo. This argument remains tentative and more studies are needed to confirm that the artifacts from El Fin del Mundo are in fact made on vitrified basalt from El Bajío or another source and that the lithologies of the chert varieties at both sites are the same.

On a broader regional scale, El Fin del Mundo and El Bajío provide new information about the early human occupation of northwest Mexico and the southwestern United States. In a regional context, the Sonoran sites are distinctive because most of the Clovis sites in the San Pedro River Valley resulted from short-term, single events: hunting and possible butchering of large game. Murray Springs is the only site in Arizona that has yielded a lithic component comparable in size and tool type diversity to those identified at El Fin del Mundo and El Bajío. At the same time, El Fin del Mundo and Murray Springs contain a rare association of a Clovis megafauna kill and an associated campsite. These sites offer the possibility of evaluating Clovis landscape use related to long-term or repeated use of one area.

How Does the Sonoran Clovis Record Compare to the San Pedro River Valley?

As mentioned in the introduction of this chapter, in the archaeological research of foraging societies the term land-use refers to three main aspects of human behavior: (1) settlement patterns and intensity of site occupation, (2) exploitation of resources, and (3) human mobility. In the previous sections of the discussion and in the description of the lithic assemblages the first two aspects (intensity of site occupation including on-site activities and exploited lithic raw materials) were partially addressed. The following section focuses on settlement patterns, the similarities and differences of the lithic technology from the Sonoran sites and the San Pedro River Valley, and mobility.

Settlement Patterns and Site Configurations

The Clovis sites in north-central Sonora and the San Pedro River Valley differ in their configuration. As previously mentioned, with the exception of Murray Springs and Lehner, the sites in the San Pedro River Valley are single-locality sites that resulted from single short-term events whereas El Fin del Mundo and El Bajío comprise multiple localities that seem to represent repeated use and/or a long-term occupation. This does not mean, however, that the sites resulted from different strategies of land-use, rather they represent different facets of the same adaptation.

With few exceptions, Clovis sites are generally small and dispersed. Known sites extending over a large area containing abundant Clovis artifacts, such as Gault in Texas (Waters, Pevney, and Carlson 2011), Mockingbird Gap in New Mexico (Hamilton and others 2013: Holliday and others 2009), Thunderbird in Virginia (Gardner 1977), Carson-Conn-Short in Tennessee (Broster and Norton 1993), and Topper in South Carolina (Goodyear 2000; Goodyear and Steffy 2003) are rare. In general, these sites were located at or near a permanent natural resource (e.g., sources of water and/or lithic raw material) and probably represent aggregations of sites where groups of families camped together on a few occasions, or where single families camped repeatedly over time (Haynes 2002). El Fin del Mundo, El Bajío, and Murray Springs represent comparable extensive sites with ideal settings where one or several resources including lithic raw materials, water, and/or game were available. Thus, these three sites likely functioned as temporary residential campsites that were used in multiple episodes. In the case of El Fin del Mundo and El Bajío the sites were also used as lithic workshops and raw material quarries. Lehner seems to represent a series of hunting events over a short period of time, with an associated ephemeral logistical camp (Haury, Sayles, and Wasley 1959). In contrast, Naco, Leikem, Navarrete, and Escapule exclusively represent megafauna hunting localities. Extensive residential campsites may have been associated with these hunting localities but have yet to be found or have been eroded away.

Similarities in site configuration and location between El Fin del Mundo and Murray Springs reveal more about the settlement patterns of Clovis foragers of this region. At El Fin del Mundo, the gomphothere kill in Locus 1 is preserved in wetland deposits overlain by a white diatomite layer, and the campsite better represented at Loci 5 and 10 is located approximately 600 m in the uplands to the south. At Murray Springs, the mammoth and bison kills (Areas 3 and 4) are preserved under a "black mat" that indicates a wetland environment (Haynes 2007), and the campsite at Areas 6 and 7 is located 50 to 150 m to the south on higher elevation ground. Both sites contain at least one megafauna kill situated in wet lowland environments and an associated campsite in the better-drained uplands surrounding the ponds or marshes where the animals were taken down and processed.

Similarities and Differences in Lithic Technology

The presence of large, good-quality lithic raw material sources at El Fin del Mundo and El Bajío is one of the most important differences between the Sonoran and San Pedro River Valley sites and is reflected in the configuration of the stone tool assemblages. The lithic assemblages from both Sonoran sites are dominated by the local raw materials. However, the overall lithic tool type composition of the assemblage from El Fin del Mundo is similar to that from Murray Springs. Both stone tool assemblages indicate significant reliance on biface production. Based on the number of bifacial tools and biface thinning debitage at Murray Springs, Huckell (2007) concluded that bifaces were the mainstay of Clovis lithic technology in the San Pedro Valley. Biface thinning debitage has not been collected at El Fin del Mundo, but the numerous complete and fragmentary bifacial implements recovered from the site indicate that biface production was central to the organization of Clovis lithic technology. In contrast to Murray Springs, no primary bifaces have been recovered from El Fin del Mundo, and the utilization of bifaces as cores has yet to be demonstrated. This is likely due to the archaeological collection strategy at the site that excluded debitage and bifacial artifacts with no obvious Clovis diagnostic features such as overshot flaking, basal fluting, lanceolate morphology, and a large size comparable to that of the complete Clovis points from the site.

Despite an incomplete archaeological record from El Fin del Mundo, differences in the biface assemblages from Murray Springs and El Fin del Mundo might be expected due to the differences in the contexts of the two sites. The closest high-quality raw material outcrop to Murray Springs is the St. David chalcedony source about 35 km from the site (Huckell 2007), a distance that would have encouraged the production of cores and blanks at the chalcedony source for more efficient transportation to the campsite. In contrast, the rhyolite outcrop (Locus 25) at El Fin del Mundo, though not as high quality as chalcedony, was an immediately available source of raw material that might have reduced the need for bifacial cores or other easy-to-transport prepared sources of flakes and blades compared to Murray Springs. Bifaces made of chert at El Fin del Mundo reflect more similarities between the two lithic assemblages. At both sites, the use-lives of chert and chalcedony bifaces and other implements were maximized through heavy resharpening and reworking, exemplified by extensively refurbished and repaired Clovis points and highly reduced end scrapers, suggesting that the sources of these raw materials were not as readily available as others.

Blade production seems to have been more common in the Sonoran sites compared to the San Pedro River Valley. The blade technology that is abundantly evident at El Fin del Mundo and El Bajío is much more reminiscent of the Southern High Plains region than it is to the surrounding area (Sánchez and Carpenter 2012). Sánchez (2016) concluded that tool manufacture and production of blanks, preforms, and transportable cores at El Bajío must have been more common than in any other Clovis site of the region due to the presence of the vitrified basalt source at the site. This resulted in the large number of blade cores, byproducts of blade production, and prismatic blades that were used as blanks for end scrapers. A similar pattern is observed at El Fin del Mundo, where abundant blades and blade production byproducts have been found in and around the local source of rhyolite. Other Clovis sites with abundant evidence of blade production are also located near sources of high-quality tool stones, including Gault in central Texas, where lithic workshops and camps have been excavated in close proximity to sources of Edwards chert (Waters, Pevney, and Carlson 2011). Easy access to good quality raw materials may have resulted in frequent blade production at these sites.

In the San Pedro Valley sites Huckell (2007) identified a number of Clovis point repair strategies related to four basic categories of damage: tip damage, basal damage, lateral damage, and various combinations of those three (Figure 5.14; Huckell 2007:199–201). Similar patterns of basal and tip damage are observed in the specimens from El Fin del Mundo. Lateral damage such as the burin breaks displayed by Clovis points from Naco, Lehner, and Murray Springs are not seen at El Fin del Mundo. Conversely, at El Fin del Mundo one Clovis point (# 59604) displays a heavy reworked body, which is markedly beveled; this beveling resulted from alternate retouch on the lateral margins of the distal half of the point. Beveling of Clovis points is not seen on the San Pedro Valley specimens, but as mentioned previously it is common on later Paleoindian projectile points such as Dalton/Golondrina (Jennings, Smallwood, and Greer 2016; Walthall and Holley 1997; Wiant and Hassen 1985).

Lastly, many of the projectile points associated with the multiple bison kill (Area 4) at Murray Springs display more extensive and complex damage than any of the points from Naco, Lehner, or El Fin del Mundo. Huckell (2007) suggests that the different breakage patterns may have resulted from different procurement methods related to the targeted animal. Huckell (2007) also raises the possibility that bison were hunted with throwing spears, while thrusting spears were used to hunt proboscideans due to differences in size. The lower accuracy resulting from a throwing spear, coupled with a smaller prey with a smaller vital target area, may have resulted in more extensive damage to the Clovis points from the bison kill at Murray Springs (Huckell 2007:200–201). The Clovis points associated with the gomphotheres at El Fin del Mundo are complete or nearly complete (Figure 5.4a–g). The only fragment is a distal portion that snapped in a clean, transverse fashion. Thus, the sample of points from this kill is consistent with those associated with proboscideans at Naco and Lehner (Figure 5.9e–y).

Mobility

The surface nature of the assemblage from El Fin del Mundo and the consequently incomplete and biased Clovis record of the site limits interpretations about patterns of group mobility. In addition, the lack of data on the location of the sources of exogenous raw materials used by the Clovis foragers of the area, such as chert and obsidian, pose an even more important caveat to understanding group movements and the diversity of landscapes that were utilized. Nonetheless, the available evidence provides insights into Clovis mobility patterns and raw material economy in Sonora. Chert constitutes the second most common raw material used for stone tool production at El Fin del Mundo (34.6% of total lithic assemblage), only surpassed by the local rhyolite (35.3%), but the proportion difference is minimal. Procurement and use of high-quality cryptocrystalline raw materials such as chert for manufacturing stone tools is one common feature in many Paleoindian lithic assemblages thought to indicate highly mobile strategies (Andrefsky 2005; Kelly and Todd 1988). This is because raw materials that are more easily controlled in the knapping process produce tools with longer and potentially more variable use-lives. This is ideal for highly mobile groups using landscapes where access to good quality raw materials is limited or cannot be anticipated. In addition, in the case of El Fin del Mundo, the exogenous nature of these high-quality tool stones and the diversity of chert varieties observed in the assemblage potentially indicates that these artifacts were made in different places, carried, and eventually deposited at the site. Other exogenous raw materials of high flaking quality present at El Fin del Mundo include vitrified basalt and obsidian. Together, the exogenous high-quality tool stones at El Fin del Mundo represent 40 percent of the total lithic assemblage (Figure 5.15). In addition, if the vitrified basalt is, in fact, the same that is available at El Bajio, then this could potentially indicate movements across a territory at least 40 km wide between the landscapes of the valleys of Rio Zanjón and Rio Bacoachi.

Heavy refurbishing and repair of implements made on chert and other high quality exogenous materials suggest that these were valuable and not easy to replace items that were used and carried around for long periods of time, possibly because of the significant distance to a quarry (Andrefsky 2009). Most of the end scrapers from El Fin del Mundo were made of chert (83% of end scrapers), and several show shortening of their length and abrupt working edge angles due to refurbishing of the working distal end, suggesting that these artifacts were deemed unusable and discarded. In addition, three Clovis points

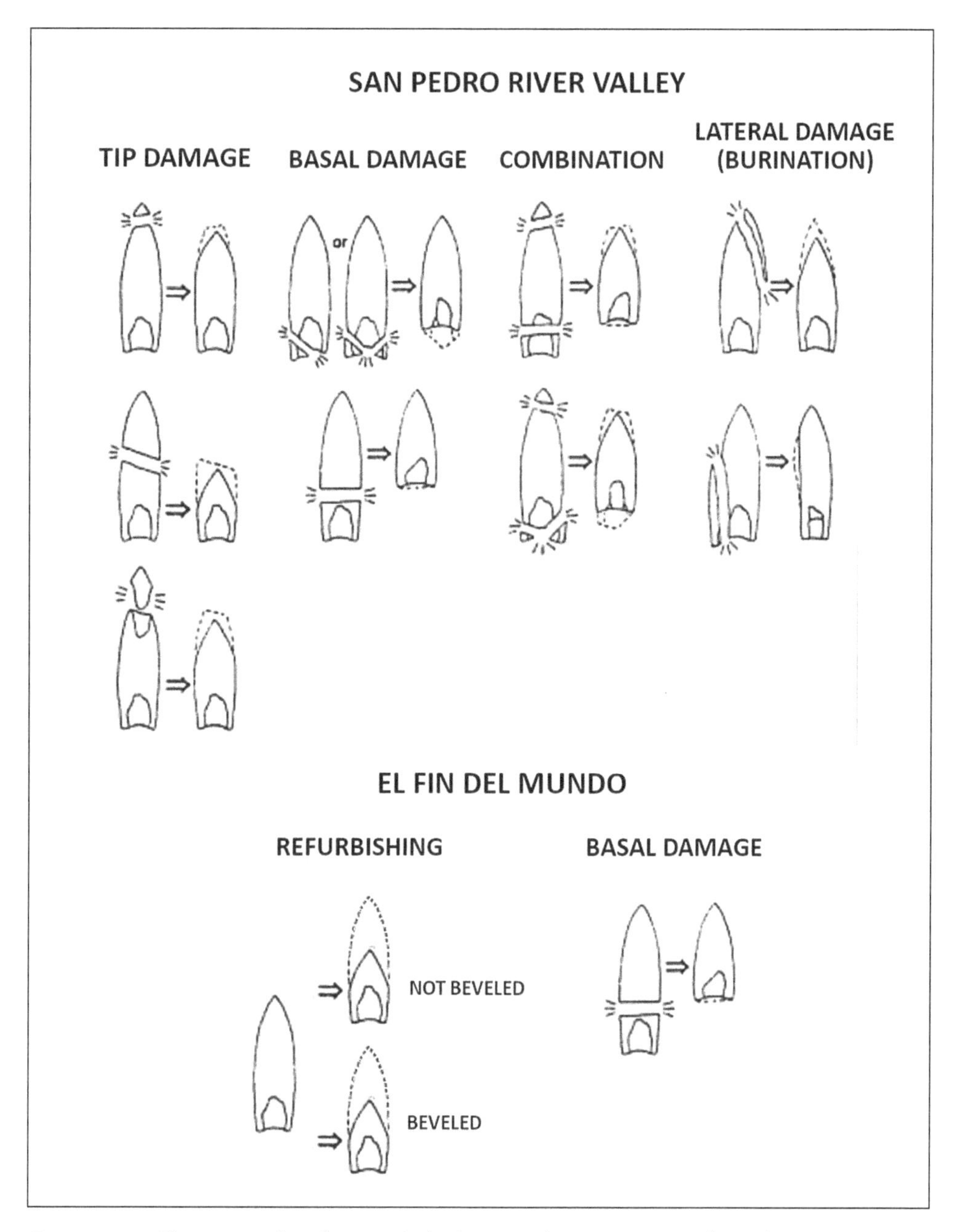

Figure 5.14. Clovis point breakage, refurbishing, and repair patterns from the San Pedro River Valley and El Fin del Mundo. Modified from Huckell 2007:200, Figure 8.9. Prepared by Ismael Sánchez-Morales.

made on chert (# 62943 [Figure 5.4c], 63008 [Figure 5.4b], and 59604 [Figure 5.4k]), were repaired and/or heavily refurbished, resulting in their small size, short body and/or beveling. The complete Clovis points from Locus 1 made on chert may represent unrecovered specimens that were lost during the hunt or during the butchering process. A somewhat different pattern is observed on the bifaces made on rhyolite. Only two specimens, Clovis point #59727 (Figure 5.4i) and preform #59287 (Figure 5.3h), show similar use-life maximization patterns, whereas most rhyolite bifaces do not exhibit reworking or recycling and seem to have been easily discarded. This seems to be a logical behavior when the source of that raw material is locally available and can be procured when needed.

The predominance of noncryptocrystalline volcanic raw materials in the Clovis points from Sonora may in part be

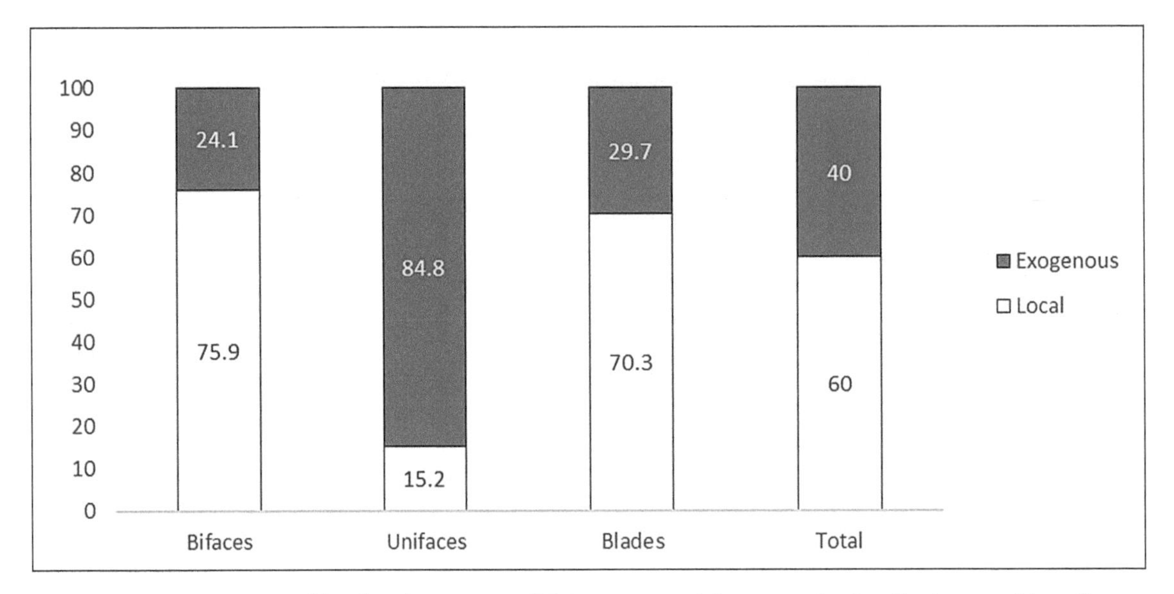

Figure 5.15. Proportions of local and exogenous lithic raw materials present in the Clovis assemblage from El Fin del Mundo. Total refers to the proportions of the whole assemblage. Local materials are dacite, jasper, quartz, quartz crystal, quartzite, rhyolite, and not identified (sedimentary and igneous). Exogenous materials are chert, obsidian, and vitrified basalt.

the result of raw material constraints in a region dominated by volcanic geology (Gaines, Sánchez, and T. Holliday 2009). The greater number of artifacts made of rhyolite and vitrified basalt in the assemblages from El Fin del Mundo and El Bajío respectively is logical, as the sources of these materials are located in the sites. Chert was the second most often utilized raw material for tool production at El Fin del Mundo, however, and most of the Clovis points from this site are made of chert. Also, at El Fin del Mundo and El Bajío there is a striking predominance of chert in tool types other than bifaces such as end scrapers. These patterns indicate that the Clovis groups that inhabited the region relied heavily on these presumably exogenous tool stones.

The blades from El Fin del Mundo provide more insights into Clovis mobility in the region. Blades obtained from chert cores are smaller compared to the larger specimens made on locally available rhyolite and quartzite. In order to visually compare both groups of blades (chert vs rhyolite/quartzite), a scatter plot was created showing the distribution of blades by width and thickness (Figure 5.16). These measurements were plotted even for incomplete specimens, given that width and thickness are more likely to remain constant through the length of a blade (Collins 1999). Most of the chert blades group together under 13 mm thickness and 32 mm in width. In contrast, most of the rhyolite and quartzite blades are scattered above 13 mm in thickness and 32 mm in width. The descriptive statistics of both groups (blades made on chert vs. blades made on rhyolite or quartzite) also indicate that chert blades are smaller (Table 5.7). This is confirmed by the results of Kruskal-Wallis tests comparing maximum width and maximum thickness, which indicate that chert blades ($n = 12$) are statistically narrower (test statistic= 8.367, $df = 1$, $p < 0.01$) and thinner (test statistic= 8.799, $df = 1$, $p < 0.01$) than blades made on rhyolite and quartzite ($n = 28$).

Assuming that chert is an exogenous raw material, the small size of the chert blades might be due to (1) the small size of naturally available sources of chert, perhaps in the form of small nodules, (2) the intentional size of either prepared cores or finished products, which needed to be small in order to be transported from long-distance sources (Kelly 1988; Kuhn 1994), or (3) highly reduced cores that produced small blades. The large size of Clovis points made on chert recovered from the site suggests that the size of the naturally available sources of chert was not a likely factor limiting blade size. Thus, blades may have been intentionally produced as small items as a strategy to maximize the utility of the transported load while minimizing its weight, indicative of high mobility (Kuhn 1994), or because of core reduction during long-distance transport.

It has been argued that increasing regionalization resulting from permanence in staging areas is a possible

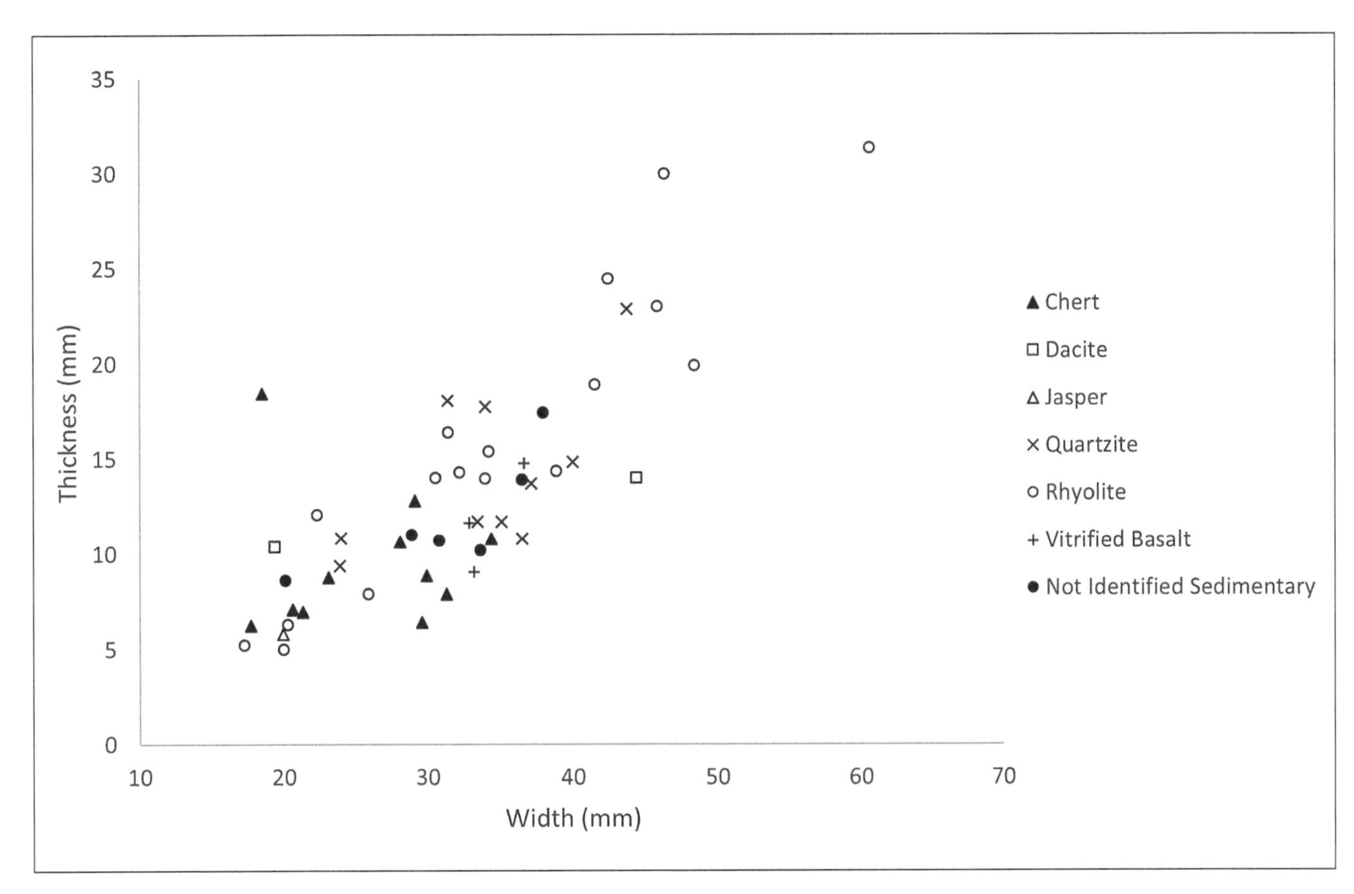

Figure 5.16. Scatter plot showing width and thickness of cortical and noncortical blades from El Fin del Mundo.

Table 5.7. Descriptive Statistics of the Chert (Exogenous) and Rhyolite and Quartzite (Local) Blades from El Fin del Mundo

		Maximum Width (in mm)	*Maximum Thickness (in mm)*
Chert Blades	Number	12	12
	Minimum	17.7	6.3
	Maximum	34.4	18.5
	Median	28.0	8.5
	Mean	26.0	9.4
	Standard Deviation	5.5	3.5
Rhyolite and Quartzite Blades	Number	28	28
	Minimum	17.3	5.0
	Maximum	120.5	59.1
	Median	34.1	14.3
	Mean	37.6	16.9
	Standard Deviation	19.0	10.6

Table 5.8. Clovis Points with No Evidence of Reworking/Refurbishing or Damage Affecting Their Maximum Length, Maximum Width, and Maximum Thickness

Artifact Number	*Site*	*Condition*	*Raw Material*	*Maximum length (in mm)*	*Maximum width (in mm)*	*Maximum thickness (in mm)*
—	El Bajío	Complete	Vitrified basalt	88	29	7
59569	El Fin del Mundo	Complete	Chert	52.4	17.7	6.4
59342	El Fin del Mundo	Complete	Quartz crystal	49.2	23.1	5.7
46023	El Fin del Mundo	Complete	Chert	88.5	28.5	7.9
63177	El Fin del Mundo	Complete	Rhyolite	95.5	30.3	8.3
A-31232	Escapule	Semicomplete	Chert	90.3	34	6.8
A-31231	Escapule	Semicomplete/tip fracture	Chalcedony	97.4	30.2	8.9
A-12683	Lehner	Complete	Quartz crystal	30.8	16.6	4.7
A-12681	Lehner	Complete	Quartz crystal	46.8	20.6	6.7
A-12682	Lehner	Complete	Chalcedony	55.7	24.6	7.2
A-12677	Lehner	Complete	Chert	74.2	28.3	7.6
A-12685	Lehner	Complete	Chert	96.3	30.3	8.2
A-12676	Lehner	Complete	Chalcedony	82.5	29.2	9.8
A-24127	Leikem	Complete	Chert	94.6	31.7	10.2
A-33116	Murray Springs	Complete	Obsidian	41.8	22.3	6.4
A-32718	Murray Springs	Complete	St. David chalcedony	56.8	28.3	9
A-32992	Murray Springs	Complete	Chert	71.8	33.3	9.6
A-11913	Naco	Complete	Chert	97	30.5	9.7
A-10899	Naco	Complete	Chert	70.8	25.8	7.8
A-10900	Naco	Nearly complete/one broken corner	Felsite	67.8	27.1	7.8
A-10901	Naco	Nearly complete/tip fracture	Felsite	80.8 (83)	29.8	7.9
A-10902	Naco	Complete	Chert	67.7	31.3	7.5
A-10903	Naco	Nearly complete/one corner is missing	Chert	116.9	34	9.5
A-10904	Naco	Complete	Chert	57.8	27.8	7.3
A-11914	Naco	Nearly complete/tip fracture	Chert	81.5	32	10

Note: Values in parenthesis are estimated total measurements.

implication of a place-oriented and less residentially mobile model of Clovis land-use (Anderson 1996; Smallwood 2012). This regionalization should be reflected in stone tool assemblage variation from region to region. Smallwood (2012) tested this hypothesis through the analysis of Clovis stone tool assemblages from three different regions in the southeastern United States. She concluded that the morphology of finished Clovis points varies significantly from one region to another, resulting in different length, width, and thickness ratios. At the same time, the biface reduction process in each assemblage varies, further supporting the idea of regionalization (Smallwood 2012). In order to look for evidence of variation between the Clovis assemblages from Sonora and the San Pedro River Valley that might be suggestive of regionalization, metrics of the unbroken/unrefurbished Clovis points from both regions (Table 5.8) are shown in two scatter plots (Figures 5.17, 5.18). The objective was to determine if the Clovis points from the Sonoran sites and the San Pedro River Valley sites form two separate groups based on length/width and width/thickness. The sample of unbroken and unrefurbished Clovis points from the Sonoran sites of El

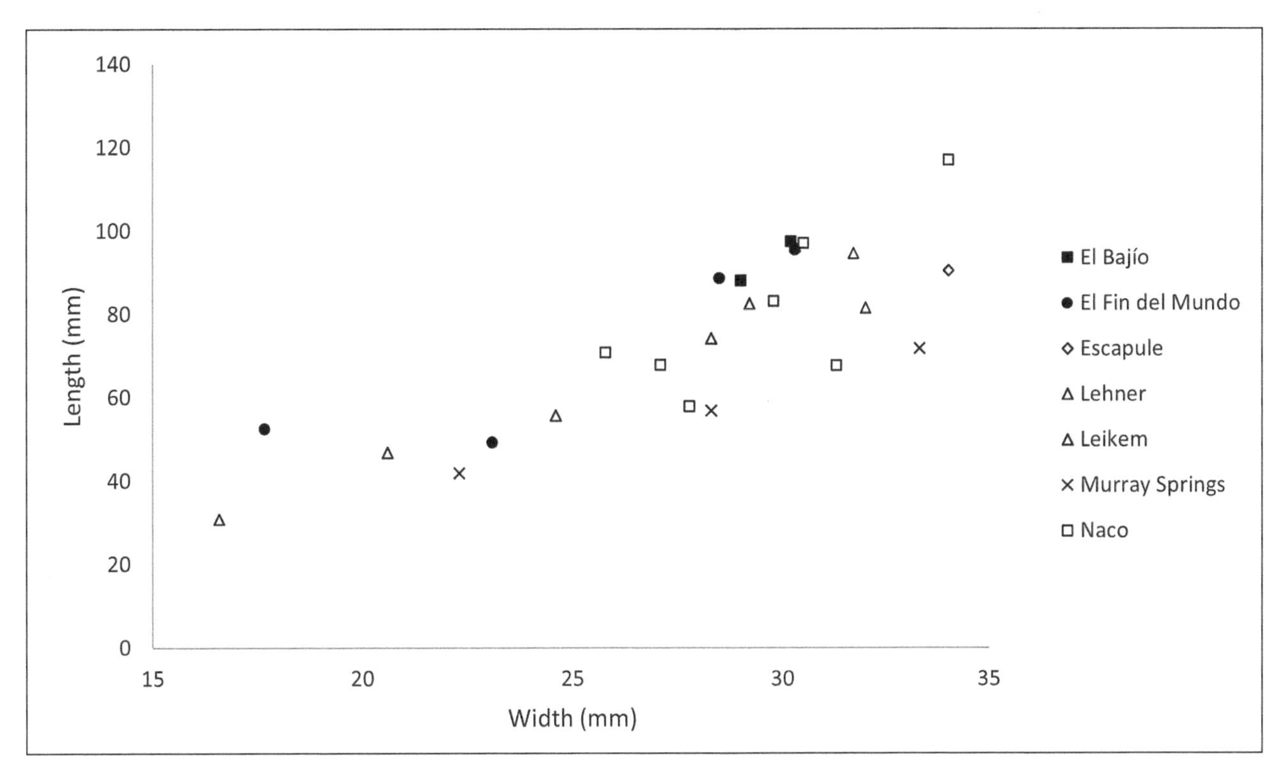

Figure 5.17. Scatter plot showing maximum length and maximum width of unbroken or unrefurbished Clovis points from El Fin del Mundo, El Bajío, and the San Pedro River Valley sites.

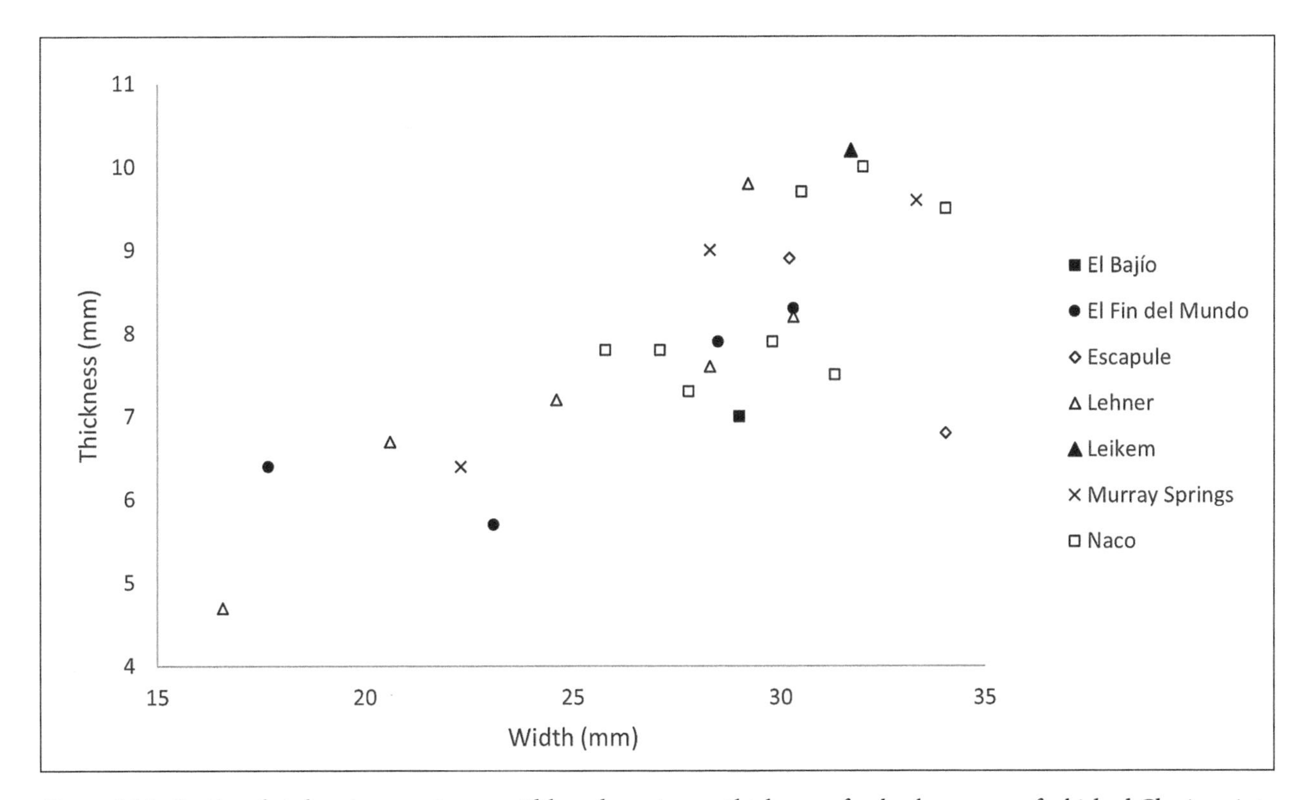

Figure 5.18. Scatter plot showing maximum width and maximum thickness of unbroken or unrefurbished Clovis points from El Fin del Mundo, El Bajío, and the San Pedro River Valley sites.

Table 5.9. Descriptive Statistics for Maximum Length, Maximum Width, and Maximum Thickness of the Complete, Semicomplete, and Unrefurbished Clovis Points from Sonoran and San Pedro River Valley Sites

		Maximum Length (in mm)	*Maximum Width (in mm)*	*Maximum Thickness in mm)*
Clovis points from Sonora	Number	5	5	5
	Minimum	49.2	17.7	5.7
	Maximum	95.5	30.3	8.3
	Median	88.0	28.5	7.0
	Mean	74.7	25.7	7.1
	Standard Deviation	22.1	5.3	1.1
Clovis Points from the San Pedro River Valley	Number	20	20	20
	Minimum	30.8	16.6	4.7
	Maximum	116.9	34.0	10.2
	Median	73.0	29.5	7.9
	Mean	74.1	28.4	8.1
	Standard Deviation	21.7	4.6	1.4

Bajío and El Fin del Mundo consists of five specimens, whereas the sample from the San Pedro River Valley Clovis sites consists of 20. These Clovis points have no evidence of extensive reworking-refurbishing or damage that could affect their maximum length, maximum width, and maximum thickness.

Descriptive statistics (Table 5.9) and the scatter plots (Figures 5.17 and 5.18) suggest that there is no notable difference in maximum length, maximum width, and maximum thickness of the Clovis points from both regions. The points, however, seem to cluster in two groups: points with a maximum width under 25 mm, and points that are wider than 28 mm. Both clusters include points from both regions. When the same data is plotted by raw material, the raw material clearly is the limiting factor of the size of the narrower, shorter, and thinner points (Figure 5.19). Three of these points are made on quartz crystal, two of which are from Lehner and one from El Fin del Mundo, and another one is the obsidian Clovis point from Murray Springs. Haury, Sayles, and Wasley (1959) noted that the small size of the quartz crystal points from Lehner is a consequence of the fact that this raw material occurs locally in the form of small nodules. The size of quartz crystal in its natural form was also the likely limiting factor for the specimen from El Fin del Mundo, because this material is observed only as small prisms and nodules at the site and the nearby primary source at Cerro del Cuarzo.

The fourth scatter plot (Figure 5.20) shows the basal width and basal concavity of Clovis points from both regions. In this dispersion graph basal fragments and reworked or refurbished points that retain their basal portion were included, resulting in a sample of 13 specimens from El Fin del Mundo and 30 from the San Pedro River Valley sites (Table 5.10). These metrics were plotted because they may represent important dimensions of the hafting element of the points that might indicate variation in hafting strategies. As seen in Figure 5.20 these metrics are so variable that no clusters are observed and the points from both regions are scattered with no clear pattern. However, the Clovis points from the San Pedro River Valley show more variation of basal concavity, perhaps due to the larger sample size (Table 5.11).

In order to evaluate whether there are significant differences in the five metrics plotted between the Clovis points from the Sonoran and the San Pedro River Valley sites, a series of Kruskal-Wallis tests were run. The results show that there are no significant differences in maximum length (test statistic= 0.000, df = 1, $p > 0.05$), maximum width (test statistic= 1.11, df = 1, $p > 0.05$), maximum thickness (test statistic= 2.038, df = 1, $p > 0.05$), and basal concavity (test statistic= 1.150, df = 1, $p > 0.05$). The statistical test for basal width (test statistic= 5.730, df = 1, $p < 0.05$) indicates that there is a significant difference between samples for this variable. However, the difference in means is only about 3

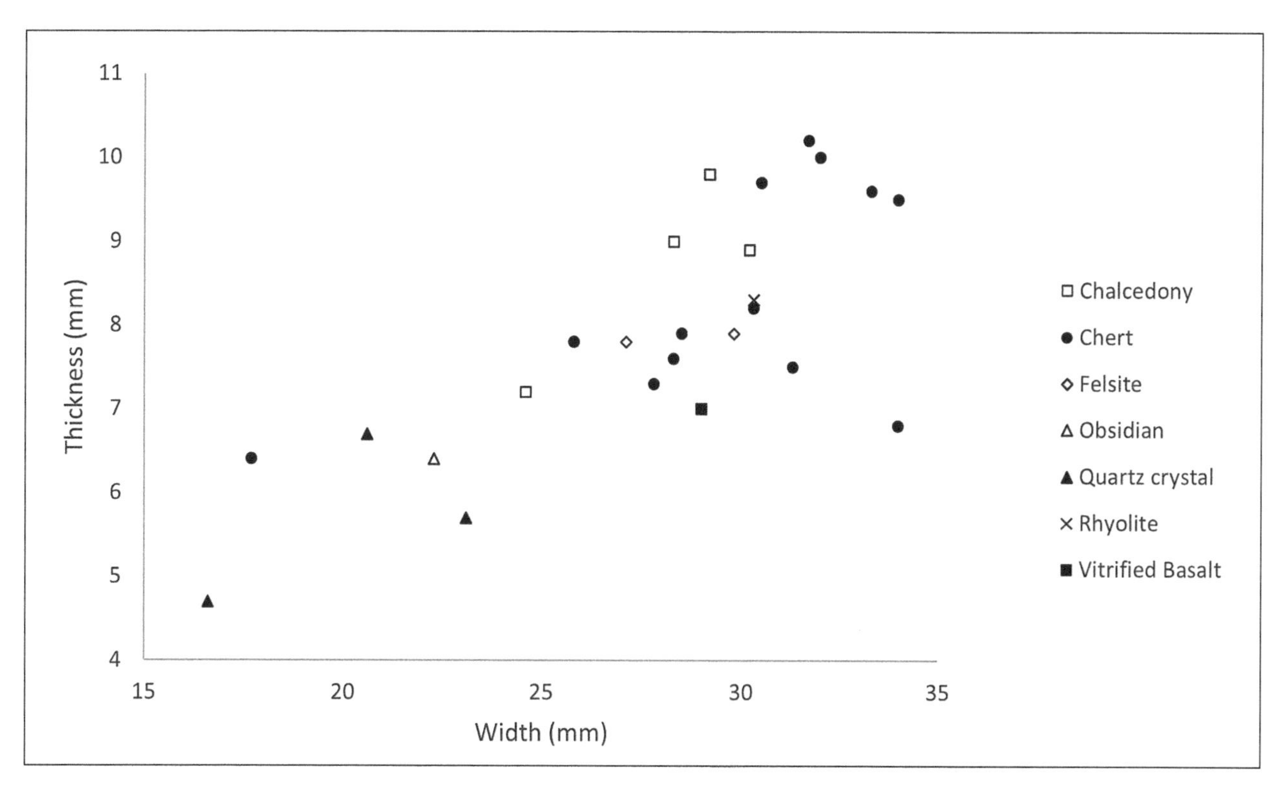

Figure 5.19. Scatter plot showing maximum width and maximum thickness of the unbroken or unrefurbished Clovis points from El Fin del Mundo, El Bajío, and the San Pedro River Valley sites by raw material.

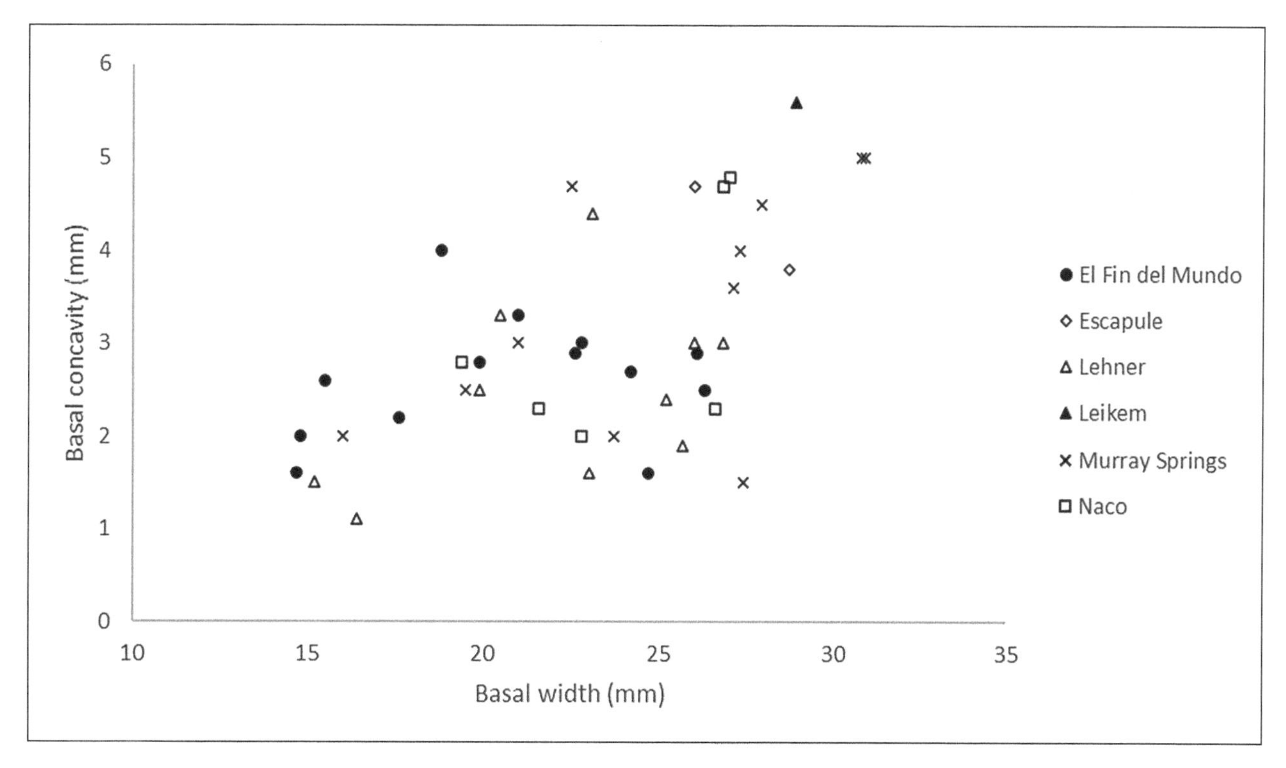

Figure 5.20. Scatter plot showing basal width and basal concavity of Clovis points from El Fin del Mundo and the San Pedro River Valley.

Table 5.10. Clovis Points from El Fin del Mundo and San Pedro River Valley Sites, Including Basal and Reworked Fragments, that Preserve Their Basal Width and Basal Concavity

Artifact No.	*Site*	*Condition*	*Raw Material*	*Basal width (in mm)*	*Basal Concavity (in mm)*
46023	El Fin del Mundo	Complete	Chert	24.7	1.6
59082	El Fin del Mundo	Proximal	Chert	21	3.3
59342	El Fin del Mundo	Complete	Quartz crystal	19.9	2.8
59569	El Fin del Mundo	Complete	Chert	17.6	2.2
59593	El Fin del Mundo	Proximal	Chert	22.6	2.9
59603	El Fin del Mundo	Proximal	Chert	26.3	2.5
62943	El Fin del Mundo	Complete/reworked	Chert	14.7	1.6
63008	El Fin del Mundo	Complete/reworked	Chalcedony	15.5	2.6
63177	El Fin del Mundo	Complete	Rhyolite	26.1	2.9
63220	El Fin del Mundo	Proximal	Chert	24.2	2.7
36516	El Fin del Mundo	Proximal	Rhyolite	18.9	4
33487	El Fin del Mundo	Proximal with catastrophic impact fracture	Quartz crystal	14.8	2
33030	El Fin del Mundo	Proximal	Chert	22.8	3
A-31231	Escapule	Semicomplete/tip fracture	Chalcedony	26	4.7
A-31232	Escapule	Semicomplete	Chert	28.7	3.8
A-12675	Lehner	Complete/reworked	Chert	19.9	2.5
A-12676	Lehner	Complete	Chalcedony	26.8	3
A-12677	Lehner	Complete	Chert	23.1	4.4
A-12678	Lehner	Nearly complete/tip fracture	Quartz crystal	15.2	1.5
A-12679	Lehner	Complete	Chert	25.2	2.4
A-12681	Lehner	Complete	Clear Quartz	20.5	3.3
A-12682	Lehner	Complete	Chalcedony	23	1.6
A-12683	Lehner	Complete	Clear Quartz	16.4	1.1
A-12685	Lehner	Complete	Chert	26	3
A-12686	Lehner	Complete	Jasper	25.7	1.9
A-24127	Leikem	Complete	Chert	28.9	5.6
A-32718	Murray Springs	Complete	St. David chalcedony	23.7	2
A-32992	Murray Springs	Complete	Chert	30.9	5
A-33109	Murray Springs	Nearly complete	Chert	21	3
A-33110	Murray Springs	Complete	Chert	22.5	4.7
A-33116	Murray Springs	Complete	Obsidian	16	2
A-33917	Murray Springs	Proximal	St. David chalcedony	27.1	3.6
A-33922	Murray Springs	Proximal	Chert	27.9	4.5
A-47139	Murray Springs	Proximal	Obsidian	19.5	2.5
A-47140	Murray Springs	Proximal	St. David chalcedony	27.4	1.5
A-47162	Murray Springs	Proximal	Chert	27.3	4
A-47283	Murray Springs	Proximal	Chert	30.8	5
A-10899	Naco	Complete	Chert	19.4	2.8
A-10901	Naco	Nearly complete/tip fracture	Felsite	27	4.8
A-10902	Naco	Complete	Chert	26.8	4.7
A-10904	Naco	Complete	Chert	21.6	2.3
A-11912	Naco	Complete	Chert	22.8	2
A-11913	Naco	Complete	Chert	26.6	2.3

Table 5.11. Descriptive Statistics for Basal Width and Basal Concavity of Clovis Points from El Fin del Mundo and the San Pedro River Valley, including Basal and Reworked Fragments, with No Evidence of Reworking-Refurbishing or Damage Affecting Original Basal Width and Basal Concavity

		Basal Width (in mm)	*Basal Concavity (in mm)*
Clovis Points from Sonora	Number	13	13
	Minimum	14.7	1.6
	Maximum	26.3	4.0
	Median	21.0	2.7
	Mean	20.7	2.6
	Standard Deviation	4.2	0.7
	Variance	17.4	0.4
Clovis Points from the San Pedro River Valley	Number	30	30
	Minimum	15.2	1.1
	Maximum	30.9	5.6
	Median	25.5	3.0
	Mean	24.1	3.2
	Standard Deviation	4.2	1.3
	Variance	18.0	1.7

mm and likely can be attributed to factors including (a) the relatively high representation in the sample from El Fin del Mundo of repaired points (which were not included in the comparisons of maximum length, maximum width, and maximum thickness) and points made on raw materials for which the size of the natural package is a likely factor limiting the size of the finished points and (b) the relatively high representation of "pristine" specimens with no evidence of repair or refurbishing from Naco, Escapule, and Leikem. The Clovis point collections from Murray Springs and Lehner are more similar to the one from El Fin del Mundo as they contain several specimens that were repaired and/or made on raw materials such as quartz crystal and obsidian, which may have limited the size of the finished points. In fact, when the statistical test includes only these two sites from the San Pedro Valley there is no significant difference between samples (independent-samples Kruskall-Wallis test; test statistic= 3.463, df =1, p= 0.063).

Finally, the assessment of possible variation in biface reduction processes between the Sonoran sites and the San Pedro River Valley is not possible at the moment. In-depth analysis of biface production at El Fin del Mundo that includes the collection of biface production byproducts in the field is needed for this objective. The comparison of metrics and morphology of the Clovis points from both regions suggests that no strong differences are present.

The analysis of the overall Clovis lithic technology at El Fin del Mundo and its comparison with the stone tool assemblages from El Bajío and the San Pedro River Valley sites suggests that the Clovis groups in both regions had similar landscape use patterns. These consisted of

1. positioning of repeatedly used campsites on better drained uplands adjacent to large game hunting locations in wetland areas;
2. high reliance on high-quality, cryptocrystalline lithic raw materials from exogenous sources for the manufacture of stone tools, and the exploitation of lower quality tool stones when readily available at or near the campsites, which is also suggestive of
3. high residential, wide ranging mobility. However, the potential distances traveled by the Clovis groups that occupied the Sonoran sites remains uncertain given the unknown locations of utilized exogenous raw material sources.

Evaluating the possibility of direct links between the Clovis sites in north-central Sonora and those in the San Pedro River Valley is difficult. The approach based on similarities in raw materials among collections has proven to be insufficient because the only information available at the moment is the macroscopic descriptions of the raw materials. No petrographic or geochemical data is available. Sánchez (2010: 273) mentions that one of the end scrapers from El Bajío (# 37549b) is made on red chert that appears to be identical to the raw material used for the manufacture of one of the end scrapers from Murray Springs. After the re-examination of these artifacts this assessment based on texture and color descriptions was not corroborated. At the moment, only speculations based on raw material source locations in Arizona and the geography of the area in between north-central Sonora and the San Pedro Valley can be made.

Based on the sources of the raw materials present in the lithic assemblage from Murray Springs, Huckell (2007) identified a movement route that Clovis foragers may have followed prior to their arrival at the site. This route may have started in the upper Little Colorado River basin where petrified wood is available, approximately 225 km north of the site. One hundred seventy-five km to 250 km south from that area is the Cow Canyon obsidian source, where obsidian was obtained to manufacture artifacts including two Clovis points. Continued movement to the west, possibly along the Gila River, and south into the San Pedro Valley would have led to the St. David area where chalcedony was largely exploited, and this was possibly the last place visited by Clovis foragers before arriving in the Murray Springs area as suggested by the abundant debitage of this raw material at the site (Figure 5.12). Based on this analysis, Clovis foragers of southeastern Arizona covered distances well over 200 km in their group movements. Thus, these groups could have moved to and from the San Pedro River Valley into north-central Sonora 250 km to the south where several Clovis sites, including El Fin del Mundo, El Bajío, El Aigame, and El Gramal, have been identified. In addition, an obvious physiographic connection between the Clovis sites in the San Pedro River Valley and those in north-central Sonora is clear. The headwaters of the San Pedro River and the headwaters of the Río Sonora, which flows into the Gulf of California, are separated only by a low rolling to flat grassy plain. Moving between these two areas would have been a relatively easy matter for mobile hunter-gatherer groups (Gaines, Sánchez, and Holliday 2009).

CONCLUSIONS

El Fin del Mundo has produced one of the largest assemblages of Clovis stone tools in Mexico. Most of the stone tools have been recovered from surface scatters of lithic materials in uplands surrounding an in situ Clovis gomphothere kill. The great number of Clovis lithic artifacts and the ample variety of tool types, including bifaces, unifaces, and blades, and the presence of tool manufacturing byproducts from the site are indicative of a campsite. This evidence suggests long and/or repeated occupations of the upland loci. The stone tools and their contexts indicate that a variety of tasks were performed at the site including resource processing, stone tool manufacture and refurbishing, retooling, and discard of exhausted and damaged lithic implements. In addition, El Fin del Mundo and El Bajío contain lithic raw material sources that were extensively exploited by Clovis foragers. These sites have produced the largest Clovis lithic assemblages south of the U.S.-Mexico border. Lastly, Clovis foragers from north-central Sonora and the San Pedro River Valley of southeastern Arizona seem to have had similar landscape use patterns, characterized by (1) the positioning of campsites on well drained uplands near wetland hunting locations where large game were available, (2) high reliance on exogenous high-quality lithic raw materials for stone tool production and exploitation of lower quality tool stones when available at or near the campsites, and (3) a wide-ranging and highly mobile lifestyle.

Acknowledgments

Special thanks to Centro INAH Sonora and the Arizona State Museum for granting access to the archaeological collections from Sonora and Arizona respectively and for providing the facilities for data collection on these materials. I am indebted to Guadalupe Sánchez for providing the metric and qualitative data on the Clovis artifacts from El Bajío and for her support while I was conducting this research. Thanks to Vance Holliday for facilitating access to the Clovis point cast collection from the Naco site and for providing feedback on this manuscript. Identification of raw materials with the binocular microscope was conducted with the assistance of Geologist Alejandra Gómez at INAH Sonora. I am grateful to Steven Kuhn, Kayla Worthey, and John Olsen for their commentary and suggestions on the original manuscript of this chapter. This research was funded by the Argonaut Archaeological Research Fund, the University of Arizona School of Anthropology, and CONACYT.

CHAPTER SIX

The Faunas of the Upper and Lower Bonebeds, Locus 1

Kayla B. Worthey and Joaquín Arroyo-Cabrales

The faunas of El Fin del Mundo offer a rare window into human-animal interactions and environment during the terminal Pleistocene of Sonora. In addition to preserving an association between gomphotheres (*Cuvieronius* sp.) and Clovis artifacts in the Upper Bonebed of Locus 1, El Fin del Mundo is unique in representing the northernmost locality with *Cuvieronius* sp. remains dating to the terminal Pleistocene yet known. Furthermore, the older, Lower Bonebed of the site contains unusual associations of gomphothere (*Cuvieronius sp.)* and mastodon (*Mammut americanum*) remains. The additional presence of a third proboscidean, mammoth (*Mammuthus* sp.), at the site suggests a unique paleoenvironmental and paleoecological history of the region surrounding El Fin del Mundo.

Inferences about the nature of human-gomphothere associations in the Upper Bonebed and proboscidean species associations in the Lower Bonebed depend upon detailed contextual information about the depositional histories and taphonomic processes that produced and modified the bonebed faunal records. Determining, for example, whether the El Fin del Mundo bonebeds formed in place (autochthonous) or as redeposited secondary accumulations (allochthonous) is important for reconstructing the role humans had in the formation of the Upper Bonebed and the paleoecological significance of the Lower Bonebed. This chapter compares and contrasts the formation processes of the Upper and Lower Bonebeds of El Fin del Mundo Locus 1 by discussing the contents of each accumulation, the spatial distributions of bone elements, and the post-depositional processes affecting the faunal remains.

THE UPPER BONEBED

Locus 1 of El Fin del Mundo consists of a landform isolated by arroyo cutting (Figure 1.7; Chapter 1). The sediments of the landform preserve two stratigraphically separate bonebeds, defined following Behrensmeyer (1991, 2007) as relative concentrations of skeletal remains representing more than one individual within a stratum. The Upper Bonebed contains faunal remains in association with Clovis lithic artifacts concentrated largely within the upper 15 cm of Stratum 3B and capped by or locally encased within a thin layer of diatomite (Figure 4.4; Chapters 2, 4). The bones of the Upper Bonebed are distributed into two concentrations in the horizontal plane, each approximately 6 m^2 in area and located in the south and north portions of the excavation area respectively (Figure 6.1). The faunal material around and between the concentrations consists mostly of small, unidentifiable bone fragments, with only small number of identifiable elements located outside the main concentrations. Artifacts are not spatially clustered in any particular area of the excavation and are instead found scattered in and around the bone concentrations (Figure 4.2).

Bioturbation mixed some sediments and skeletal elements (layers 4 to the top of 3A) in the easternmost portion of the landform over an approximately 20 m^2 area east of the excavation's 161E coordinate gridline (Figure 6.1). In the following description of the Upper Bonebed contents, elements derived from this portion of the excavation area are excluded due to their insecure provenience. This section reports on Upper Bonebed faunal remains excavated in the 2007, 2008, 2010, and 2020 field seasons.

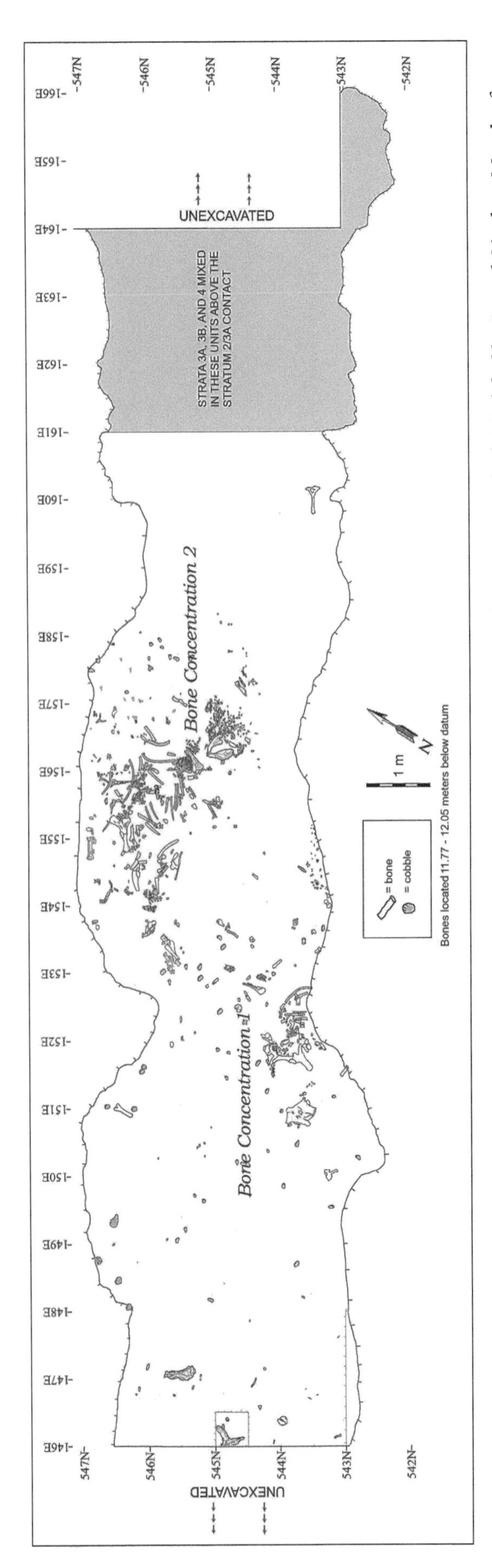

Figure 6.1. Map of Locus 1 indicating spatial distribution of skeletal elements recovered from the Upper Bonebed. Modified by Ismael Sánchez-Morales from original map by Edmund Gaines.

Taxonomic Representation

The Upper Bonebed is best described as monotaxic, with the vast majority of identifiable skeletal elements attributed to gomphothere (*Cuvieronius* sp.). One hundred fifty-nine faunal specimens from undisturbed contexts in the Upper Bonebed were identified to element and taxon. Of these, 149 are either identifiable to the genus level as *Cuvieronius* sp. or are identifiable to the order Proboscidea more generally (Appendix D). We therefore conclude that nearly all of the faunal material present in the Upper Bonebed of the site is gomphothere (*Cuvieronius* sp.).

Only 10 specimens in the Upper Bonebed are identified as nonproboscidean, all small in size. These include 1 rodent humerus, 1 tortoise or turtle plastron fragment, and 8 dermal ossicles belonging to Harlan's ground sloth (*Paramylodon harlani*) (Appendix D). As is typical of monotaxic bonebeds, the minor component of faunal remains from other species is hypothesized to have a distinctly separate depositional history, representing "background" accumulation of vertebrate skeletal material (Behrensmeyer 2007). This is supported by the presence of rodent finds in the sediments both above and below the Upper Bonebed (n=4), as well as tortoise or turtle (n = 5) and *P. harlani* (n = 17) finds in the sediments between the Upper and Lower Bonebeds.

Gomphothere Individuals, Element Distributions, Age, and Sex

Multiple lines of evidence point to the presence of two semi-articulated gomphothere (*Cuvieronius* sp.) skeletons in the Upper Bonebed of the site. The minimum number of gomphothere individuals (MNI) in the total assemblage is two. Furthermore, evidence drawn from (1) distributions of paired elements, (2) the presence of semi-articulated elements, and (3) states of molar eruption and bone fusion suggest that the two bone concentrations present in the southern and northern areas of the excavation respectively each derive from individual gomphotheres.

Bone concentration 1 (BC1; in the south area of the excavation block) contains molar enamel and tusk fragments, a scapula, ribs, an innominate and fragments, sacral and caudal vertebrae, a tibia, two fibulas, carpals and tarsals, a metatarsal, and phalanges (Table 6.1). Bone concentration 2 (BC2; north area of the excavation block) contains a mandible, molar enamel fragments, cranial fragments, stylohyoids, a scapula, ribs, an innominate and fragments, lumbar and caudal vertebrae, a radius, a tibia, carpals and tarsals, and phalanges (Table 6.2). A small number of bones are not clearly grouped into either bone concentration. These

Table 6.1. Identified Contents of Bone Concentration 1 (BC1)

Element	*NISP*	*MNE*	*MNI*
Cranial			
Molar	2	1	1
Tusk	1	1	1
Axial			
Scapula	1	1	1
Rib	11	1	1
Thoracic or lumbar vertebra	1	1	1
Innominate	3	2	1
Sacral vertebra	2	2	1
Sacral or caudal vertebra	1	1	1
Caudal vertebra	8	8	1
Vertebra (unidentified)	7	2	1
Front Limb			
Magnum	1	1	1
Scaphoid	1	1	1
Hind Limb			
Tibia	1	1	1
Fibula	2	2	1
Calcaneus	1	1	1
Metatarsal	1	1	1
Unassigned Appendicular			
First phalanx	3	3	1
First or second phalanx	1	1	1
Sesamoid	2	2	1
Totals	**50**	**33**	**1**

Key: NISP = number of identified specimens; MNE = minimum number of elements; MNI = minimum number of individuals.

Table 6.2. Identified Contents of Bone Concentration 2 (BC2)

Element	*NISP*	*MNE*	*MNI*
Cranial			
Cranium	3	1	1
Stylohyoid	2	2	1
Mandible	1	1	1
Molar	11	1	1
Tusk	2	1	1
Axial			
Scapula	1	1	1
Rib	29	7	1
Thoracic vertebra	4	4	1
Innominate	1	1	1
Lumbar vertebra	5	5	1
Caudal vertebra	4	4	1
Vertebra (unidentified)	14	4	1
Front Limb			
Radius	1	1	1
Hind Limb			
Tibia	2	2	1
Cuboid	2	2	1
Navicular	1	1	1
Unassigned Appendicular			
Phalanx	2	2	1
Third phalanx	1	1	1
Long bone	2	1	1
Totals	**88**	**42**	**1**

Key: NISP = number of identified specimens; MNE = minimum number of elements; MNI = minimum number of individuals.

include molar enamel fragments, a rib, thoracic vertebra, ischium fragment, humerus, and pedal bones (Table 6.3).

Paired Elements

Five element pairs consisting of bones from the right and left sides of the body of the same size, shape, and in the same state of epiphyseal fusion were identified as likely to have come from the same individual. Identified pairs include paired fibulas located in BC1 (Figure 6.2E) and paired tibias, stylohyoids, and cuboids in BC2 (Figure 6.3A–C). Mixing of elements between the bone concentrations is assumed to be limited or absent, given that paired elements are located within their same respective concentrations. Linear distances between paired elements range from 0.6 to 3.5 meters, with the average distance between paired elements in BC2 being 1.2 m and the distance between paired fibulas in BC1 being 3.5 m (Figure 6.4).

Semi-articulated Elements

In addition to the presence of paired elements, semi-articulated elements in the two bone concentrations support the interpretation that each concentration represents a single individual gomphothere. A concentration of seven caudal vertebrae in BC1 indicate the presence of a semi-articulated tail, and the pelvis is articulated at the pubic symphysis. The left tibia, fibula, and calcaneus are also

Table 6.3. Identified Specimens Not Clearly Grouped into Either Bone Concentration

Element	*NISP*	*MNE*	*MNI*
Cranial			
Molar	4	1	1
Axial			
Rib	1	1	1
Thoracic vertebra	1	1	1
Ischium	1	1	1
Front Limb			
Humerus	1	1	1
Unassigned Appendicular			
Phalanx	1	1	1
Carpal or tarsal	1	1	1
Long bone	1	1	1
Total	**11**	**8**	**1**

NISP = number of identified specimens; MNE = minimum number of elements; MNI = minimum number of individuals.

located in close proximity (Figure 6.4). A concentration of semi-articulated ribs and vertebrae make up the center of BC2. Cranial elements tend to occur in the eastern portion of BC2, while distal limb elements occur in the western portion (Figure 6.4), suggesting that the animal was deposited with its head facing east. The original orientation of the individual represented by BC1 is less certain because cranial bones are absent. Fragments of enamel and tusk dentine suggest that its cranial skeletal elements may have originally been deposited east of the concentration but have since been eroded from the landform.

Age and Sex

States of molar eruption and epiphyseal fusion are consistent within each bone concentration. They indicate that the gomphothere comprising BC1 was older than the individual comprising BC2. Information on epiphyseal fusion schedules of proboscidean bones is limited. Since no information on the bone fusion or molar eruption schedules specific to *Cuvieronius* sp. or other gomphothere taxa is yet available, we use data from modern African elephants (*Loxodonta* sp.) to approximate the ages of the gomphotheres from El Fin del Mundo, following Haynes (1991). It was not possible to make definitive sex determinations based on pelvis morphology due to incomplete ilia of the innominates in BC1 and BC2. However, given that male and female proboscideans have different epiphyseal fusion schedules (Haynes 1991), sexes of the individuals were inferred from the combination of age markers present in each bone concentration. Given the reliance of age and sex determinations on African elephant data, determinations are considered estimates and subject to revision if more data becomes available on estimated epiphyseal fusion schedules, molar eruption schedules, and sexual differences in skeletal ontogeny from more closely related extinct proboscidean taxa.

Bone Concentration 1 contains three skeletal markers that are useful for determining age and for which there are African elephant reference data: an innominate, a sacral vertebra, and a distal tibia. The pubis and ilium bones of the innominate are partially fused, indicating an age of less than eight African elephant equivalent years (AEY) if the individual is male, and an age of eight AEY or a few years older if the individual is female. The sacral vertebra of BC1 is unfused. This marker is not useful for determining age in male elephants since they fuse late in life but indicate an age less than 19 AEY if female. The distal tibia of BC1 is partially fused, which in female elephants fuses completely by age 20. Given the state of epiphyseal fusion among the three age markers in BC1, we conclude that the gomphothere individual was a young female aged between approximately 8 and 19 AEY.

Specimens used to age the individual in BC2 include an intact mandible preserving a record of molar eruption and wear, an innominate, a proximal radius, and two tibias. Dental eruption schedules are more reliable age indicators than epiphyseal fusion schedules, since they vary less between individuals and have been the subject of more study. As such, the absolute age estimation of the gomphothere in BC2 follows Haynes (1991), based on tooth eruption data from modern elephants. However, the eruption schedules of gomphotheres are expected to vary from those of more recent proboscidean lineages including modern elephants and mammoths, which usually have only one or two molariform teeth in wear at a time. Gomphothere tooth eruption follows a pattern similar to that of mastodons, whereby multiple teeth are typically in wear concurrently. Therefore, as an independent measure of age we also use tooth eruption schedules of mastodons to determine a relative age class for the gomphothere in BC2, following Green (2002) and Green and Hulbert (2005).

The mandible in BC2 contains articulated right and left third and fourth premolars (dP3, dP4) and first molars (M1). These are the second, third, and fourth molariform teeth to erupt in the life of a proboscidean. The dP3s are in an advanced stage of wear, the right dP3 is fractured, and

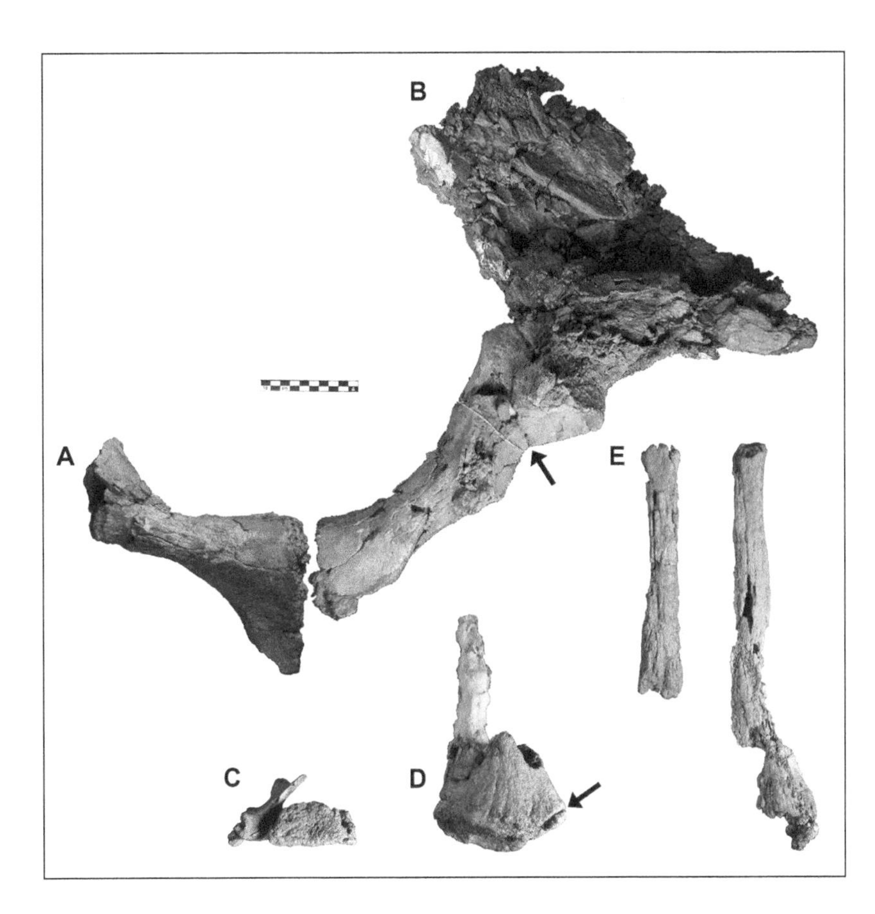

Figure 6.2. Selected identified elements from Bone Concentration 1: (A) left pubis (posterior view) articulated with (B) right innominate with partially fused pubis and ilium (posterior view); (C) unfused sacral vertebra (cranial view); (D) left partially fused distal tibia (anterior view); and (E) right and left unfused fibulas (anterior view). Arrows indicate partially fused epiphyseal lines. Scale = 10 cm. Photographs by Kayla B. Worthey.

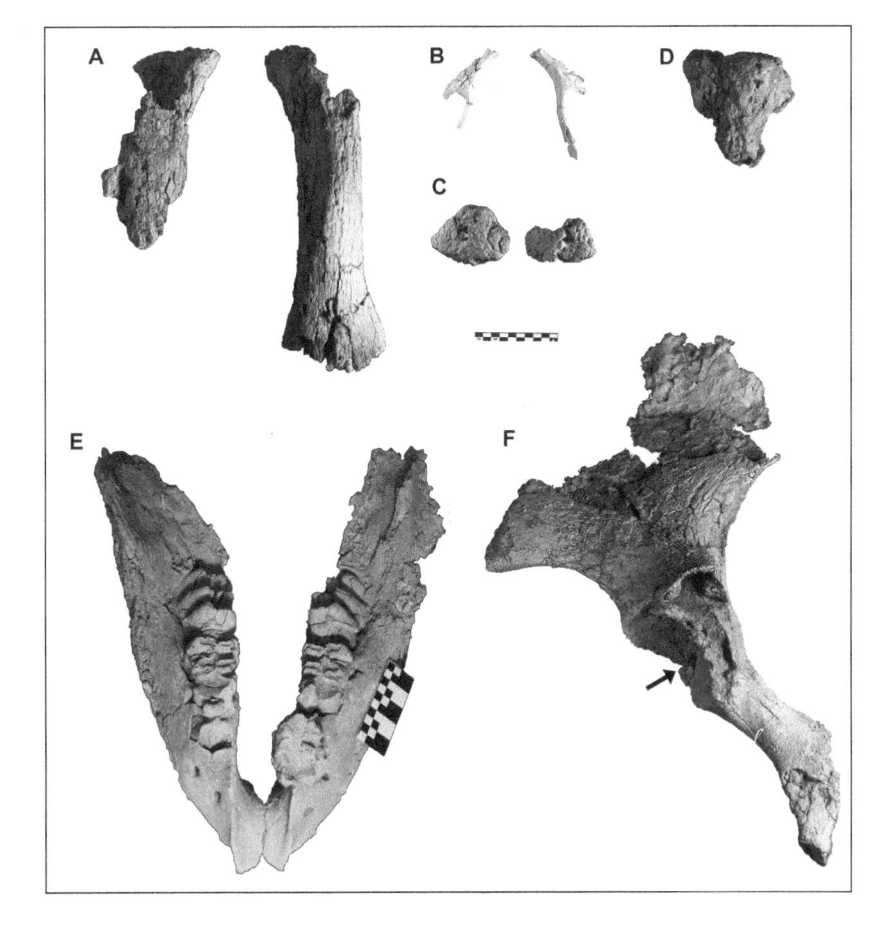

Figure 6.3. Selected identified elements from Bone Concentration 2: (A) right and left unfused tibias (anterior view); (B) right and left stylohyoids (lateral views); (C) left and right cuboids (dorsal view); (D) right unfused proximal radius (anterior view); (E) mandible with articulated dP3s, dP4s, and M1s; and (F) left innominate with partially fused pubis and ilium (posterior view). Arrow indicates partially fused epiphyseal line. Scale = 10 cm. Photographs A–D and F by Kayla B. Worthey. Photograph E modified from Sánchez and others (2014). Scales 10 cm.

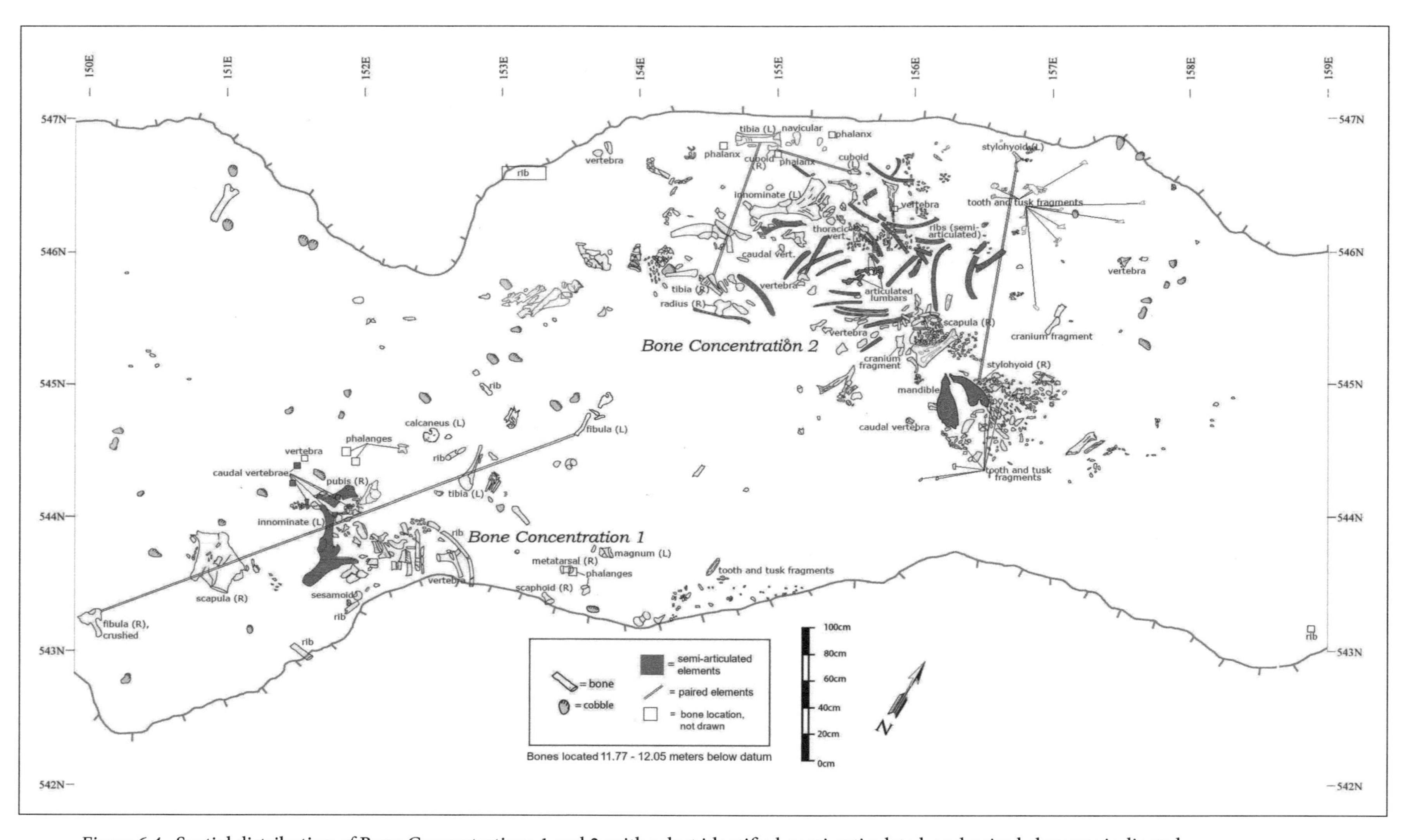

Figure 6.4. Spatial distribution of Bone Concentrations 1 and 2, with select identified, semi-articulated, and paired elements indicated. Modified from original map by Edmund Gaines.

both appear to be on the verge of replacement by the dP4s. The dP4s display light-medium wear, while the M1s have only just come into wear. Using modern elephant tooth eruption schedules (Haynes 1991), the second gomphothere individual is estimated to be two to eight years old (AEY). This estimate is consistent with an age class system based on the tooth eruption schedules of *Mammut americanum* (Green 2002; Green and Hulbert 2005), which would classify the gomphothere as an older juvenile.

States of epiphyseal fusion in BC2 are consistent with an age of two to eight AEY. The innominate in BC2 is partially fused, while the proximal radius, proximal tibias, and distal tibia are all unfused. The similar states of fusion in the innominates of BC1 and BC2 may indicate that the individuals were of different sexes since male pelvises fuse earlier than those of females (Haynes 1991), indicating that the younger gomphothere individual from BC2 is more likely to be male.

Bone Alterations

Weathering

The bones of the Upper Bonebed are well weathered, with some specimens requiring consolidation before extraction from the excavation block. Following Behrensmeyer's (1978) weathering stage system, all bones are in at least weathering stage 3, displaying fibrous outer surfaces and desiccation cracks (43%). An additional 48 percent of the specimens are in weathering stage 4 with coarsely fibrous surfaces and severe cracks or splinters. About 10 percent of the assemblage is in weathering stage 5, characterized by extremely fragile bone that has occasionally been broken beyond recognition (Appendix D). Advanced weathering is expected when bones are exposed to fluctuating temperature and humidity on the surface or near-surface environments, typically for at least several years (Behrensmeyer 1978). Periods of subaerial exposure and wet-dry cycles that could have contributed to bone weathering are evidenced by moderate pedogenesis and desiccation cracks in Stratum 3B (Chapters 2, 3).

The two bone concentrations display similar degrees of weathering, although BC1 has a greater percentage of specimens in weathering stage 3 (59%) relative to stage 4 (32%), whereas BC2 contains a lower percentage of specimens in weathering stage 3 (33%) and more in stage 4 (57%; $X^2(2, N=114) = 8.0, p<0.05$). In both concentrations, 9 to 10 percent of bones are in weathering stage 5. Differences in weathering may result from the different ages of the individuals in each concentration or differing amounts of time that the bone concentrations experienced subaerial exposure. Significant heterogeneity in weathering exists among paired elements from the same bone concentration, however, perhaps the result of horizontal variation in sediment properties or uneven distributions of vegetation after deposition. For example, the paired right and left cuboid bones in BC2 are located only 0.6 meters apart and are in very different states of weathering, with one cuboid complete with moderately fibrous surfaces and some small cracks, and the other broken with evidence of dissolution and highly fibrous surfaces (Figure 6.3C).

Other Damage

Carnivore damage on the bones is light, occurring on only 0.02 percent of specimens in the form of puncture and drag marks. There are no cuts or striations observed on the bones that could be definitively attributed to trampling marks or anthropogenic cut marks. Due to the advanced stages of weathering presented on the majority of the assemblage, however, it is likely that if originally present, carnivore damage, trampling damage, and shallow cut marks on the bone surfaces were obscured. Heavy cuts or chop marks on limb bones potentially indicative of human scavenging of stiff carcasses (Saunders 2007; Saunders and Daeschler 1994) are not present.

Element Orientations

The orientations of 86 elongated skeletal elements were measured from the Upper Bonebed using site maps and a protractor as degrees relative to grid north (30.7° west of true north). Elongated elements were defined as being at least twice as long as they are wide. Orientation data was taken for all elongated bones that were plotted in the site maps regardless their size or whether they were identified, including many specimens less than 20 cm long. Rose diagrams for visual representation of data were constructed in R using 10° bins. Diagram petal lengths are scaled to the square root of relative frequencies within each bin (Figure 6.5). A Rayleigh Test of Uniformity was carried out in R to determine if bone orientations are more likely to be random or unidirectional.

The Rayleigh test produced a p-value of 0.36, indicating that the radial orientations of elongated bone elements in the Upper Bonebed is statistically random. This indicates that water action is unlikely to have influenced the final orientations of bones in the Upper Bonebed.

Depositional History of the Upper Bonebed

The Upper Bonebed of Locus 1 is best described as autochthonous, meaning that the skeletal elements preserved

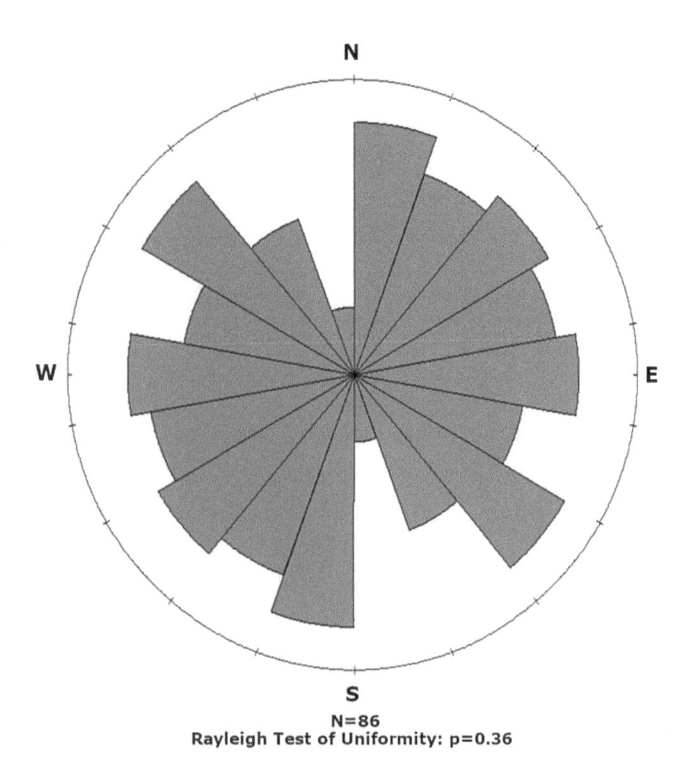

Figure 6.5. Rose diagram of elongated skeletal element orientations relative to grid north in 10° bins, with frequencies square-root scaled. The Rayleigh statistical test p-value is 0.36, indicating statistically random orientations of elements.

in the bonebed have not moved substantially from where they were deposited at death or shortly thereafter. The presence of semi-articulated and paired elements, the spatial distribution of bones in rough anatomical order, the preservation of two separate bone concentrations, each containing elements from an individual gomphothere, and the nondirectional orientation of elongated bones all indicate that the remains of two young gomphotheres were deposited in place with soft tissue attachments present. The gomphotheres likely were killed and butchered by Clovis peoples in Locus 1 as suggested by the association of the bones with complete diagnostic Clovis points, a Clovis point distal fragment, and small flakes possibly indicating tool re-sharpening (Figure 4.2). Evidence of late-access scavenging, such as deep cut marks or chop marks at joints is absent (Haynes and Klimowicz 2015; Saunders 2007; Saunders and Daeschler 1994).

A degree of post-depositional movement of bones did occur, however. Not all skeletal elements are articulated, and there are a number of elements notably absent from the bone concentrations, such as femora, some cranial and front limb elements in BC1, and some front limb elements in BC2. A number of these missing elements likely eroded from the landform prior to excavation, although some may have also been removed during carcass processing for transport or due to scavenger activity. Fragments of a pelvis and humerus at the western edge of the excavation area and a thoracic vertebra at the eastern edge may indicate movements of a few elements from their original bone concentrations approximately 4 to 6 m. The completeness and degree of element articulation among the gomphotheres from El Fin del Mundo is comparable to that observed at other Clovis kill sites including the Clovis site (aka Blackwater Draw Locality No. 1; Saunders and Daeschler 1994), Murray Springs (Hemmings 2007), and Naco (Haury 1953).

The processing of gomphothere carcasses for meat consumption is not necessarily expected to have resulted in thorough disarticulation, displacement, or removal of skeletal elements. Ethnographic and experimental work has shown that meat can be stripped from proboscidean carcasses with limited disarticulation of elements (Haynes and Klimowicz 2015). Among elephant cull butchers in Zimbabwe during the 1980s, Gary Haynes (1991) documented only the disarticulation of front limbs for access to meat below the scapula and disarticulation of hind limbs to aid in turning over the carcass. Meanwhile, of four ethnographically documented occasions of elephant butchering by Ituri Forest people, three were characterized by the disarticulation and removal only of the humeri and femora after meat-stripping, for the purposes of transport to campsites for marrow extraction (Duffy 1984; Fisher 1992).

Similar degrees of bone weathering and spatial dispersal of elements in BC1 and BC2 may indicate that the two gomphothere individuals were killed during a single hunting event. If so, and if gomphothere social organization was analogous to that of modern elephants, it would indicate the individuals were likely kin from the same herd. That the two individuals were a mother-offspring pair is unlikely, but possible if the gomphothere in BC1 was in the older end of her age estimate and the gomphothere in BC2 was in the younger end of his. It is also possible that the gomphothere carcasses were deposited at different times, as has been proposed for other Clovis proboscidean kill sites including Dent (Brunswig 2007; Fisher and Fox 2007; Haynes 1987; Hoppe 2004), Miami (Holliday and others 1994, Hoppe 2004), and Lehner (Haury, Sayles, and Wasley, 1959; Haynes 1987). One individual may have been hunted and butchered by Clovis peoples while the second died at a separate time of natural causes. Alternatively, Locus 1 may have been a repeatedly used hunting site for Clovis hunters and the two gomphotheres were hunted and butchered at different times. The abundant lithics recovered from the upland camp locality at El Fin del Mundo is indicative of repeated site visits by Clovis peoples (Sánchez-Morales 2018; Chapter 5), so it is reasonable to expect hunting localities in the surrounding lowlands to have been used repeatedly as well. The erosional truncation of Locus 1 by surrounding arroyos would have erased evidence of additional gomphothere individuals in this location, if originally present.

The gomphothere remains were exposed long enough for weathering, breakage, and some movement of bone to occur, but not long enough for complete deterioration or dispersal of bone elements. For the site to have been preserved in this state, it likely became buried relatively soon (within a few years) after the butchering of the gomphotheres. Most of the bonebed became buried by Stratum 3B sediments, while some bones remained exposed and were capped by a thin white diatomite layer, indicating the initiation of a shallow lacustrine environment in this location (Chapter 2).

THE LOWER BONEBED

The Lower Bonebed lies within Stratum 3A of Locus 1. Approximately 40 to 50 cm of sediment vertically separate the base of the Upper Bonebed from the top of the Lower Bonebed. The highest concentration of bones in the Lower Bonebed is located at the base of Stratum 3A just above an abrupt erosional contact with underlying Stratum 2, while a lower density of bones occurs above the contact.

The top of Stratum 3A in the portion of the Locus 1 landform east of the 160E excavation gridline was disturbed by bioturbation and mixed with overlying layers, and as such is excluded from analysis. The base of Stratum 3A was intact, however, and materials from this context are included in the analysis of the Lower Bonebed. Faunal remains found on the surface in proximity to Locus 1 have affinity to the taxa present in the intact deposits of the Lower Bonebed and are thought to have eroded from this context. Due to their likely association with the Lower Bonebed, these are included in the analysis of taxonomic representation and minimum numbers of individuals (MNI) but excluded from spatial analyses. Since including some specimens from surface contexts potentially introduces a minor component of remains from the Upper Bonebed into the Lower Bonebed assemblage, all elements from disturbed contexts are indicated in tables and NISP estimates (Table 6.4).

This chapter presents an analysis of approximately 60 percent of the total faunal assemblage from the Lower Bonebed excavated in the 2007, 2008, 2010–2011, 2012, and 2014–2015, 2018, and 2020 field seasons, focusing primarily on point-plotted specimens. Materials recovered from screens were targeted for study from the 2020 field season only.

Taxonomic Representation

The Lower Bonebed is dominated by megafauna remains but, unlike the Upper Bonebed, the assemblage is multitaxic with a moderately diverse array of large-bodied species. The most abundant identified species in the Lower Bonebed include *Cuvieronius* sp. (number of identified specimens (NISP = 21; all in situ; Figure 6.6), *Mammut americanum* (NISP= 8; all in situ; Figure 6.7), *Equus* sp. (NISP = 6; 3 in situ; Figure 6.8F–I), and *Paramylodon harlani* (NISP = 27; 26 in situ; Figures 6.8E, 6.9). Represented by one or a few specimens are *Mammuthus* sp. (NISP = 2; not in situ; Figure 6.8A and B), *Tapirus* sp. (NISP = 1; in situ; Figure 6.8D), family Camelidae (NISP = 1; not in situ; Figure 6.8C), and family Leporidae (NISP =1; 1 in situ). In addition to specimens with taxonomic identifications to the species, genus, or family level, the analyzed faunal assemblage includes order Rodentia (NISP = 2; not in situ), order Testudines (NISP = 11; in situ), and 64 specimens (62 in situ) identified to the order Proboscidea (Table 6.4, Appendix D).

Minimum Numbers of Individuals (MNI)

The minimum number of gomphothere (*Cuvieronius* sp.) and mastodon (*Mammut americanum*) individuals represented in the Lower Bonebed was determined through the examination of molars, the best represented element type among these taxa. Identification of molars, degree of wear, and inferred molar eruption schedules all contributed to the MNI determinations. Schedules of *Mammut americanum* tooth eruption and wear, such as those developed in Green (2002) and Green and Hulbert (2005; Tables 6.5, 6.6), can be used for approximating relative age categories in mastodons as well as gomphotheres, which are expected to have comparable tooth eruption schedules. At least four gomphothere individuals of different age categories are represented in the Lower Bonebed: one youth with a lightly worn M2/, one young adult with a moderately worn M/2, one adult with light- to moderately worn M/3s, and one adult late in life with advanced M3 wear. Multiple mastodon individuals are represented in the Lower Bonebed as well, indicated by molar eruption and wear schedules. At least

Table 6.4. Analyzed Contents of the Lower Bonebed

Taxon and Element	*NISP*	*MNE*	*MNI*
***Cuvieronius* sp.**			
Molar (incl. articulated)	8	8	4
Tusk	1	1	1
Mandible	2	2	1
Cervical vertebra	1	1	1
Innominate	1	1	1
Radius	2	2	1
Ulna	1	1	1
Femur	1	1	1
Tibia	1	1	1
Fibula	3	3	3
Totals	**21**	**21**	**4**
Mammut americanum			
Molar	6	5	3
Tusk	2	1	1
Totals	**8**	**6**	**3**
***Mammuthus* sp.**			
Molar	(2)	1	1
***Equus* sp.**			
Molar, premolar & incisor	2 (3)	5	1
Radioulna	1	1	1
Totals	**3 (3)**	**6**	**1**
Paramylodon harlani			
Molariform tooth	(1)	1	1
Dermal ossicle	26	26	1
Totals	**26 (1)**	**27**	**1**
***Tapirus* sp.**			
Molar	1	1	1
Camelidae			
Astragalus	(1)	1	1
Leporidae			
Humerus	1	1	1
Order Proboscidea			
Molar	19 (1)	2	1
Tusk	21	1	1
Thoracic vertebra	2	2	1
Rib	4 (1)	1	1
Scapula	1	1	1
Humerus	1	1	1
Carpal or tarsal	4	4	1
Phalanx or metapodial	3	3	1
Vertebra (unidentified)	1	1	1
Long bone (unidentified)	3	1	1
Flat bone (unidentified)	3	1	1
Totals	**62 (2)**	**18**	**1**
Order Rodentia			
Cranium	(1)	1	1
Mandible	(1)	1	1
Totals	**(2)**	**2**	**1**
Order Testudines			
Carapace	6	1	1
Plastron	9	1	1
Totals	**15**	**2**	**1**

Key: NISP = number of identified specimens; MNE = minimum number of elements; MNI= Minimum number of individuals.

NISP values in parentheses indicate the number of specimens found eroded from the landform or from disturbed areas of the site. NISP values without parentheses indicate the number of specimens found in situ.

three mastodon individuals are present, including one youth with a lightly worn M2, one young adult with a moderately worn M2, and one adult late in life with heavily worn M3s. The remaining taxa represented in the Lower Bonebed are less well represented and have an MNI of 1 (Table 6.4).

Element Distributions and Orientations

For spatial analysis of the Lower Bonebed, materials were divided according to stratigraphic context (Figures 6.10 and 6.11). The majority of faunal materials were recovered from a dense concentration of bones along the contact between Strata 3A and 2. These specimens were considered separately from the bones in the upper portion of Stratum 3A, which were more sparsely distributed and may have had a different depositional history.

In both the Stratum 3A/2 contact zone and upper Stratum 3A, bones from various species are intermixed, with no spatial concentrations of bones belonging to a particular taxon (Figures 6.12–6.15). Most bones are not articulated or paired, with a few exceptions at the 3A/2 contact: (1) left and right portions of a *Cuvieronius* sp. mandible located 1.2 m apart (IDs #33491 and #33493; Figure 6.13) and (2) a *Cuvieronius* sp. tibia and adjacent fibula (IDs #62629 and #62631; Figure 6.14).

The orientations of elongated bones in the Lower Bonebed were measured using the same methods described for the Upper Bonebed. Site maps and a protractor were used to measure degrees of elongated skeletal elements relative to grid north (30.7° west of true north). Orientations for bones concentrated at the contact between

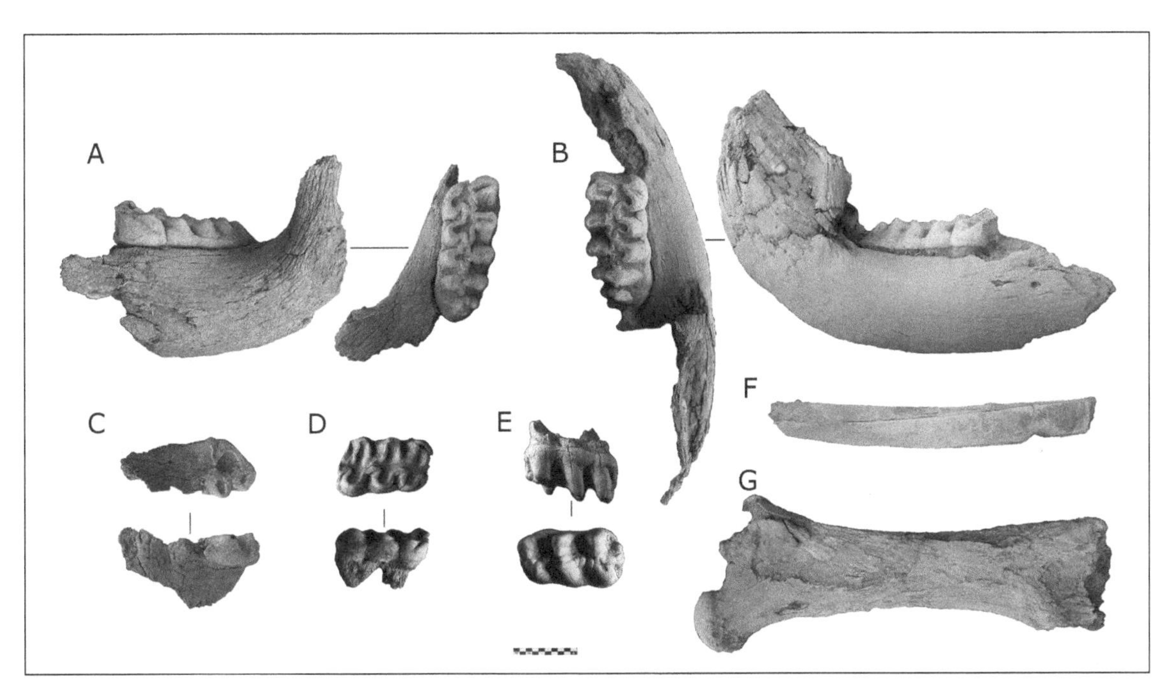

Figure 6.6. Selected *Cuvieronius* sp. specimens from the Lower Bonebed: (A) left mandible with articulated M/3 (ID #33491); (B) right mandible with articulated M/3, which is likely paired with A (ID #33493); (C) highly worn upper or lower M3 (ID #74566A); (D) right M/2 (ID #62628); (E) right M2/ (ID #62542); (F) medial tusk fragment with enamel band (ID #62630); and (G) right femur, posterior view (ID #74560A). All pictured specimens were recovered from the 3A/2 contact except for the femur (G). Scale = 10 cm. Photographs A, B, and G by Ismael Sánchez-Morales; C–E by Kayla B. Worthey; F by El Fin del Mundo Project 2008.

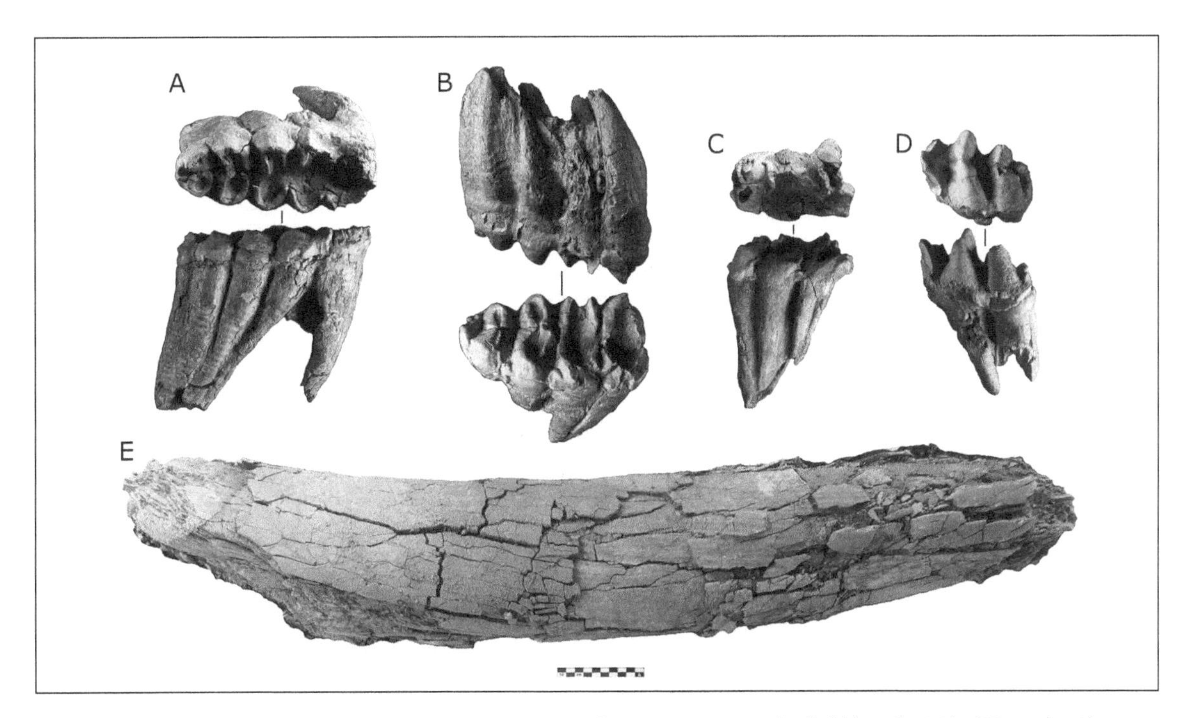

Figure 6.7. Selected *Mammut americanum* specimens from Lower Bonebed: (A) right M/3 (ID #59178); (B) right M2/ (no ID); (C) M/3 fragment (ID #62623); (D) upper or lower M2 (ID #62622); (E) medial tusk fragment (ID #74564A). All pictured specimens were recovered from the 3A/2 contact. Scale = 10 cm. Photographs by Kayla B. Worthey.

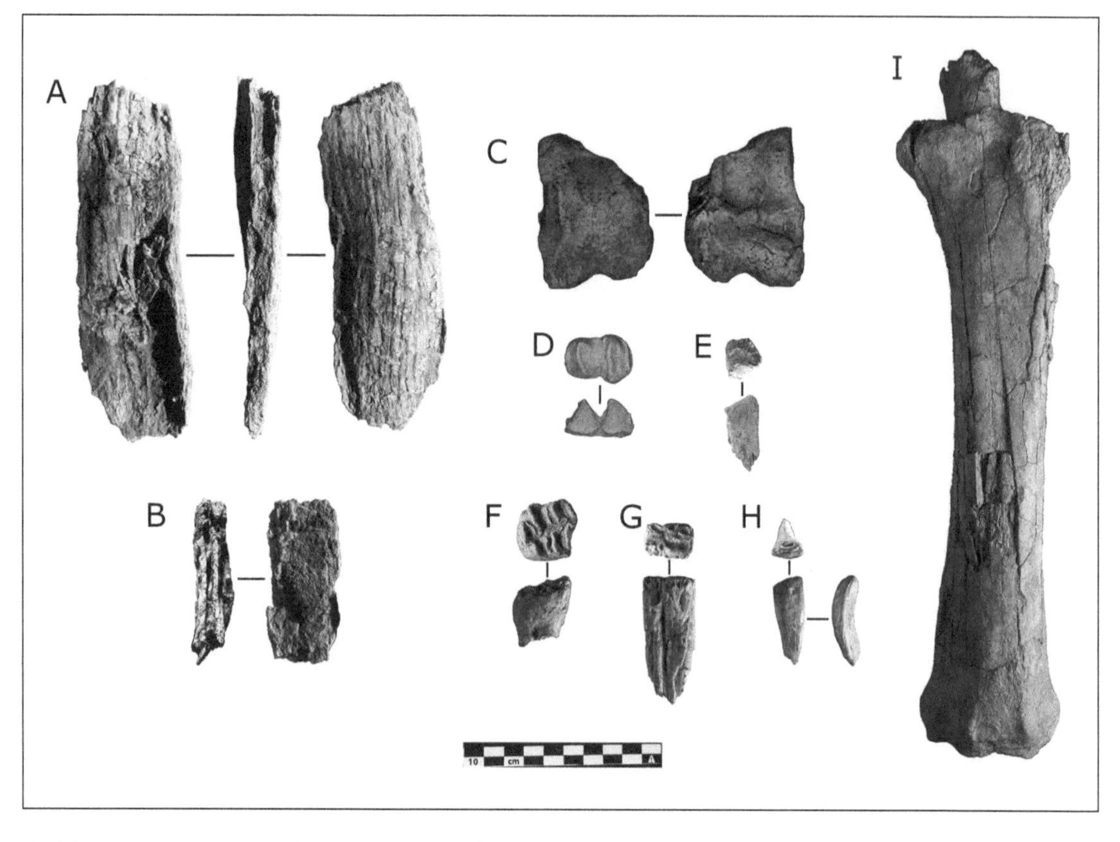

Figure 6.8. Selected mammoth, camel, tapir, Harlan's ground sloth, and horse specimens from Lower Bonebed. (A and B) *Mammuthus* sp. molar plate fragment, showing parallel enamel plates separated by cementum and dentine. Specimens found eroded from the Locus 1 landform (A) and 140m west of the landform (B) (no IDs); (C) camelid astragalus found eroded from the Locus 1 landform (ID #59345); (D) tapir (*Tapirus haysii*) molar recovered from 3A/2 contact (ID #46256); (E) *Paramylodon harlani* molariform tooth fragment recovered from the south profile of stratum 3A (ID #74573A); (F) *Equus* sp. upper molar or premolar recovered from 3A/2 contact (ID #62611); (G) *Equus* sp. lower molar or premolar recovered from a bioturbated context (ID #46027); (H) *Equus* sp. incisor recovered from 3A/2 contact; I: *Equus* sp. right radioulna recovered from 3A/2 contact. Scale = 10 cm Photographs A, B, E–I by Kayla B. Worthey, C and D by Alejandro Jiménez.

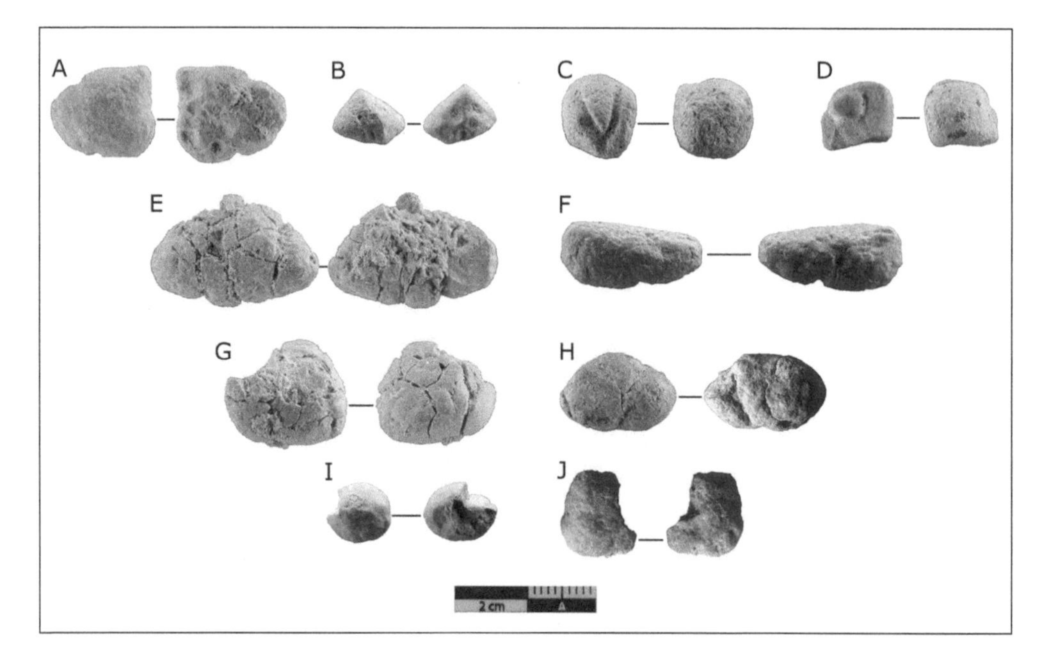

Figure 6.9. Selected *Paramylodon harlani* (Harlan's ground sloth) dermal ossicles (osteoderms) from Lower and Upper Bonebeds. Diagnostic features observed on the ossicles include pitting (A and B), grooves (C, D, E, F, H), foraminifera (B), and an interior structure indicating concentric growth of bone around a nucleus (I) (Brambilla and others 2019; McDonald 2018; Merriam 1906; Moreno and Woodard 1899). (A) ID #93915A, (B) ID #93932A, (C) ID #59892, (D) ID #63057, (E) ID #93946, (F) ID #74933A, (G) ID #93942A, (H) #ID 63043, (I) #ID 74596A, and (J) ID #93915A. All pictured ossicles were found in the Lower Bonebed except specimens D and H, which were recovered from the Upper Bonebed. Scale = 2 cm. Photographs by Kayla B. Worthey.

Table 6.5. Description of Mastodont Tooth Wear Stages

Wear Stage	*Description*
0	Crown formed, not erupted (in bone)
0+	Crown erupted, not worn (in bone)
0/+	Crown unworn, eruption state unclear (isolated)
1	Wear on protoloph/id only.
2	Light wear on all loph/ids; protoloph/id may show adjacent enamel figures.
2+	Protoloph/id and metaloph/id showing adjacent enamel figures.
3	Extensive wear but pattern still clear; all loph/ids with adjacent enamel figures.
4	Severe wear. Pattern partly or wholly obliterated; enamel figures contiguous.

Note: Wear stages follow Green and Hulbert. (2010), Saunders (1977), and Simpson de Paula Couto (1957). This system is applicable only to proboscidean taxa with bunodont tooth morphology, including *Mammut americanum* and *Cuvieronius* sp.

Table 6.6. Schedule of Tooth Eruption and Wear in *Mammut americanum*

Age	*Tooth Eruption and Wear*					
Birth	dP2 (0+)	dP3 (0+)				
Juvenile (younger)	dP2 (1 - 4)	dP3 (1 – 3)	dP4 (0+ - 2)			
Juvenile (older)		dP3 (4)	dP4 (2 – 3)	M1 (0+ - 2)		
Youth			dP4 (4)	M1 (2 – 3)	M2 (0+ - 2)	
Young adult				M1 (4)	M2 (2 – 3)	M3 (0+ - 2)
Adult					M2 (4)	M3 (2 - 3)
Adult late in life						M3 (4)

Schedule follows Green (2002) and Green and Hulbert. (2005). The teeth listed in each age category include only those that are expected to have partially or wholly erupted at that age. The expected range of potential wear stages for each tooth and age category is indicated in parentheses.

Stratum 3A and 2 (n =112) are reported separately from bones recovered from above the contact (n = 25). Statistical tests and plots were carried out using R. Rose diagrams for visual representation of data were constructed using 10° bins and the square root of relative frequencies within each bin (Figures 6.16 and 6.17).

The Rayleigh Test of Uniformity produced a statistically significant p-value of <0.05 for bones at the Stratum 3A/2 contact, indicating nonrandom, unidirectional patterns in bone orientations consistent with water transport (Figure 6.16). Meanwhile, orientations of bones located above the Stratum 3A/2 contact appear random when considering all specimens (p=0.30), but nonrandom when considering only small (<20cm) specimens (p<0.05; Figure 6.17). Relative to grid north, bones are generally oriented northwest to southeast at the stratum 3A/2 contact and west-east among small bones in the upper portion of Stratum 3A, indicating changes in flow direction during the deposition of Stratum 3A, perhaps due to channel migration.

Bone Alterations

Weathering

Degrees of weathering among bones in the Lower Bonebed are similar to those observed in the Upper Bonebed, with weathering stages defined by Behrensmeyer (1978) ranging from 3 (fibrous surface texture and cracks present) to 5 (highly friable). The majority of elements are in weathering stages 3 (47%) and 4 (coarsely fibrous surface texture with open cracks; 40%), with a minor component in weathering stage 5 (13%). The strong similarities between the weathering and fragility of bones from the Upper and Lower Bonebeds are likely to have resulted from periods

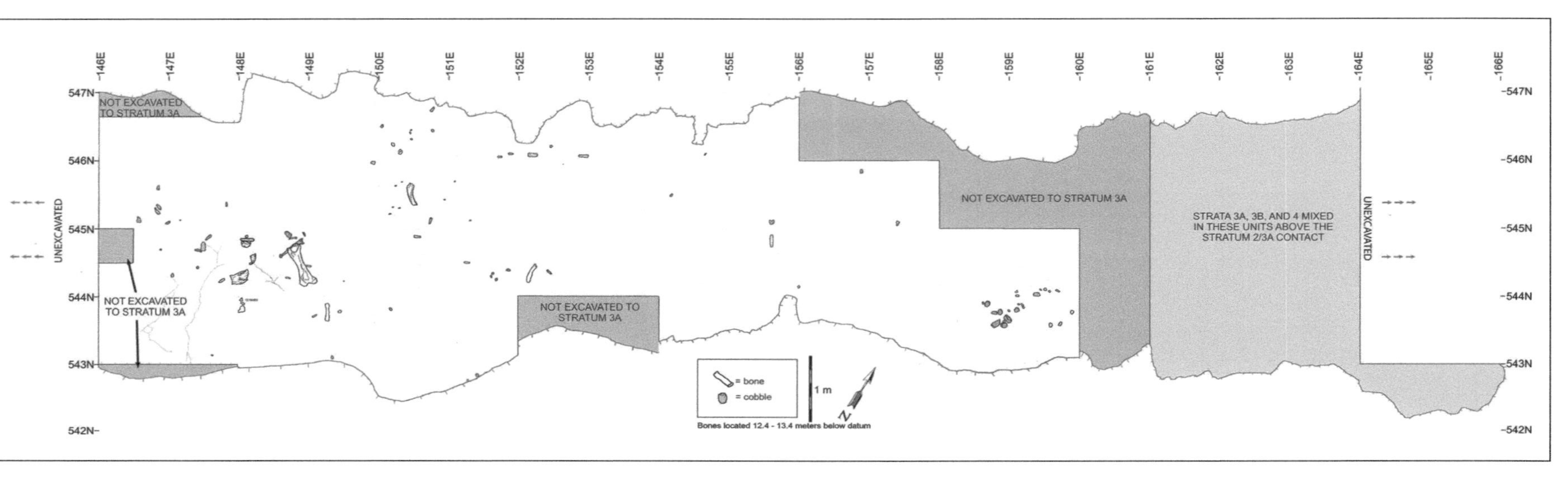

Figure 6.10. Spatial distribution of skeletal elements recovered from excavated portions of Lower Bonebed in upper portion of Stratum 3A. Modified from original maps by Ismael Sánchez-Morales and Edmund Gaines.

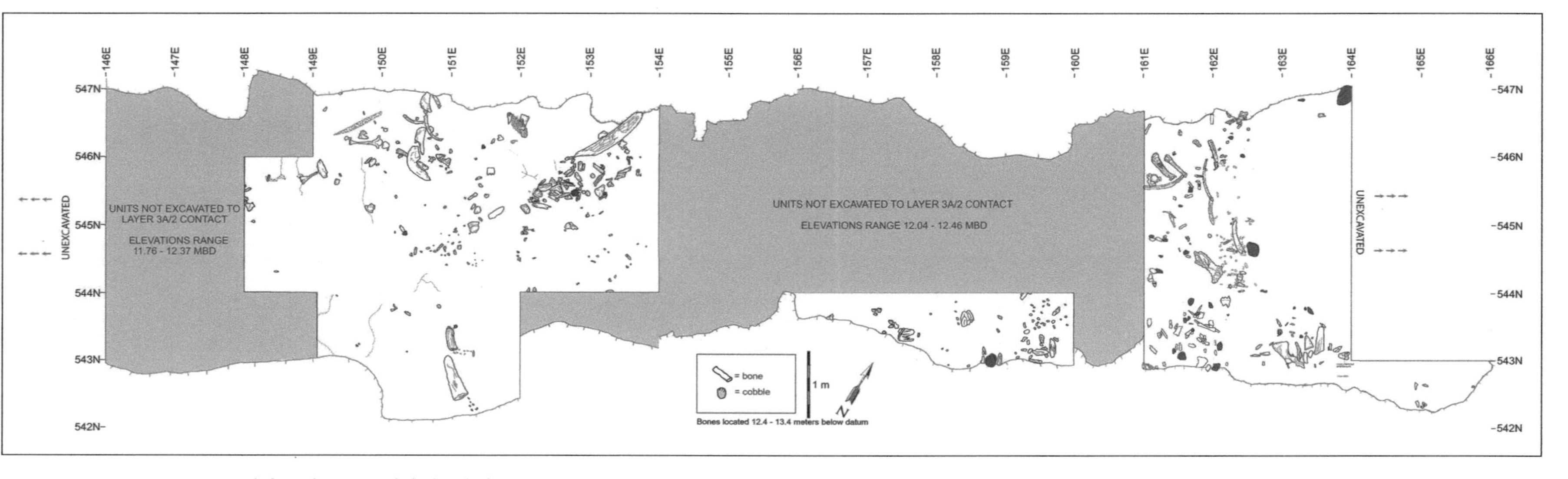

Figure 6.11. Spatial distribution of skeletal elements recovered from the excavated portions of the Lower Bonebed at the Stratum 3A/2 contact. Modified from original maps by Ismael Sánchez-Morales, Edmund Gaines, and Kayla B. Worthey.

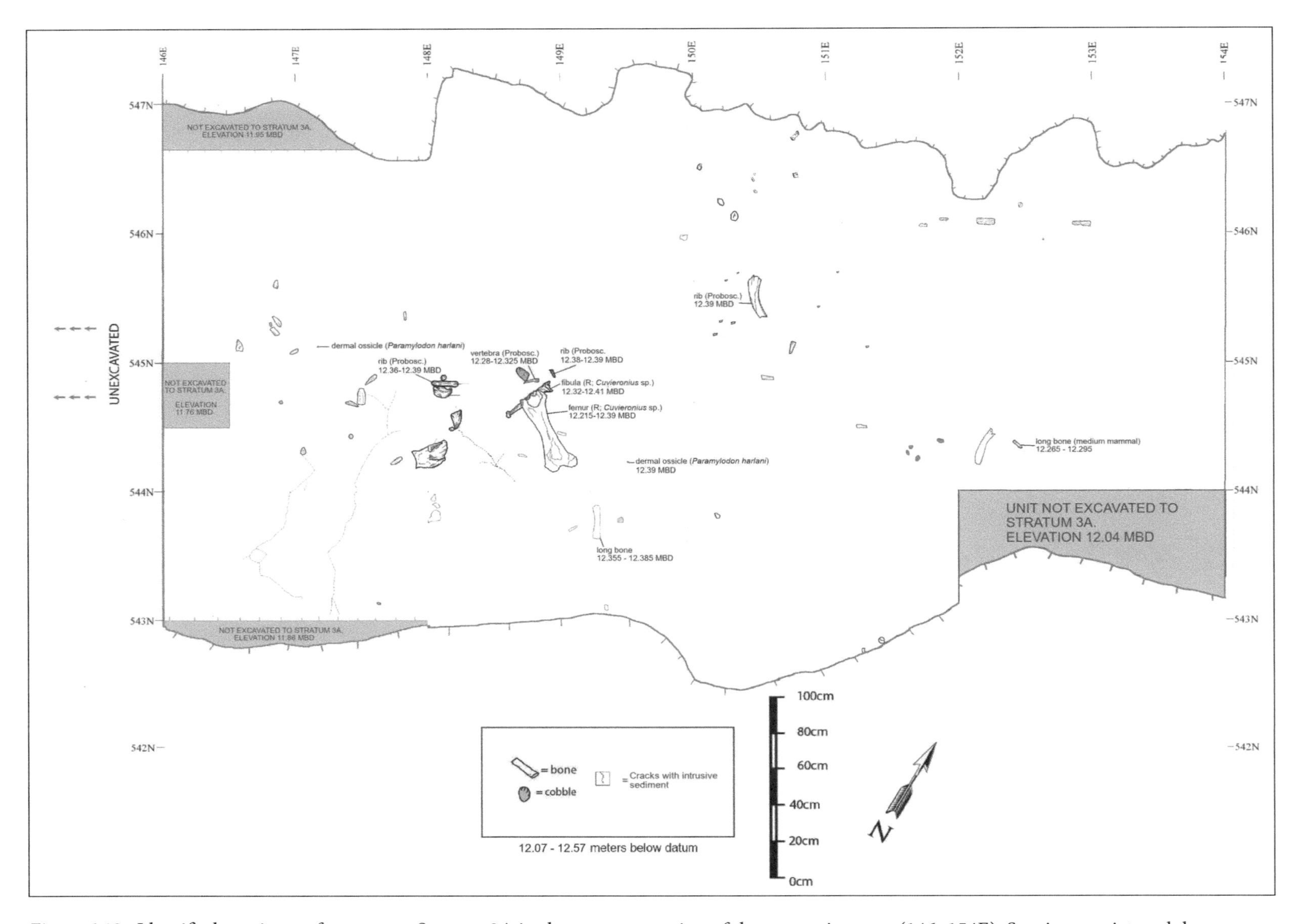

Figure 6.12. Identified specimens from upper Stratum 3A in the western portion of the excavation area (146–154E). Specimens pictured do not rest directly on the Stratum 3A/2 contact. Modified from original map by Ismael Sánchez-Morales.

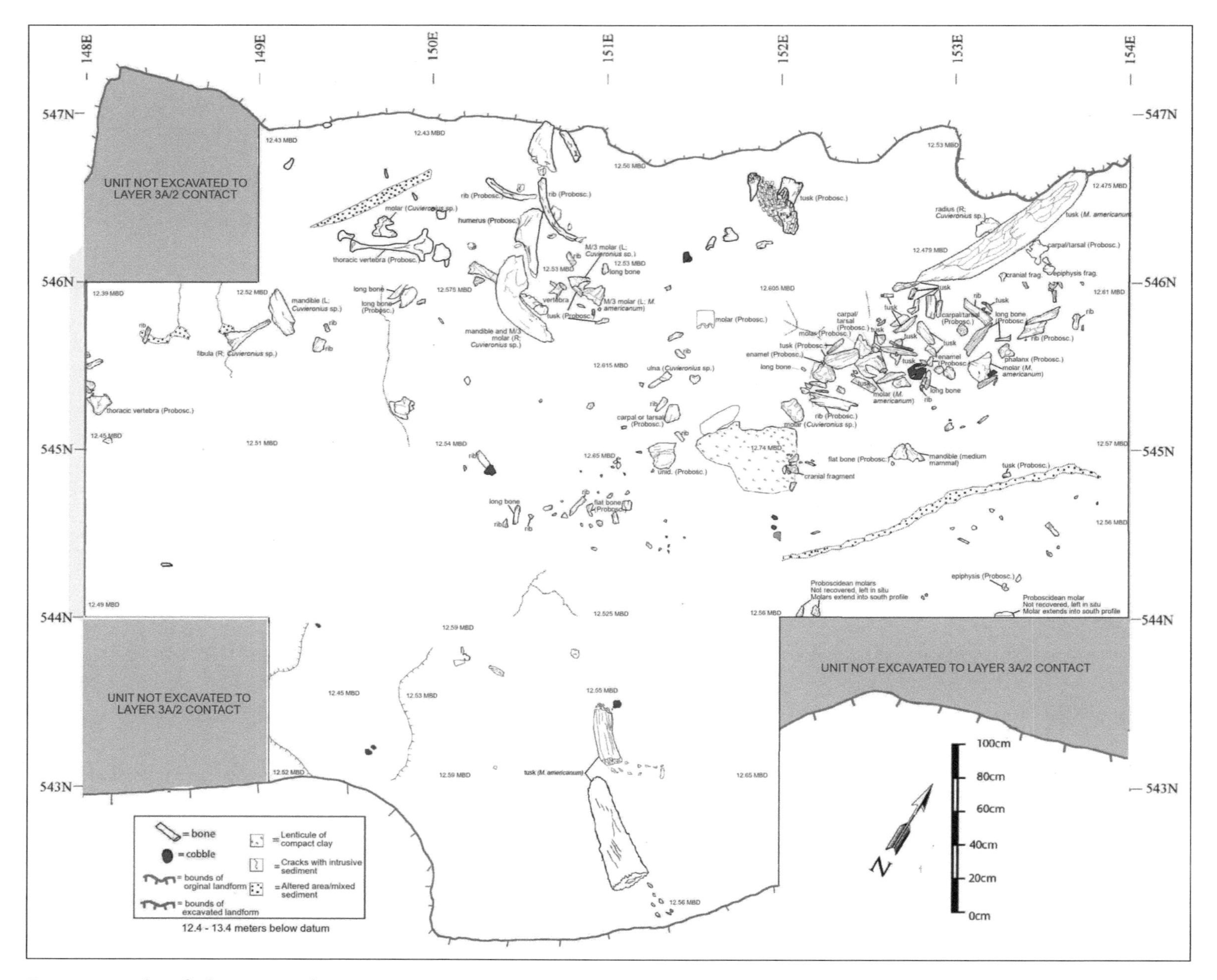

Figure 6.13. Identified specimens from Stratum 3A/2 contact in the western portion of the excavation area (148–154E). Specimens pictured are situated at the base of Stratum 3A and rest directly on Stratum 2. Modified from original map by Ismael Sánchez-Morales.

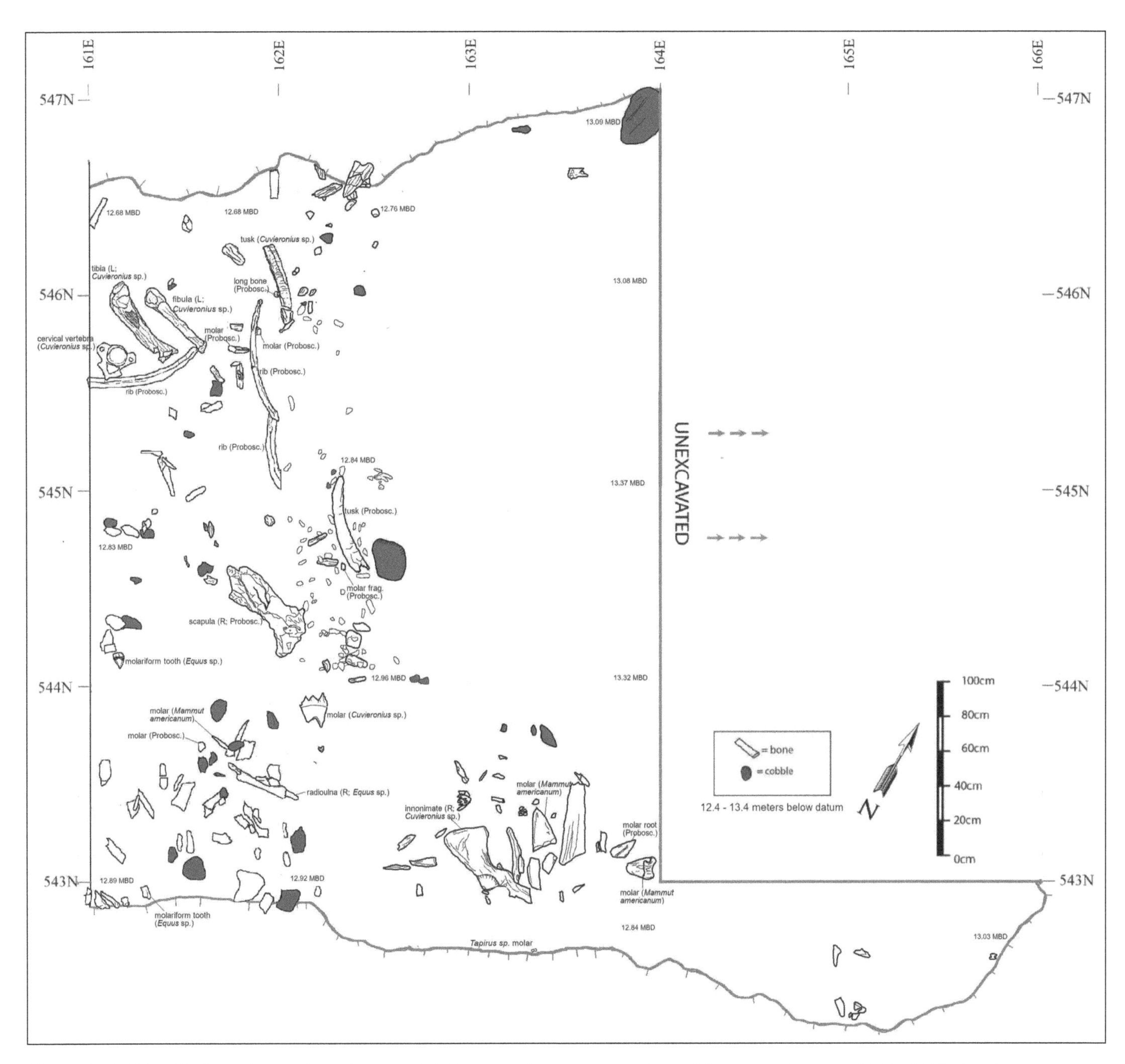

Figure 6.14. Identified specimens from Stratum 3A/2 contact in eastern portion of the excavation area (161– 66E). Specimens pictured are situated at the base of Stratum 3A and rest directly on Stratum 2. Map by Kayla B. Worthey.

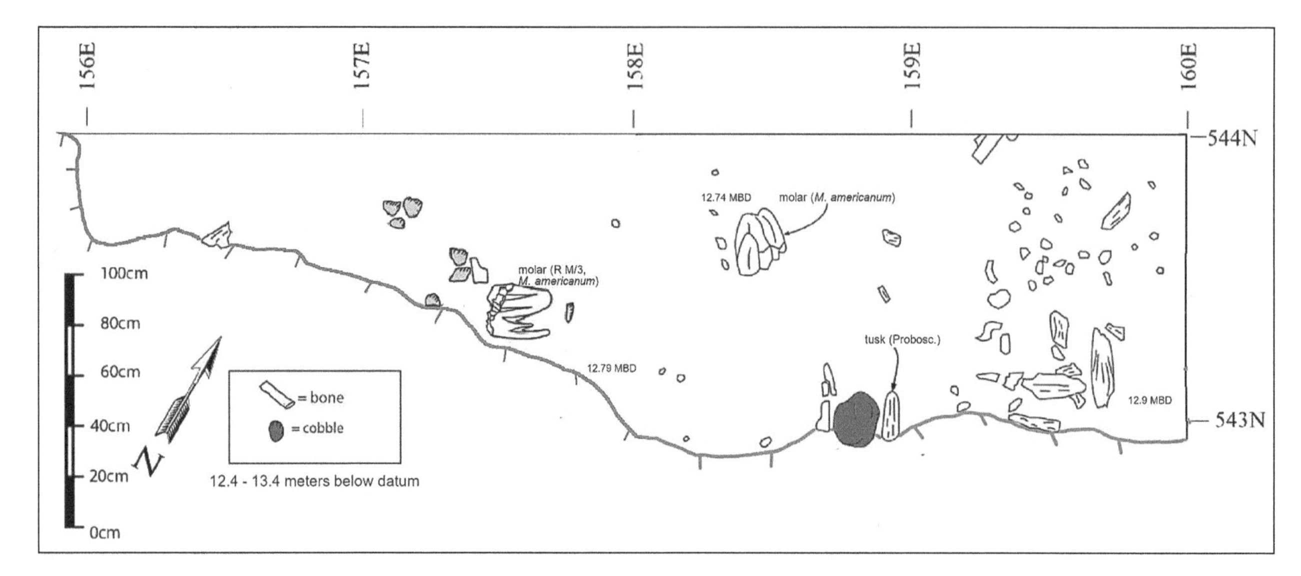

Figure 6.15. Identified specimens from Stratum 3A/2 contact in south-central portion of excavation area (156–160E). Specimens pictured are situated at the base of Stratum 3A and rest directly on Stratum 2. Modified from original map by Edmund Gaines.

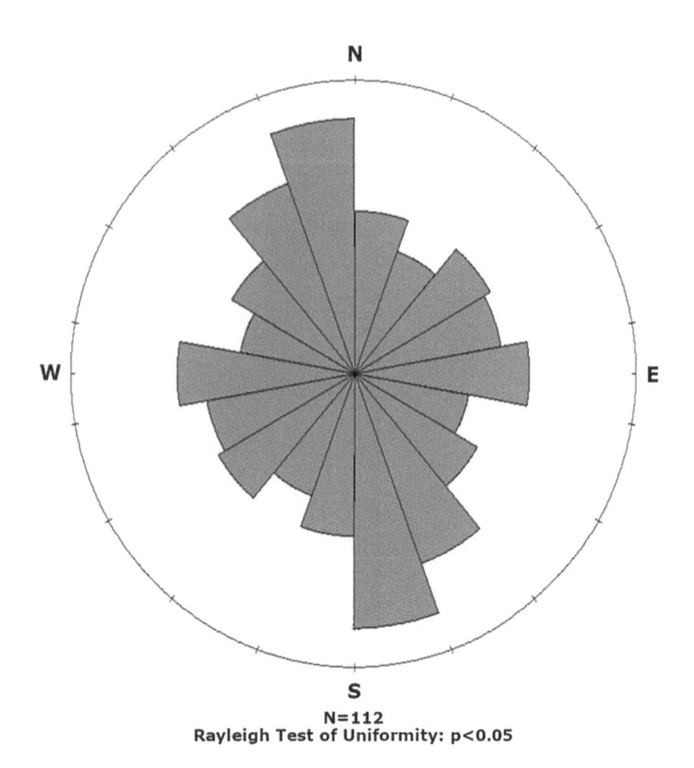

Figure 6.16. Orientations of elongated skeletal elements at Stratum 3A/2 contact of Lower Bonebed. Rose diagram is relative to grid north in 10° bins, with frequencies square-root scaled. The Rayleigh statistical test p-value is <0.05, indicating statistically nonrandom (directional) orientations of elements along a northwest to southeast axis relative to the grid.

of exposure and fluctuating wet and dry conditions affecting Stratum 3, evidenced by moderate pedogenesis and desiccation cracking within this stratum (Chapters 2, 3).

Other Damage

As in the Upper Bonebed, damage to bones in the form of carnivore tooth and gnaw marks or trampling marks is largely absent, possibly erased or obscured due to the advanced stages of weathering that dominate the assemblage. Several specimens, however, do display polishing or edge-rounding consistent with water transport. Waterworn specimens consist of small, unidentifiable bone fragments.

Depositional History and Paleoecology of the Lower Bonebed

Evidence is consistent with an allochthonous origin of the Lower Bonebed, meaning that the bones were transported from elsewhere before being deposited at Locus 1. The variety of species and numbers of individuals represented in the assemblage, the paucity of articulated skeletal elements, evidence for unidirectionality among elongated bones, and the sedimentary setting of the bonebed within channel deposits all suggest that the Lower Bonebed accumulated via the fluvial transport and deposition of disarticulated skeletal elements at a point of lower energy along a stream, such as a point bar on the inside bend of the channel or another obstruction (Aslan and Behrensmeyer 1996). In addition, there are several small-scale cut-and-fill features within Stratum 3A associated with the deposition of bone

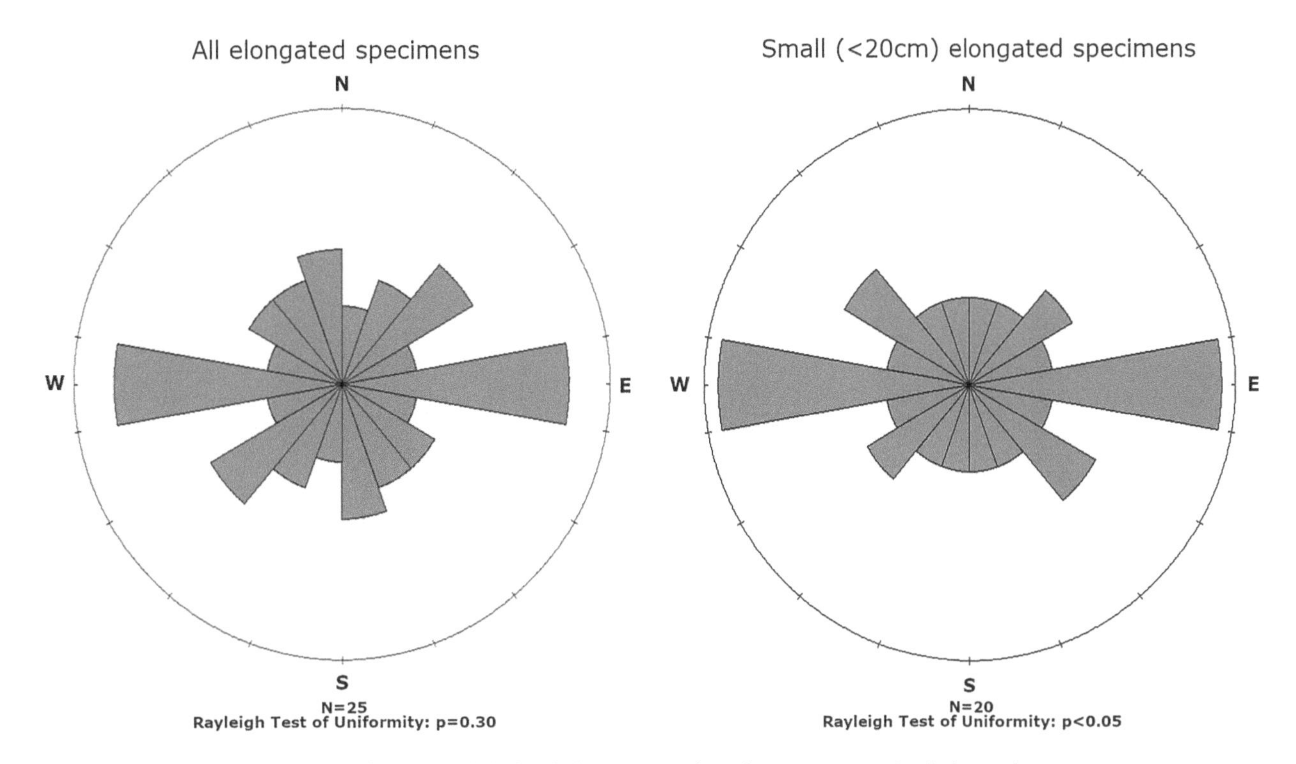

Figure 6.17. Orientations of elongated skeletal elements within the Lower Bonebed above the Stratum 3A/2 contact. Rose diagrams of all elongated specimens (left) and those with a maximum dimension smaller than 20 cm (right) are plotted relative to grid north in 10° bins, with frequencies square-root scaled. The Rayleigh statistical test p-value for all specimens is 0.30, indicating statistically random orientations of elements. When considering smaller specimens only, however, the Rayleigh test p-value is <0.05, indicating statistically non-random (directional) orientations of elements along an east-west axis relative to the grid.

and gravel, suggesting that variable stream discharge contributed to bone deposition (Figure 6.18).

The erosional contact between Strata 3A and 2 is associated with a dense concentration of bones resting on and just above the contact. Bones from large and medium-bodied mammals dominate, although variance in the skeletal element sizes recovered from the bonebed indicates that the hydraulic sorting was relatively poor. In multiple areas along the contact, bone elements are associated with large cobbles as well (Figure 6.18), suggesting a high-energy depositional environment.

We hypothesize that a high-energy current eroded the top of Stratum 2 as well as bone-bearing sediments along the banks upstream, redepositing bones at Locus 1 due to an obstruction or obstructions in the stream bed and promptly burying them as the current slowed. Many, though not all, bones were disarticulated prior to burial. After the burial of the bone concentration at the Stratum 3A/2 contact, a lower density of bones was deposited in the same location during smaller scale cut-and-fill events. Small-sized elements in these sediments are aligned to a different axis indicating a change in flow direction, possibly due to channel migration.

An alternative explanation for the concentration of bones at the layer 3A/2 contact is the exhumation of overlying bone-bearing sediments via erosion, which could have concentrated bones on an erosional surface as finer sediments were winnowed away and heavier bones remained in place (Rogers and Kidwell 2007). However, rounded edges and surfaces of bones due to the abrasion of water, while present, are rare in the assemblage.

Regardless of the exact mechanism that led to the accumulation of bones in the Lower Bonebed, the contents likely represent a time-averaged assemblage. The densest concentration of bones at the Stratum 3A/2 contact is bracketed by radiocarbon dates on Succinid shells, which date to 21,530 ± 80 (Stratum 2, AA-94053) and 12,890 ± 50 ^{14}C years (Stratum upper 3A, AA-94052). In addition, charred plant fragments from lower Stratum 3A, closest to the bone concentration, produced a date of 21,200 ±1600 ^{14}C years (Table 2.2). These dates are best considered maximum ages for the deposition of the strata due to evidence

Figure 6.18. Photographs of the Stratum 3A/2 contact during excavation of the Lower Bonebed. Arrows indicate large cobbles in association with the bones: (A) example of a cut-and-fill feature that resulted in the deposition of gravel and small bone fragments; (B) horse radioulna associated with large cobbles; and (C) concentration of bones and teeth including molars belonging to two different proboscidean taxa: *Cuvieronius* sp. and *Mammut americanum*. Photograph A and B by Edmund Gaines; Photograph C by Ismael Sánchez-Morales.

for the presence of older, reworked plant remains and shell in the deposits (Chapter 2). That said, the bones themselves could also be older than the date of sediment deposition. While some of the individuals represented in the bonebed may have died relatively soon before being incorporated into the deposit, erosion and redeposition of isolated bones or other bonebeds from the banks and floodplain deposits upstream are also expected. Bones eroded from banks upstream could potentially represent long and diverse timescales of skeletal element accumulation and reworking. Without more information, resolving the amount of time represented by the bones of the Lower Bonebed is difficult to accomplish with certainty. Other studies of the taphonomy of fluvial bonebed deposits estimate that most bonebeds represent a span of time on the scale of 10^2 to 10^4 years (Behrensmeyer 1982).

The Lower Bonebed of El Fin del Mundo is unique in that it preserves an unusual association of proboscidean taxa. Mastodons (*Mammut americanum*) and gomphotheres (including *Cuvieronius* sp., *Stegomastodon* sp.) have opposing distributions in Mexico during the Pleistocene, suggesting that they were parapatric taxa (Arroyo-Cabrales and others 2007), but they co-occur in close association at El Fin del Mundo. This is the first documented occurrence of *Mammut americanum* in Sonora (for a review, see Mead, Arroyo-Cabrales, and Swift 2019). In addition, mammoth molar plates were found eroded from the east side of Locus 1 from Stratum 3 and are thought to be associated with Stratum 3A, while another concentration of mammoth molar plate fragments was recovered from the surface approximately 140m west of Locus 1. The close association between mastodons and *Cuvieronius* sp. in the Lower Bonebed of El Fin del Mundo as well as the presence of mammoth remains at the site is an anomaly, with only two other known paleontological sites in Mexico (Tequixquiac in Estado de México and Valsequillo in Puebla) containing the remains of mastodons, gomphotheres, and mammoths (Alberdi and Corona-M. 2005; Arroyo-Cabrales and others 2003; Hibbard 1955; Pérez-Crespo and others 2012; Polaco and others 2001).

Given that the Lower Bonebed is expected to include skeletal elements from animals who lived centuries or even millennia apart in time, the association between mammoths, mastodons, and gomphotheres does not necessarily indicate that these species existed contemporaneously on the landscape. And if they did, their association may indicate that the region contained multiple microhabitats that could support diverse proboscidean taxa. The extreme rarity of Pleistocene fossil sites in the region preserving the remains of three proboscidean families despite presumably similar degrees of time averaging at other localities suggests a unique paleoenvironmental and paleoecological history of the area surrounding El Fin del Mundo.

The Lower Bonebed faunas of El Fin del Mundo include those frequently associated with open habitats (*Equus* sp., *Mammuthus* sp., *Paramylodon harlani*), closed habitats (*Mammut americanum, Tapirus* sp.), and those associated with either or mixture of the two (*Cuvieronius* sp., Camelidae). Stable isotope analyses of *Cuvieronius* sp., *Mammut americanum,* and *Tapirus* sp. teeth from El Fin del Mundo indicate the past presence of both C_3 plants and C_4 grasses on the landscape, with *Cuvieronius* sp. consuming a mixed diet. In contrast, mastodon and tapir individuals consumed a diet strongly dominated by C_3 plants (Pérez-Crespo, Morales-Puente, Arroyo-Cabrales, and Ochoa-Castillo 2019; Pérez-Crespo, Prado, Alberdi, Arroyo-Cabrales, and Johnson 2020). This either indicates the past presence of a habitat mosaic in the region or changes in habitat composition through time. Nearby paleontological localities dating to the late Pleistocene are dominated by faunas typical of open grassland habitats (i.e., La Playa; Mead and others 2010), although some other late Pleistocene sites in Sonora such as Térapa (ca. 43 to 40 ka) contain co-occurring taxa with affinities to open grasslands (bison, horse, mammoth, peccary) and to wooded or mixed (deer, gomphothere, ground sloth, llama) and marshy habitats (capybara, crocodile, tapir; Mead and others 2006; Mead, Arroyo-Cabrales, and Swift 2019; Nunez and others 2010).

BONEBED COMPARISONS AND CONCLUSIONS

Despite their stratigraphic proximity, the Lower and Upper Bonebeds of El Fin del Mundo Locus 1 have very different depositional histories. The Lower Bonebed is composed of the disarticulated elements of at least 10 different taxa with evidence that skeletal elements were transported by water prior to burial. Bones may have been derived from upstream floodplain or other bonebed deposits and likely represent a time-averaged assemblage. In contrast, the Upper Bonebed is monotaxic and is represented by two semi-articulated gomphothere (*Cuvieronius* sp.) skeletons. The gomphotheres were likely a male and female aged approximately between 2 and 8 and 8 and 19 years old (AEY: African elephant equivalent years) respectively and were found in close association with Clovis lithic artifacts. The Upper Bonebed likely represents a short-duration event in which Clovis peoples butchered one or both gomphothere individuals, or alternatively represents two butchering events spaced in time. Some artifacts were left behind in the process, and at least a portion of the gomphothere skeletal elements remained in place until burial by a shallow lake or marsh.

Although the bonebeds differ in their nature, each contains information indicating a unique paleoecological history of the region. The Lower Bonebed contains the remains of three proboscidean taxa: *Cuvieronius* sp., *Mammut americanum,* and *Mammuthus* sp., making El Fin del Mundo anomalous among paleontological localities in northern Mexico and the southwestern United States. The habitats to support these diverse proboscideans, as well as the other taxa represented in the Lower Bonebed, either existed contemporaneously in the area surrounding El Fin del Mundo or developed and diminished in succession, bringing different proboscidean species in and out of the area through time. Furthermore, the Upper Bonebed provides evidence that *Cuvieronius* sp. persisted in Sonora until the terminal Pleistocene. Elsewhere in North America, this genus of gomphothere is thought to have been locally extinct by the time of the early Rancholabrean (Lucas 2008) with some populations persisting to approximately 24 to 17 thousand years ago (Lundelius and others 2019), perhaps due to interspecific competition with mammoths or other megaherbivores (Smith and DeSantis 2020). El Fin del Mundo, however, indicates that the paleoecology of north-central Sonora was somehow different in allowing *Cuvieronius* sp. to persist until Clovis times, despite evidence from the Lower Bonebed of the prior presence of mammoths and mastodons in the area.

Acknowledgments

The authors thank Alejandro Jiménez, Eileen Johnson, Maria Teresa Alberdi, Aurelio Ocaña, and Iván Alarcón Duran for their work on faunal identifications and bone curation at the start of this project, and Ismael Sánchez-Morales for contributing maps, reports, and data from the site excavations, as well as helpful comments on this manuscript. Thank you to Mary Stiner and two anonymous reviewers whose comments helped to improve this text. Luciano Brambilla, Gary Takeuchi, and Greg McDonald helped to confirm identifications of sloth osteoderms at the site. We are especially grateful to Guadalupe Sánchez Miranda, Vance Holliday, and all former and current field directors, students, staff, and volunteers of El Fin del Mundo excavations. Faunal analysis of the Upper and Lower Bonebeds was funded by grants from the University of Arizona School of Anthropology, University of Arizona Graduate and Professional Student Council, and the Argonaut Archaeological Research Fund. Thank you in particular to donors Dennis Fenwick and Martha Lewis for their support of the project.

CHAPTER SEVEN

Palynological Evidence of Paleoenvironmental Changes

Carmen Isela Ortega-Rosas and Thanairi Gamez

Pollen is widely used for paleoenvironmental research in archaeological and non-archaeological contexts, although it may be poorly preserved in arid environments. But pollen recovery may be enhanced in paleo-wetlands that are now arid. Pollen is also useful in reconstructing plants used by ancient hunter-gatherers. Paleoenvironmental investigations of terminal Pleistocene and early Holocene setting are rare in northern Mexico. A recent study from the La Playa archaeological site (Ibarra-Arzabe and others 2018) shows the potential to reconstruct paleoenvironmental conditions in Sonora using pollen.

The pollen and related paleoenvironmental studies (Chapters 8, 9, 10) at El Fin del Mundo provide the first picture of terminal Pleistocene and early Holocene vegetation and related environmental conditions in northern Sonora based on paleobotany. This interdisciplinary research is a unique opportunity to better understand the environmental setting of Clovis hunters that once occupied the region and to reconstruct the subsequent environment into the earliest Holocene. The diversity of local environments promoted by climate change during the Pleistocene-Holocene transition in northwestern Mexico has been little studied. Most work on late Pleistocene and early Holocene environments in Mexico focused mainly on the center and southeast of the country, as summarized by Cruz-y-Cruz and others (2016). Very broadly, regional patterns of climate change in northwestern Mexico appear to contrast with global trends. Globally, the Last Glacial Maximum (LGM) was marked by drier conditions in most of the subtropical and tropical belt, with the subsequent climate transition to the Holocene marked by increased humidity (Street-Perrott and others 1989). In contrast, Arroyo-Cabrales, Carreño, and Lozano-García (2008) indicate that at the beginning of the Holocene temperatures gradually rose as ice masses decreased. Regional patterns of rainfall also progressively decreased to its current configuration due to the displacement of the Intertropical Convergence Belt to the North, with the consequent increase in summer rains over much of the south of the country. In northern Mexico the decrease in rainfall caused a decrease in lacustrine levels, and, in general, arid conditions in the northern altiplano and northwestern Sonora and Baja California. This study from northern Sonora and other paleovegetation studies along the U.S.-Mexican borderlands provide high resolution records to test those models (e.g., Holmgren, Peñalba, Rylander, and Betancourt 2003; Holmgren, Betancourt, and Rylander 2006; Holmgren, Norris, and Betancourt 2007.

The modern plant communities at El Fin del Mundo are represented by a mixture of Arizona Upland and Lower Colorado plant communities of the Sonoran Desert biome (Turner and Brown 1994). Most important species in the area are *Fouquieria splendens* (ocotillo), *Larrea tridentata* (creosote bush), *Ambrosia deltoidea* (bursage), *Muhlenbergia porteri* (summer grass), *Stenocereus thurberi* (pitahaya; organ pipe), *Pachycereus schottii* (senita), *Carnegiea gigantea* (saguaro), and several species of *Agave*; the dominant tree is *Olneya tesota* (ironwood).

METHODS

Soil, pollen, and diatom samples were taken from the complete stratigraphic sequence at the western end of the excavations in Locus 1 (08-1; Figure 4.3; Chapter 2). No samples for pollen analysis were collected from Stratum 5 or the calcareous facies of upper Stratum 4de (surface

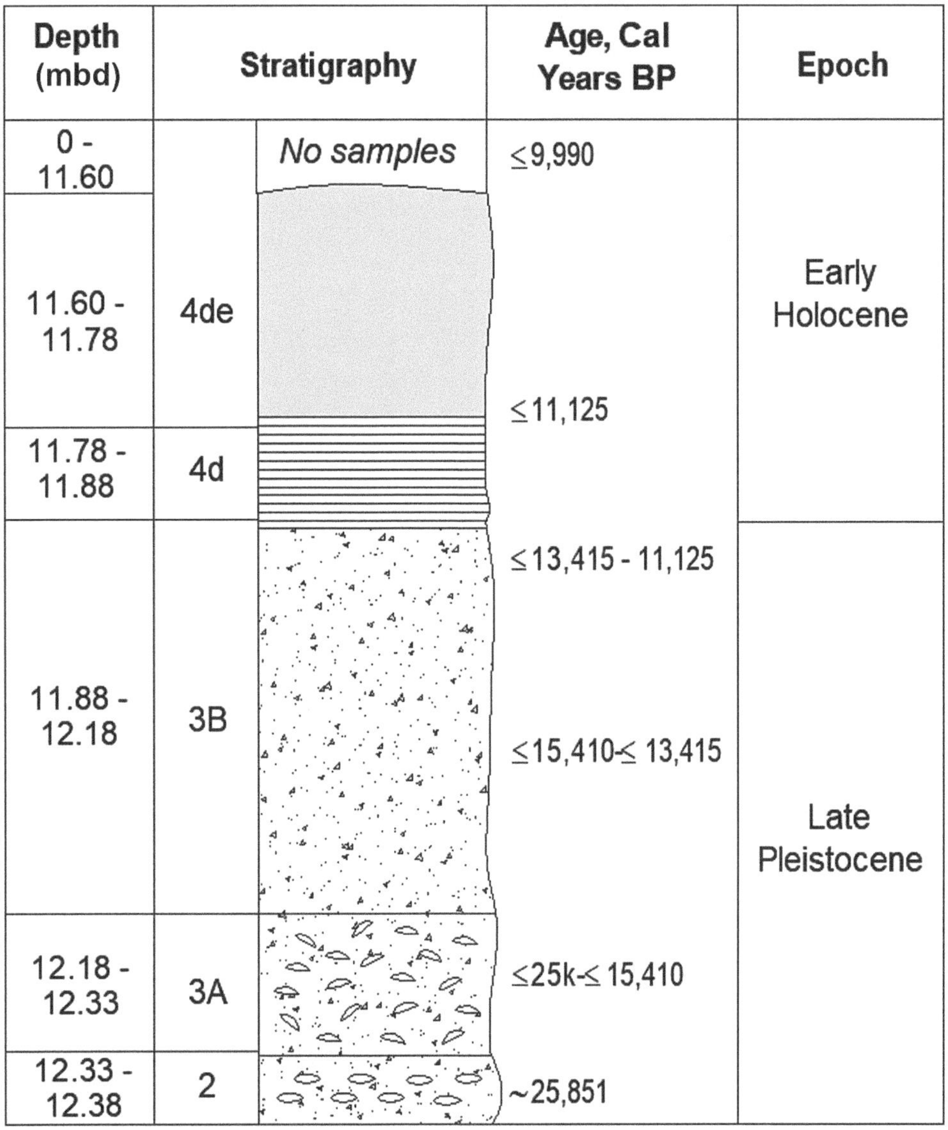

Depth (mbd)	Stratigraphy		Age, Cal Years BP	Epoch
0 - 11.60		No samples	≤9,990	Early Holocene
11.60 - 11.78	4de		≤11,125	
11.78 - 11.88	4d			
11.88 - 12.18	3B		≤13,415 - 11,125 ≤15,410-≤ 13,415	Late Pleistocene
12.18 - 12.33	3A		≤25k-≤ 15,410	
12.33 - 12.38	2		~25,851	

Figure 7.1. Age control and stratigraphic section (profile 08-1, Locus 1) for pollen sampling.

to 11.60 meters below datum, "mbd") because calcium carbonate inhibits pollen preservation. Below 11.60 mbd, samples were taken every 5 cm to obtain a representative sample of the profile (Figure 7.1, Table 7.1).

Pollen extraction in the laboratory followed Faegri and Iversen (1989) and adapted to archaeological samples (Gamez-Rascon 2016). From each level, 40 g of sediment was taken. Hydrochloric acid (HCl; diluted 10 %) was added and the solution left for 20 minutes. Distilled water was applied to stop the reaction. After 24 hours, the sample was rinsed, and hydrofluoric acid (HF) was added and left for a day. Then the sample was washed until it was acid free. Finally, a treatment with potassium hydroxide (KOH; 10%) was employed in a water bath. The final residue of the chemical treatment was stored in glycerin.

Pollen analysis was conducted with an optical microscope at 40X and 100X. We observed many slides to achieve the minimal number of pollen and spores, counting 100 to 300 grains per sample. Pollen identification was supported with pollen rain samples from the region (Ortega-Rosas, Peñalba, and Guiot 2008), several pollen atlases (Heusser 1971; Willard and others 2004), and digital pollen atlases that are available online (e.g., University of Arizona Pollen Atlas, Australasian Pollen and Spore Atlas, https://apsa.anu.edu.au/). Counting data of pollen and spores was managed in Tilia 2.0.4 software to obtain a pollen

Table 7.1. Pollen Samples

Depth, mbd	*Strata*	*Radiocarbon age cal. yrs B.P.*	*Pollen Zone*	*Pleistocene Fauna*
11.60–11.63	4 de	~9988 (Shell)[a]	4	
11.63–11.78	4de		4	
11.68–11.73	4de		4	
11.73–11.78	4de	~11,127 (Charred plant)	4	
11.78–11.83	4d		3	
11.83–11.88	3B/4d		3	Gomphothere 1 and 2
11.88–11.93	3B		3	Gomphothere 1 and 2
11.93–11.98	3B	≤13,415 (Charred plant)	3	Gomphothere 1 and 2
11.98–12.03	3B		3	Gomphothere 1 and 2
12.03–12.08	3B		3	Gomphothere 1 and 2
12.08–12.13	3B/3A		3	Gomphothere 1 and 2
12.13–12.18(1)	3B/3A		3	Gomphothere 1 and 2
12.13–12.18 (2)	3B/3A	≤15,410 (Shell)	3	Gomphothere 1 and 2
12.18–12.23	3A		2	Mastodon, gomphothere, mammoth, tapir, camel, horse
12.23–12.28	3A		2	Mastodon, gomphothere, mammoth, tapir, camel, horse
12.28–12.33	3A		2	Mastodon, gomphothere, mammoth, tapir, camel, horse
12.33–12.38	2	≤25,850 (Shell)	1	Mastodon, gomphothere, mammoth, tapir, camel, horse

Key: mbd= meters below site datum; a. This shell was collected at ~11.00 mbd.

diagram. Pollen counts are expressed in percentage with respect to total pollen and spore group sum (ferns and fungal), whereas the Pteridophyte group is expressed with respect only to the total sum of spores (ferns and fungal). In addition, the pollen zones were divided using a cluster analysis with CONISS (Grimm 2004).

RESULTS

Pleistocene-Holocene Vegetation Composition Based on Pollen Data

Pollen composition at El Fin del Mundo changed through the latest Pleistocene and into the earliest Holocene (Table 7.2). Most of the fossil pollen recorded was very deteriorated (Figure 7.2) and pollen abundance was relatively low (Table 7.3). Nevertheless, significant changes in pollen composition were recorded. We identified 23 different pollen types. Some fungal spores and Pteridophyte spores were also identified (Tables 7.2 and 7.3). The pollen diagram (Figure 7.3) shows that the most important taxa are the herbaceous ones, principally Poaceae, and Chenopodiaceae-Amaranthaceae. We identified three pollen zones based on eminent changes in the pollen diagram.

Table 7.2. Pollen and Spore Diversity

Arboreal and Shrubs Pollen	*Herbaceous Pollen*	*Pteridophyte and Fungal Spores*
Ambrosia sp.	Asteraceae	*Cicatricosisporites* sp.
Boerhaavia sp.	Chenopodiaceae-Amaranthaceae	Pteridophyte group
Cactaceae	Euphorbiaceae	Smuts group
Celtis sp.	Poaceae	Trilete group
Fabaceae	Portulacaceae	Monolete group
Oleaceae	*Potamogeton* sp.	
Parkinsonia sp.	Solanaceae	
Pinus sp.	*Tribulus* sp.	
Prosopis sp.		
Quercus sp.		
Salix sp.		

Table 7.3. Percentages of Pollen and Spores by Sample

Plant Group		*Stratum*																
		4de				*4d*	*3/4*	*3B*							*3A*			*2*
		Depth, mbd																
		11.60–11.63	*11.63–11.68*	*11.68–11.73*	*11.73–11.78*	*11.78–11.83*	*11.83–11.88*	*11.88–11.93*	*11.93–11.98*	*11.98–12.03*	*12.03–12.08*	*12.08–12.13*	*12.13–12.18 (1)*	*12.13–12.18 (2)*	*12.18–12.23*	*12.23–12.28*	*12.28–12.33*	*12.33–12.38*
Trees and Large Bushes	Cactaceae	0	0	0	0	0	0	0	0	0	0	0	0	0	0	0	0	11.1
	Celtis sp.	4.2	0	0	0	0	0	0	0	0	0	0	0	0	0	0	0	0
	Fabaceae	0	0	0	0	0	0	0	0	0	0	0	0	0	0	0	0	11.1
	Juniperus sp.	0	0	0	50	0	0	0	0	0	0	33.3	0	0	0	0	0	0
	Oleaceae	0	0	0	0	0	0	0	0	0	0	0	0	0	0	0	0	0
	Parkinsonia sp.	0	0	0	0	0	0	0	0	0	0	0	0	0	0	0	0	0
	Pinus sp.	0	25	0	0	0	0	0	16.7	0	0	33.3	0	0	0	0	0	11.1
	Prosopis sp.	4.2	0	0	0	0	0	0	0	0	0	0	0	0	0	0	0	0
	Quercus sp.	4.2	0	0	0	0	0	0	0	0	0	0	0	0	0	0	0	0
	Salix sp.	0	0	0	0	0	0	0	0	0	0	0	0	0	0	0	0	0
	Ulmaceae	0	0	0	0	0	0	0	16.7	0	0	0	0	0	0	0	0	0
	Totals	**12.5**	**25**	**0**	**50**	**0**	**0**	**0**	**33.4**	**0**	**0**	**66.6**	**0**	**0**	**0**	**0**	**0**	**33.3**
Shrubs and Herbs	*Ambrosia* sp.	0	0	0	50	0	0	0	0	0	0	0	0	0	7.7	50	2.6	0
	Asteraceae	0	0	100	0	0	0	0	16.7	0	0	0	0	0	7.7	0	2.6	0
	Boerhavia sp.	0	0	0	0	0	0	0	0	0	0	0	0	0	0	0	0	0
	Chenopodiaceae/ Amaranthaceae	8.3	25	0	0	100	0	12.5	16.7	25	0	0	25	0	7.7	50	10.5	0

continued

Table 7.3. (continued)

Plant Group		*Stratum*																
		4de				*4d*	*3/4*	*3B*							*3A*			*2*
		Depth, mbd																
		11.60–11.63	*11.63–11.68*	*11.68–11.73*	*11.73–11.78*	*11.78–11.83*	*11.83–11.88*	*11.88–11.93*	*11.93–11.98*	*11.98–12.03*	*12.03–12.08*	*12.08–12.13*	*12.13–12.18 (1)*	*12.13–12.18 (2)*	*12.18–12.23*	*12.23–12.28*	*12.28–12.33*	*12.33–12.38*
Shrubs and Herbs (continued)	Euphorbiaceae	29.2	0	0	0	0	0	0	0	0	0	0	0	0	0	0	5.3	0
	Poaceae	45.8	50	0	0	0	100	87.5	16.7	50	100	0	75	100	76.9	0	73.7	44.4
	Portulacaeae	0	0	0	0	0	0	0	0	0	0	0	0	0	0	0	0	0
	Potamogeton sp.	0	0	0	0	0	0	0	0	0	0	0	0	0	0	0	5.3	11.1
	Scrophulariaceae	0	0	0	0	0	0	0	0	25	0	0	0	0	0	0	0	0
	Solanaceae	4.2	0	0	0	0	0	0	0	0	0	0	0	0	0	0	0	0
	Tribulus sp.	0	0	0	0	0	0	0	0	0	0	0	0	0	0	0	0	11.1
	Urticaceae	0	0	0	0	0	0	0	16.7	0	0	33.3	0	0	0	0	0	0
	Totals	**87.5**	**75**	**100**	**50**	**0**	**100**	**100**	**66.7**	**100**	**100**	**33.3**	**100**	**10**	**100**	**100**	**100**	**66.7**
Fungi	Ascospores	7.7	42.9	83.3	12.5	75	0	0	0	28.6	0	0	0	20	0	0	0	0
	Smuts	0	0	0	62.5	25	0	0	53.8	14.3	10	25	14.3	60	0	25	0	0
	Cicatricosisporites sp.	0	0	0	0	0	0	0	0	0	0	0	0	0	0	0	2.5	0
	Monolete	0	0	0	0	0	0	0	0	0	0	0	0	0	0	0	0	0
	Totals	**7.7**	**42.9**	**83.3**	**75**	**100**	**0**	**0**	**53.8**	**42.9**	**10**	**25**	**14.3**	**80**	**0**	**25**	**2.5**	**0**
Ferns	Pteridophytes	0	28.6	0	0	0	0	0	0	0	0	0	0	0	0	0	0	0
	Trilete	0	0	0	0	0	0	0	0	0	0	0	28.6	0	0	25	2.5	0
	Totals	**0**	**28.6**	**0**	**0**	**0**	**0**	**0**	**0**	**0**	**0**	**0**	**28.6**	**0**	**0**	**25**	**2.5**	**0**

Note: Species at less than 5% abundance are not listed in Figure 7.3 (*Celtis* sp., Oleaceae, *Parkinsonia* sp., *Salix* sp., *Boerhavia* sp, Portulacaceae, Solanaceae, and *Cicatricosisporites* sp.)
Pollen percentages were calculated on a sum excluding Pteridophyte group. Fungi and ferns were calculated on a total sum of all pollen + fungi + ferns.

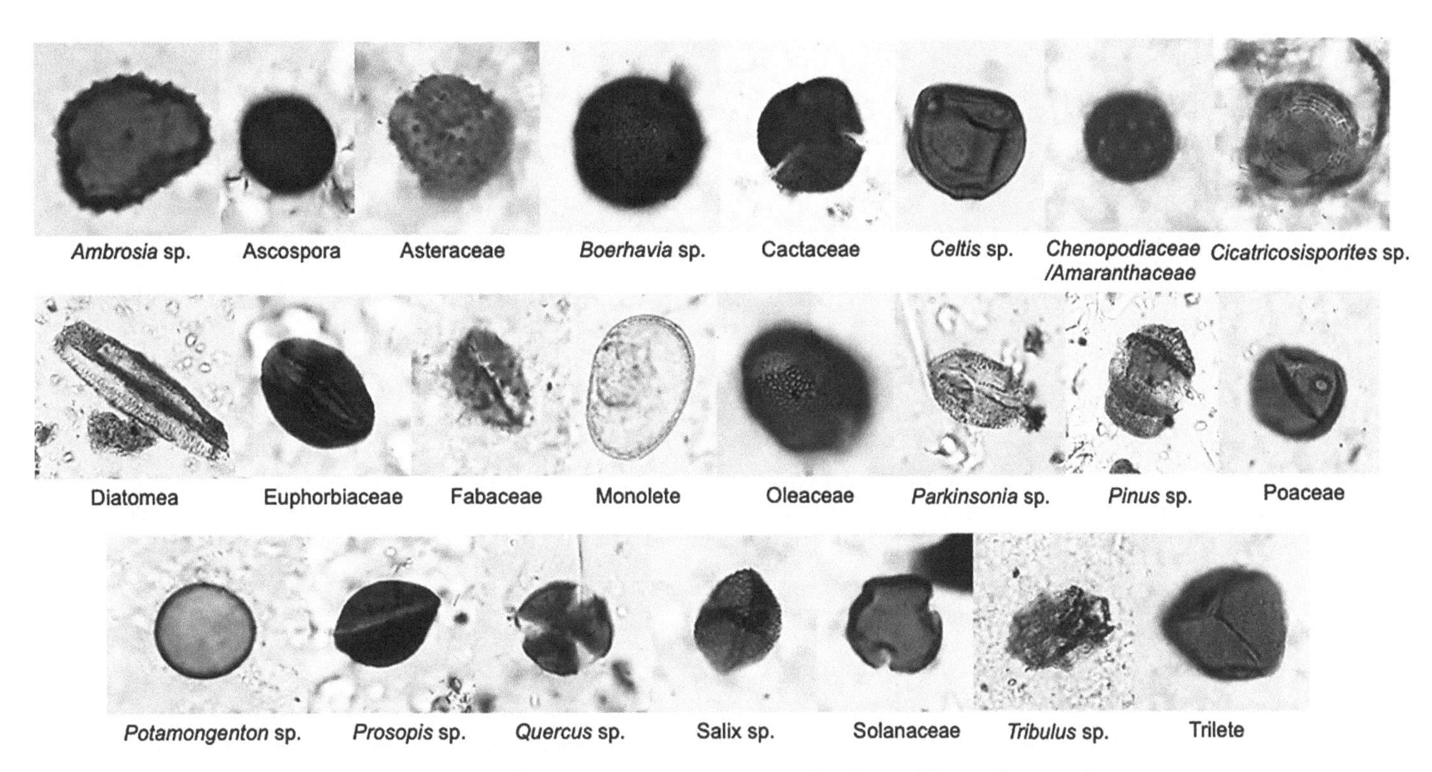

Figure 7.2. Morphological diversity of fossil pollen and spores in the El Fin del Mundo record.

Zone 1, Strata 2 and 3A

Zone 1 includes Strata 2 and 3A (12.18 to 12.33 mbd), which dates from ~25,850 to ~15,410 cal yr B.P. High percentages of Poaceae (mean of 40%), *Ambrosia* (mean of 30%), Chenopodiaceae-Amaranthaceae (10%), Asteraceae (*Tribulus sp.* type) (10%), Fabaceae (10%), and *Pinus* (10%) characterize this zone, related to a steppe-dominant landscape that included large fauna such as mammoth, gomphothere, mastodon, and tapir (Chapter 6). A significant percentage of fungal spores with smuts (30%) also was recorded (Table 7.3, Figure 7.3).

Zone 2, Stratum 3B

High percentages of non-arboreal pollen were likewise recorded for this zone, which is from 11.83 to 12.18 mbd and dates from ~15,410 to ≤13,415 cal yr B.P. Poaceae pollen had the greatest percent with a mean of 50 percent. Asteraceae pollen was present in some samples (10%), along with Chenopodiaceae-Amaranthaceae (10%). New taxa appear, such as Scrophulariaceae (10%) and Urticaceae (20%). Arboreal pollen is present in two peaks (at 12.08–12.13 and 11.93–11.98mbd) with the highest frequencies of *Pinus* pollen in the entire record (25% and 15% respectively), and a peak of *Juniperus* pollen at 30 percent (12.08–12.13mbd). Higher levels of fungal spores, smuts (8%), and ascospores (10%), were also recorded through much of 3B (Table 7.3, Figure 7.3). Upper Stratum 3B is also coeval with gomphothere and the Clovis occupation at the site.

Zone 3

This zone contains the highest diversity in pollen and spores in the record with Asteraceae reaching 80 percent in one sample. Zone 3 occurs in Strata 4d and 4de (11.60–11.83 mbd), which date to <13,415 to > 9,990 cal yr B.P.[6] Poaceae pollen occurs only in the upper two samples, with mean percentages of 40 percent. New taxa such as Euphorbiaceae (25%) appear and Chenopodiaceae-Amaranthaceae increase slightly (25%). Arboreal pollen includes one peak of *Pinus* at 20 percent (11.63–11.68 mbd) and one peak of *Juniperus* at 40 percent (11.73–11.78 mbd). Other arboreal taxa in this zone include *Celtis*, Oleaceae, *Parkinsonia*, *Prosopis*, and *Quercus* with 5 percent each. Ferns are present in higher percentages in this zone with the Pteridophyte group at 20 percent (Table 7.3, Figure 7.3). A tropical-desert vegetation dominates this period.

[6] The date of ~9990 cal yr B.P. (AA88885) was determined on shell collected within 10 cm of the surface, i.e., stratigraphically above the section sampled for pollen.

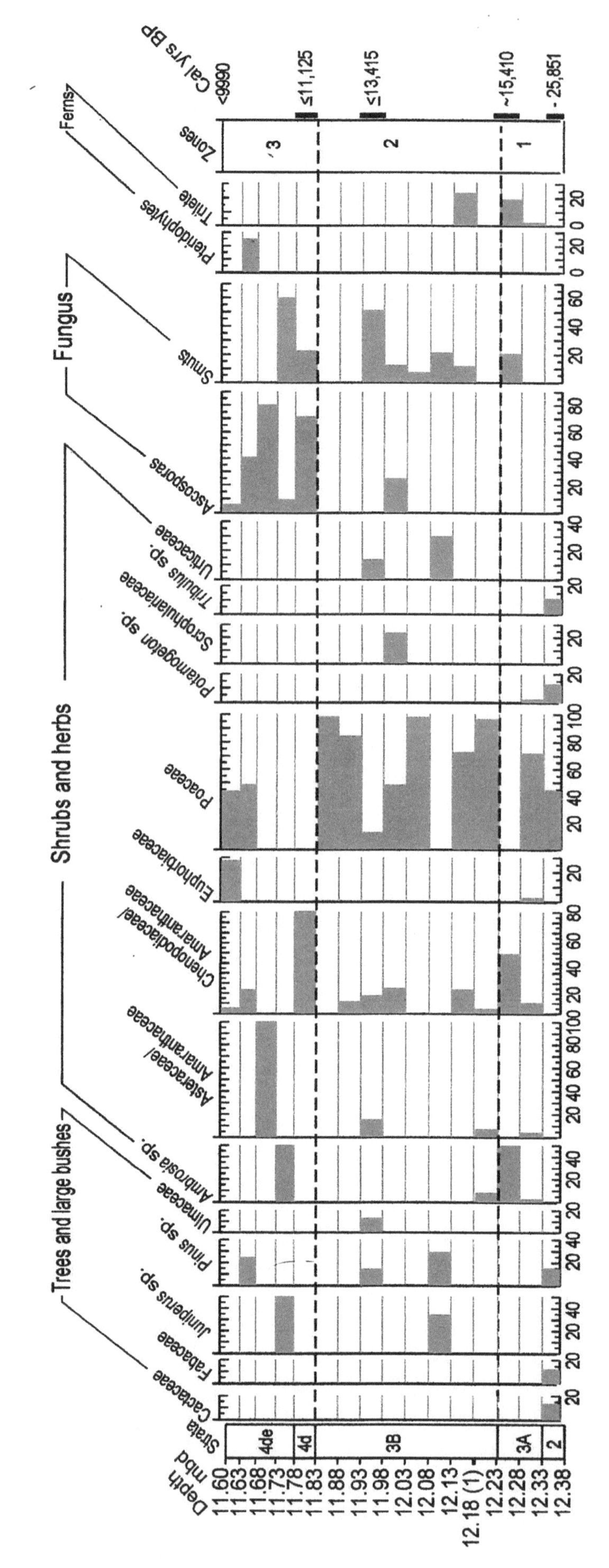

Figure 7.3. Fossil pollen diagram for the latest Pleistocene and earliest Holocene at El Fin del Mundo. Calibrated radiocarbon ages for the profile are shown to the right of the diagram. The abscissa indicates the pollen percentages for each taxon. Shading indicates presence of the taxon. Three pollen zones are shown with black horizontal dashed lines.

DISCUSSION AND CONCLUSIONS

Palynological evidence from El Fin del Mundo clearly documents significant changes in distribution and composition of vegetation communities through the end of the Quaternary. Strata 2 and 3 are separated by an unconformity that likely represents at least several thousand years. Stratum 3, particularly Stratum 3A, is characterized by cut-and-fill sequences (Chapter 2). Therefore, the Strata 2-3 sequence does not represent a continuous paleobotanical record but does provide important clues to the environment prior to and during the Clovis occupation. During the period ~25,400 cal yr B.P. (upper Stratum 2) and ~15,410 cal yr B.P. (Stratum 3A), a steppe landscape with scarce pines characterized the site area, as indicated by presence of *Pinus* pollen in some levels, along with a high percentage of grasses (Zone 1, Figure 7.3). Vegetation diversity was high at this time, and likely represents colder and more humid conditions relative to modern time, which is also indicated by the higher frequencies of fungal and pteridophyte spores.

Pollen Zone 1 corresponds to the earliest Last Glacial Maximum (LGM) and early post-glacial time. These conditions are further documented regionally based on the association of *Pinus*, *Juniperus*, and *Quercus* communities at elevations between 600 and 1,675 meters above sea level (Van Devender 1990; Thompson and others 1993; Holmgren, Peñalba, Rylander, and Betancourt 2003; Holmgren, Betancourt, and Rylander 2006; Holmgren, Norris, and Betancourt 2007). However, *Pinus* appears to disappear before the end of the LGM at FDM in contrast to its persistence in the northern Chihuahua Desert into the post-LGM late Pleistocene (Holmgren, Peñalba, Rylander, and Betancourt 2003; Holmgren, Betancourt, and Rylander 2006). El Fin del Mundo is at a lower elevation (~630m), which could account for an earlier shift to more xeric plant communities. To the east, in Texas, LGM expansion of the woodland zone is also is documented (Wells 1966). Wetter LGM conditions also are indicated in a paleolake to the east in Chihuahua (Metcalfe and others 2002). In the post-LGM late Pleistocene, vegetation shifted with the disappearance of arboreal pollen from the record and the spread of herbaceous pollen such as *Ambrosia* (Zone 1, Figure 7.3). This change indicates that climatic conditions were not favorable for the continuity of a forest community in the region. The precipitation regimen at that time likely was more related to winter rains, as suggested for the southwestern United States and northern Mexico (Metcalfe and others 2002; Roy and others 2012).

After about 15,410 years ago and through the approximate time of the Clovis occupation (≤13,400 cal yr B.P.), the Stratum 3 and particularly Stratum 3B (Zone 2, Figure 7.3) pollen record indicates an increase in arboreal pollen related to *Pinus* and *Juniperus* in some levels, suggesting that there were colder temperatures at that time around the site. Trees from the Ulmaceae family were also present. Non-arboreal pollen is less diverse and at lower frequencies compared to Zone 1. Some grasses and herbs persisted. These indicators suggest open conifer woodland vegetation with grasslands under more humid conditions, as suggested for northwestern Mexico (Ortega-Rosas Peñalba, and Guiot 2008; Ortega-Rosas, Vidal-Solano, Williamson, Peñalba, and Guiot 2017). Pine-pollen levels suggest the presence of pine forest near El Fin del Mundo site, now well distributed at elevations above 1,000 meters above sea level (Búrquez-Montijo, Martínez, and Martin 1992).

The remains of large mammals from Strata 3A and 3B further support the pollen data in terms of both their ecology and isotopic signals (Chapter 6). Tapirs and mastodons lived in wooded areas and fed on C_3 plants (pollen Zone 1) whereas gomphotheres had a mixed C_3/C_4 diet and inhabited areas that had some trees (pollen Zone 2).

Following the time of the Clovis occupation ≤13,400 to ~11,000 cal yr B.P., steppe conditions persisted as indicated by increased Asteraceae pollen (Stratum 4d, upper Zone 2, Figure 7.3). This change in vegetation probably reflects warming conditions, which have been documented for the early Holocene elsewhere in northwestern Mexico (e.g., Roy and others 2013).

During the Pleistocene-Holocene transition ~11,000 to ~10,000 cal yr B.P. (Strata 4d and lower 4de, Zone 3, Figure 7.3), the presence of *Juniperus* in the pollen record, together with an increase in fungal spores, denote wetter conditions. After this brief interval, grassland dominated. The presence of Poaceae and other herbs such as Chenopodiaceae-Amaranthaceae, Asteraceae, and Euphorbiaceae are related to warmer temperatures and climate typical of a monsoon precipitation system, because these herbs bloom after the summer rainy season.

Clovis foragers are well documented at El Fin del Mundo and across northern Sonora (Gaines, Sánchez, and Holliday 2009; Sánchez and others 2015; Sánchez 2016). At El Fin del Mundo they were supported by flowing to standing water and an open forest with grasses widely spread around the site. This information further documents the broad range of environments inhabited by Clovis groups across North America (Smallwood and Jennings, 2015).

The paleoenvironmental conditions during latest Pleistocene and earliest Holocene at El Fin del Mundo allowed establishment of heterogeneous vegetation, which served a diverse species of herbivorous mammals, with different feeding habits and habitat preferences. A temperate open forest was the setting for Clovis groups, followed by a transition to more steppe vegetation during earliest. These results correlate well with the other paleoclimatic proxies studied at this site and from the region.

Acknowledgments

We are grateful to the Consejo Nacional de Ciencia y Tecnología (CONACYT) and Secretaria de Medio Ambiente y Recursos Naturales (SEMARNAT) for the Grant to Carmen Ortega-Rosas (Project Number 263413) for equipment and laboratory material to complete this research. We thank the Laboratory of Aerobiology, State University of Sonora, for use of equipment for fossil pollen extraction.

CHAPTER EIGHT

Phytolith Analysis

Kristen Wroth

To complement the data derived from other biological indicators of past environments at El Fin del Mundo and to better understand the paleoenvironmental setting, phytoliths were analyzed in samples from Locus 1. Phytoliths are biogenic silica infillings that form in and around plant cells to support the plant with a variety of processes such as water retention, stability, and protection from pests and diseases (Piperno 2006). Phytoliths form in many parts of plants, including stems, leaves, and inflorescences, and after the plant dies the silica hardens and retains its original shape. These hardened silica bodies are then incorporated into the surrounding sediment as the organic parts decay away, creating a record of both the type of plant and the anatomical part of the plant in which the phytoliths developed. In addition, these silica microremains are very resistant to many types of diagenesis and weathering, allowing for good preservation through long periods of time and in a variety of depositional circumstances (Cabanes and Shahack-Gross 2015; Piperno 2006). Nearly all families of plants produce phytoliths, but the number of phytoliths produced in a given plant and the uniqueness of the resulting phytoliths are controlled by a variety of factors such as genetics and environmental conditions. For example, monocots like grasses produce far more phytoliths than dicotyledons like trees, but very few grass species produce phytoliths that can be identified at the species level (Piperno 2006; Albert and others 1999). In addition, each plant produces multiple types of phytoliths in its various anatomical parts. Due to this combination of many plant families producing similar phytolith morphotypes and individual plants producing a variety of phytoliths, phytolith analysts must rely on modern comparative collections, a careful understanding of the sample context, and integration with other techniques (e.g., micromorphology, charcoal analysis) to interpret phytolith assemblages (Vrydaghs, Ball, and Devos 2016).

Because phytoliths are incorporated into the sediment only after a plant dies, they tend to record more local plant dynamics, contrasting with pollen that is specifically spread through a variety of mechanisms during the plant's life cycle. This more local pattern of deposition is utilized by archaeologists and paleoecologists to reconstruct local changes in plant communities through time. In addition, grasses are not only prolific phytolith producers, but they produce well-defined morphotypes known as short cells that have been linked to specific classes of grass that grow in distinct environmental conditions (Twiss, Suess, and Smith 1969). This pattern can help phytolith specialists understand both how the local area may have changed as well as link those changes with larger scale climatic patterns (Piperno 2006). Phytoliths can also help archaeologists understand what types of plants people brought to a site, as silica is much more resistant to physical degradation than many other types of plant remains such as unburnt seeds or fruits.

Phytolith analysis has been applied to previous research into Paleoindian archaeological sites, primarily to categorize the past environment (Yost 2016) but also as a way to determine resources that may have been used by early populations moving through the Americas (Yost and Blinnikov 2011; Piperno 1991; Bement 2009).

METHODS

Phytoliths were recovered from samples collected at Profile 08-1 in Locus 1 (Figure 4.3). The extraction of phytoliths and other silica-based microremains from the bulk sediment

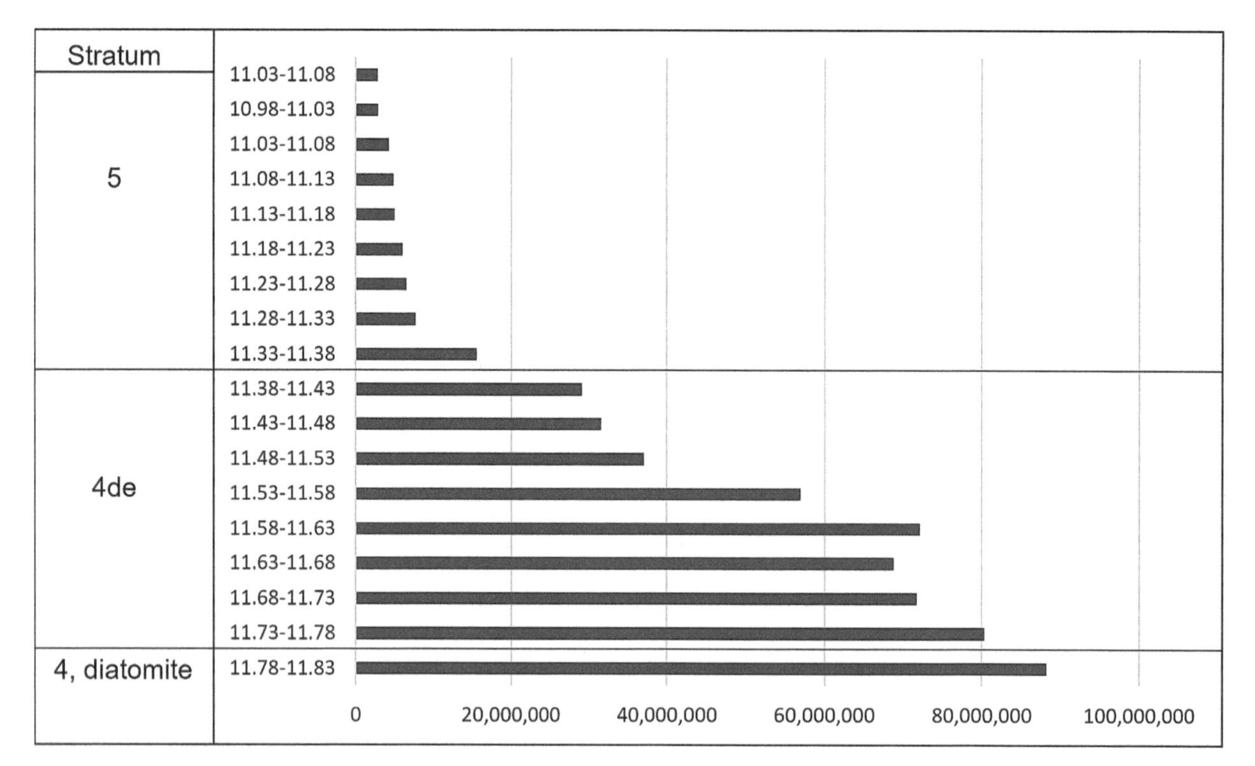

Figure 8.1. Concentration of diatoms per 1 gram of sediment. Elevations are meters below the site datum. Left column contains stratigraphic designations: Strata 3B, 4d (diatomite), 4de (diatomaceous earth), and 5.

samples followed the rapid method laid out by Katz and others (2010). An initial trial was carried out on the Fin del Mundo samples with approximately 30 mg of sediment, but the extraordinarily high density of silica microremains made identification difficult. Thus, the majority of the analysis was carried out on ~10 mg of sediment for each sample to allow for better viewing under the microscope. After the sediment was placed in a 1.5 ml Eppendorf tube, 50 µl of 1N HCl was added, and the tube was shaken periodically until any reaction had ceased. Next, 450 µl of a heavy liquid solution made from sodium polytungstate (density = 2.4g/ml) was added and the mixture was vortexed to combine. The Eppendorf tubes were placed in an ultrasonic bath and sonicated for 25 minutes to deflocculate the clay particles, and the tubes were centrifuged at 5,000 rpm for 5 minutes. The resulting supernatant was placed in a clean tube and 50 µl of the solution was placed on a glass microscope slide.

The samples were viewed at 200× magnification using a petrographic microscope and the number of phytoliths and diatoms in 16 fields of view were counted. Based on this number and the initial weight of the sediment sample, the number of silica remains in 1 g of sediment was extrapolated. The phytoliths were then identified at 200x and 500x magnification and categorized according to shape, texture, and anatomical part according to the conventions laid out in Neumann and others (2019) and Madella and others (2005). Detailed identification of the diatoms is presented by Palacios-Fest in Chapter 6 of this volume.

The phytoliths and diatoms present in each sample were quantified separately according to the guidelines outlined for phytoliths by Katz and others (2010), allowing the concentration of each to be compared. The concentrations of the diatoms and phytoliths may be underestimated as there were many fragments of silica that were unidentifiable in the background of each sample. Unidentifiable silica fragments can include weathered or broken diatoms and phytoliths, as well as volcanic glass, which is present in the underlying bedrock of the area. Only silica fragments that could clearly be assigned as either diatom or phytolith were counted in the totals presented here.

RESULTS

As expected, the concentration of diatoms is the highest in the samples collected from the layer of diatomite, Stratum 4d, and low in Stratum 4de, ranging from ~68 to 88 million diatoms per gram of sediment (Figure 8.1, Table 8.1). The number of diatoms decreases slightly in samples collected

Table 8.1. Diatom Concentrations

Depth, mbd	*Stratum*	*Lab Number*	*IW*	*Diatoms counted per slide*	*Diatoms in 1 g of sediment*	*Assoc. samples*
11.93–11.98	5	FDM18	10.9	88	2,800,459	
10.98–11.03	5	FDM17	11	91	2,869,602	
11.03–11.08	5	FDM16	10.2	125	4,250,919	
11.08–11.13	5	FDM15	10.4	145	4,836,238	
11.13–11.18	5	FDM14	10.7	153	4,959,988	
11.18–11.23	5	FDM13	10.3	179	6,028,216	FDM-1-13, FDM-1-2
11.23–11.28	5	FDM12	10	187	6,486,563	FDM-1-2
11.28–11.33	5	FDM11	10.2	225	7,651,654	
11.33–11.38	5	FDM10	10.2	459	15,609,375	
11.38–11.43	4de	FDM9	10.3	865	29,130,765	
11.43–11.48	4de	FDM8	10.9	994	31,632,454	
11.48–11.53	4de	FDM7	10.3	1102	37,112,257	
11.53–11.58	4de	FDM6	11	1808	57,013,636	
11.58–11.63	4de	FDM5	10.5	2183	72,116,964	FDM-1-4
11.63–11.68	4de	FDM4	10.7	2122	68,791,472	FDM-1-5, FDM-1-4
11.68–11.73	4de	FDM3	10.7	2211	71,676,694	FDM-1-5
11.73–11.78	4de	FDM2	10.8	2502	80,359,375	FDM-1-6, FDM-1-14
11.78–11.83	4, diatomite	FDM1	10.6	2696	88,224,057	FDM-1-6, FDM-1-14

Key: mbd = meters below datum; IW = initial weight of the sediment sample in grams (g).

from the diatomaceous earth in upper Stratum 4de to ~29 to 57 million diatoms per gram of sediment. There is a sharp decrease in diatom abundance for the samples collected from the upper part of the profile in Stratum 5, and the concentrations continually decrease from ~15 million diatoms per gram of sediment at 11.33 to 11.38 mbd to approximately 2.8 million diatoms per gram of sediment at the very top of the sequence.

The phytoliths present a different picture. Overall, the concentrations of phytoliths from samples collected from the profile are very high and none fall below 11 million phytoliths per gram of sediment (Figure 8.2, Table 8.2). Interestingly, while the diatoms show a clear decrease in abundance from the bottom to the top of the profile, the concentration of phytoliths is consistently high, and fluctuate independently of the diatoms, with two peaks at in the middle of Stratum 4de and another in the middle of Stratum 5. Even the portion of the sequence described as diatomite contains many phytoliths.

In all the analyzed samples, monocotyledonous phytoliths are the dominant type (Figure 8.3, and see Table 8.3 for the full record of identifiable phytoliths by morphotype for each sample). Very few of the irregular or blocky type phytoliths associated with dicotyledons were identified in these samples, which may be reflective of the actual composition of the surrounding plant community or because these phytoliths are more difficult to recognize specifically when the samples have such a high density of silica-based microremains.

Overall, the analyzed samples are well preserved, with few weathered phytoliths at the base of the sequence increasing to 5 to 7 percent weathered phytoliths in the upper part of the sequence. Conversely, the number of phytoliths in anatomical connection, which is sometimes also used as a marker of preservation, is very low in the deeper samples, and increases to an average of 13 percent of the identifiable phytoliths above 50 cm. The majority of the monocotyledonous phytoliths are either part of the general monocot category or originate in the leaves or stems of the plants, but there is also a significant input of phytoliths coming from the inflorescences (Figure 8.4). Grass phytoliths are the primary component at El Fin del Mundo,

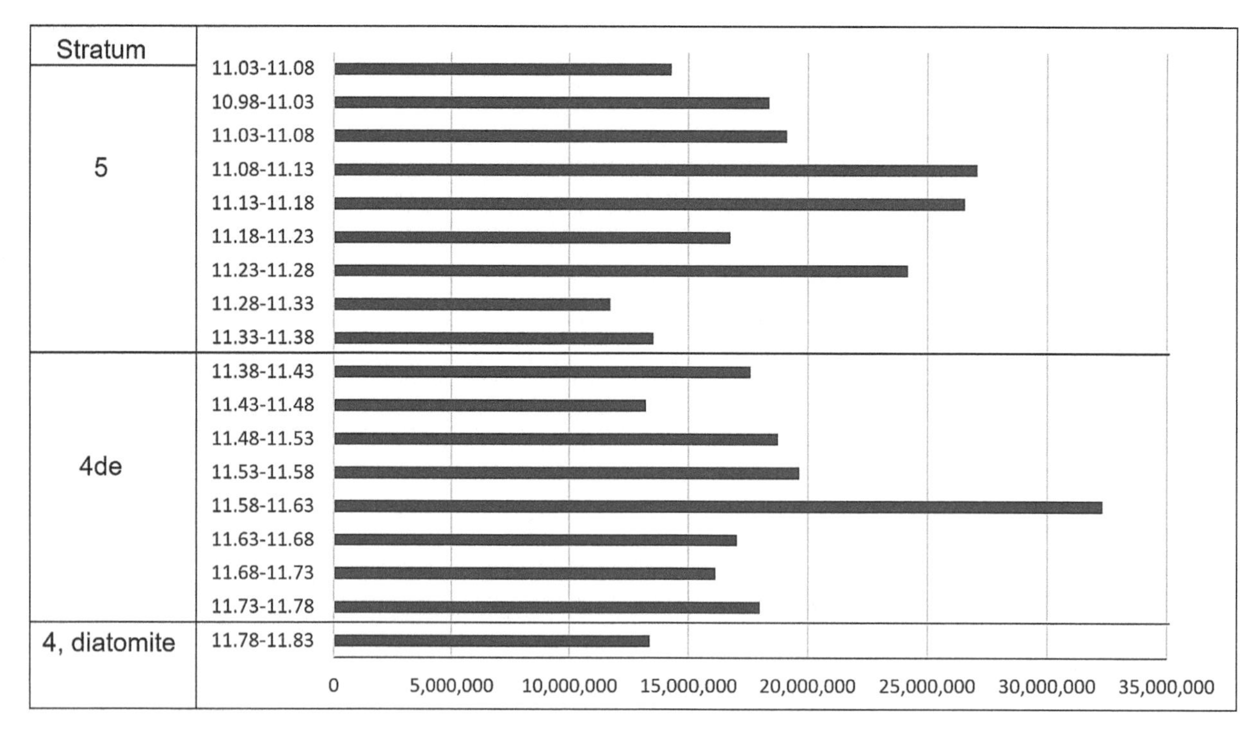

Figure 8.2. Concentration of phytoliths in 1 g of sediment. Elevations are meters below the site datum.

Table 8.2. Phytolith Concentrations

Depth, mbd	*Stratum*	*Lab Number*	*IW*	*Phytoliths counted per slide*	*Phytoliths in 1g of sediment*	*Associated Samples*
10.93–10.98	5	FDM18	10.9	449	14,288,704	
10.98–11.03	5	FDM17	11.0	583	18,384,375	
11.03–11.08	5	FDM16	10.2	562	19,112,132	
11.08–11.13	5	FDM15	10.4	812	27,082,933	
11.13–11.18	5	FDM14	10.7	820	26,582,944	
11.18–11.23	5	FDM13	10.3	497	16,737,561	FDM-1-13, FDM-1-2
11.23–11.28	5	FDM12	10.0	697	24,177,188	FDM-1-2
11.28–11.33	5	FDM11	10.2	345	11,732,537	
11.33–11.38	5	FDM10	10.2	398	13,534,926	
11.38–11.43	4de	FDM9	10.3	523	17,613,167	
11.43–11.48	4de	FDM8	10.9	415	13,206,709	
11.48–11.53	4de	FDM7	10.3	557	18,758,192	
11.53–11.58	4de	FDM6	11.0	623	19,645,739	
11.58–11.63	4de	FDM5	10.5	978	32,308,929	FDM-1-4
11.63–11.68	4de	FDM4	10.7	525	17,019,568	FDM-1-5, FDM-1-4
11.68–11.73	4de	FDM3	10.7	497	16,111,857	FDM-1-5
11.73–11.78	4de	FDM2	10.8	560	17,986,111	FDM-1-6, FDM-1-14
11.78-11.83	4, diatomite	FDM1	10.6	409	13,384,139	FDM-1-6, FDM-1-14

Key: mbd = meters below datum; IW = Initial weight of the sediment sample in grams (g).

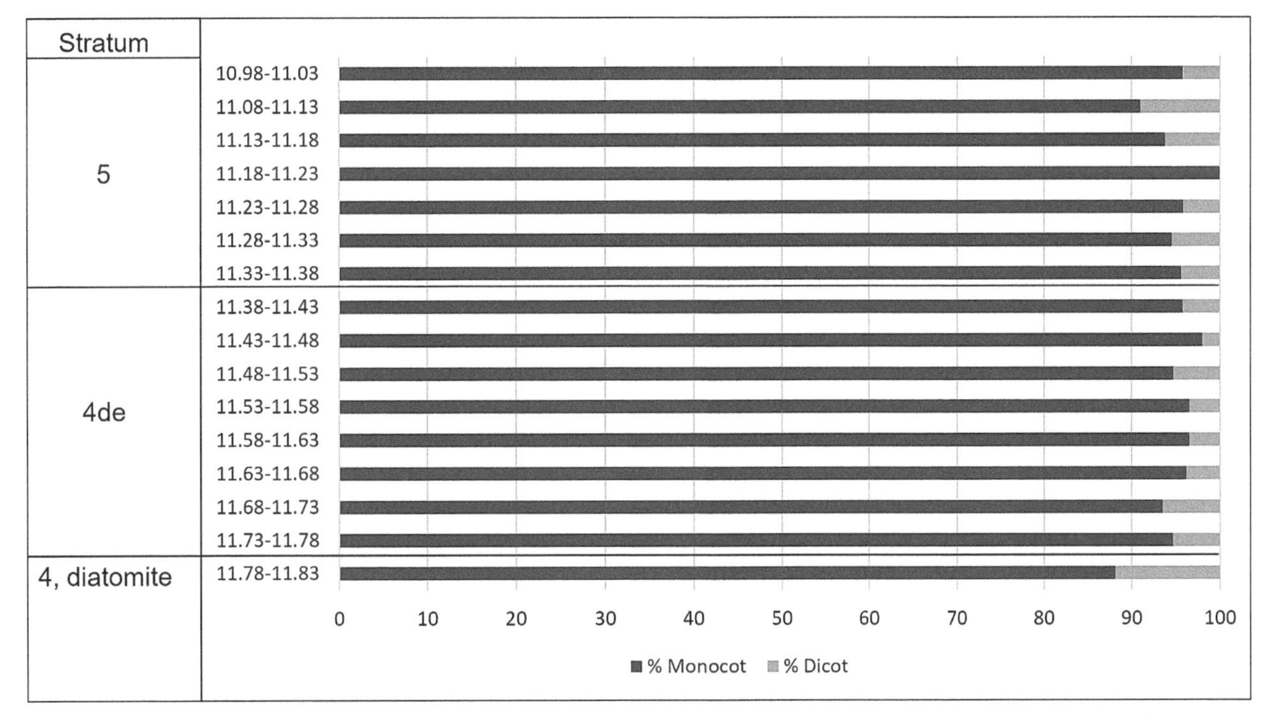

Figure 8.3. Percentage of phytoliths originating in monocotyledonous plants vs. those from dicotyledons.

but there is also a small, persistent population of the hat-shaped or conical phytoliths that originate in *Cyperaceae* plants (a family of monocot plants known colloquially as sedges). In general, all of the samples are rich in the types of phytoliths associated with aquatic or wetland vegetation, such as a high concentration of BULLIFORMS, SADDLES, and CONICAL phytoliths. Both cuneiform and parallelipedal bulliform phytoliths persist throughout the sequence but the samples from 11.33 mbd and up (FDM11-17) show the largest concentration of these cells that are also distinctly well-silicified. All the samples also have large numbers of prickle or LANCEOLATE forms that are very robust. The samples are somewhat differentiated by the type and concentration of grass short cells (Figure 8.5). The base of the sequence is predominately composed of RONDELS and SADDLES and there is a shift towards a mix of short cell types that include mostly BILOBATE and RONDEL short cells towards the middle and upper sections of the profile. Examples of the different types are provided in Figure 8.6.

DISCUSSION

Overall, the phytolith and diatom assemblages from El Fin del Mundo are very well preserved. This level of preservation is demonstrated not only by the high concentration of silica remains in general but also by the presence of many types of decorated phytolith morphotypes that tend to be more fragile and are thus more prone to weathering and fragmentation. The concentration of the phytoliths seems particularly high for an open-air locality and indicates that a large volume of plant material was incorporated into the sediment throughout the entire sequence. It may also indicate that these layers were quickly buried and the carbonate capping at the top of the profile provided protection from further weathering.

Although the overall preservation is good, there is a difference in the preservation of the lower parts of the sequence in comparison with the upper section. In the lower section, the preservation of all of the silica microremains is similar. There is little pitting or other evidence of surface modification on either the phytoliths or diatoms. The phytoliths are generally recovered as individuals, rather than occurring in anatomical connection. In the samples collected from 11.33 mbd and above, however, there appears to be more of a mixture of the types of phytolith preservation (Figure 8.7). For example, there are some phytoliths that are very robust and well-silicified, others that appear finer or more fragile but are still very well-preserved, and still others that occur as both individuals and phytoliths in anatomical connection but with surface pitting, weathering, and some type of coating or adhering sediment grains. This observation suggests that

Table 8.3 Identifiable Phytolith Morphotypes

Morphotypes	**FDM Sample Number**															
	1	*2*	*3*	*4*	*5*	*6*	*7*	*8*	*9*	*10*	*11*	*12*	*13*	*14*	*15*	*17*
Depth (mbd)	*11.78–11.83*	*11.73–11.78*	*11.68–11.73*	*11.63–11.68*	*11.58–11.63*	*11.53–11.58*	*11.48–11.53*	*11.43–11.48*	*11.38–11.43*	*11.33–11.38*	*11.28–11.33*	*11.23–11.28*	*11.18–11.23*	*11.13–11.18*	*11.08–11.13*	*10.98–11.03*
Mc Polyhedral	6													5	3	
Jigsaw puzzle	4	2												3	8	4
Tracheid (cylindric sulcate)																2
Ellipsoid rugulate											1					
Irregular psilate											3	2				2
Irregular rugulate	5	3	4	3	6	4	5	3	6	4	3	4		5	6	2
Irregular verrucate			2				1				1					
Parallelepiped blocky psilate	4	2	2	1		3	2		4	1	2	1				
Parallelepiped blocky rugulate		4	3	4	2			1		3	1	3		3	4	1
Parallelepiped thin psilate											2					
Parallelepiped thin rugulate	5		2				3			1				1		
Spheroid rugulate											1					
Cylindroid psilate											3	1				
Cylindroid rugulate	6	2			3		1	3			4	2		2		
Long cell polylobate	7	4	5	3	6	2	3	4	5	4						
Long cell sinuous	7	7	5	4	5	6	7	5		6	3			5	3	
Long cell wavy	10	5	7	6	7	5	8	6	13	12	20			6	6	
Bulliform cell cuneiform	13	17	6	8	13	11	6	9	16	12	18	36	32	20	23	22
Bulliform cell parallelipedal	11	14	12	9	21	19	16	12	9	8	29	20	18	15	11	18

Table 8.3 (continued)

Morphotypes	**FDM Sample Number**															
	1	*2*	*3*	*4*	*5*	*6*	*7*	*8*	*9*	*10*	*11*	*12*	*13*	*14*	*15*	*17*
Depth (mbd)	*11.78–11.83*	*11.73–11.78*	*11.68–11.73*	*11.63–11.68*	*11.58–11.63*	*11.53–11.58*	*11.48–11.53*	*11.43–11.48*	*11.38–11.43*	*11.33–11.38*	*11.28–11.33*	*11.23–11.28*	*11.18–11.23*	*11.13–11.18*	*11.08–11.13*	*10.98–11.03*
Parallelepiped elongate psilate	15	12	17	12	16	8	12	9	13	10	5	9	6	8	5	2
Parallelepiped elongate rugulate	20	27	32	39	48	35	33	32	45	38	52	19	16	15	15	8
Mesophyll		4			6	4	5		8			10	8	11	13	
Prickle	15	18	13	14	19	16	11	12	9	18	8	11	9	10	8	14
Stomata				5			2		5	4		10	12	13	11	12
Hair	4		1		3	1		2		2	1		1	2	2	1
Long cell dendritic	5	2		1	2		1	1	2					2		
Long cell echinate	9	6	5	6	3	5	3	9	6	4	15	12	13	15	10	
Long cell verrucate			1			2					3	4	7	11	2	
Papillae	5	3	2	5	3	5	4	3	6	2	4	6	4	3	4	
Short cell bilobate	6	5	10	11	7	6	11	11	14	12	13	27	32	44	15	34
Short cell polylobate	4	3	2	3	4	5	5	4	5	3	1	5	4	3	5	2
Short cell saddle	18	23	20	18	18	22	20	22	27	25	18	28	25	18	21	53
Short cell cross shaped		2	1	3		4	2	3	1	2	2	1	3	1	4	6
Short cells rondel	15	33	43	44	32	33	43	42	35	28	34	22	28	40	42	63
Short cell trapeziform	2	4	2	6		2	1	1	4	2	3	2	1	4	3	4
Cyperaceae (Sedges)	6	4	4	4	6	5	4	6	3	4	8	6	10	8	9	12
Weathered	3	1	4	3	1	3	4	5	3	6	12	11	15	19	16	14
Total Identifiable	**202**	**206**	**201**	**209**	**230**	**203**	**209**	**200**	**236**	**205**	**258**	**241**	**229**	**273**	**233**	**262**

Key: mbd= meters below datum.

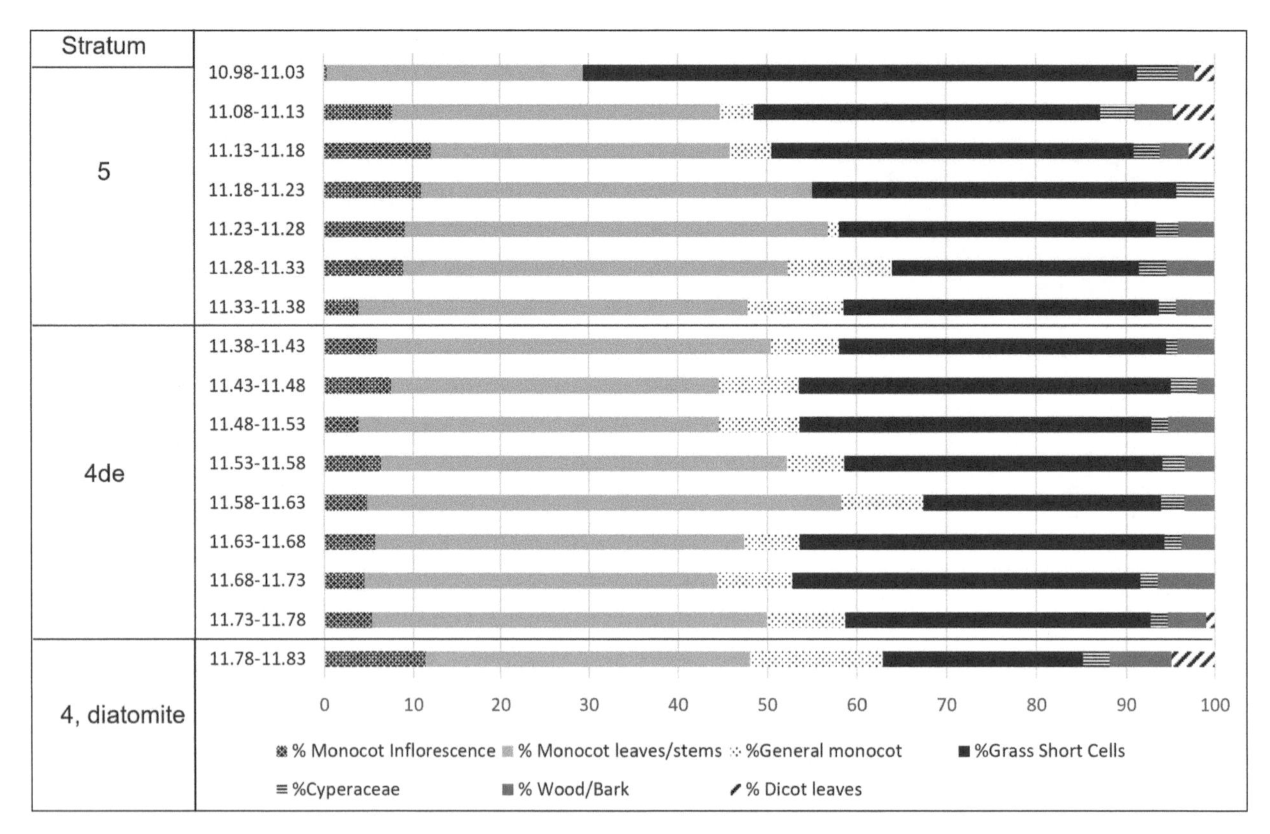

Figure 8.4. Detailed breakdown of different plant types and parts based on the morphotype identification. The overall pattern is broadly similar for all of the analyzed samples.

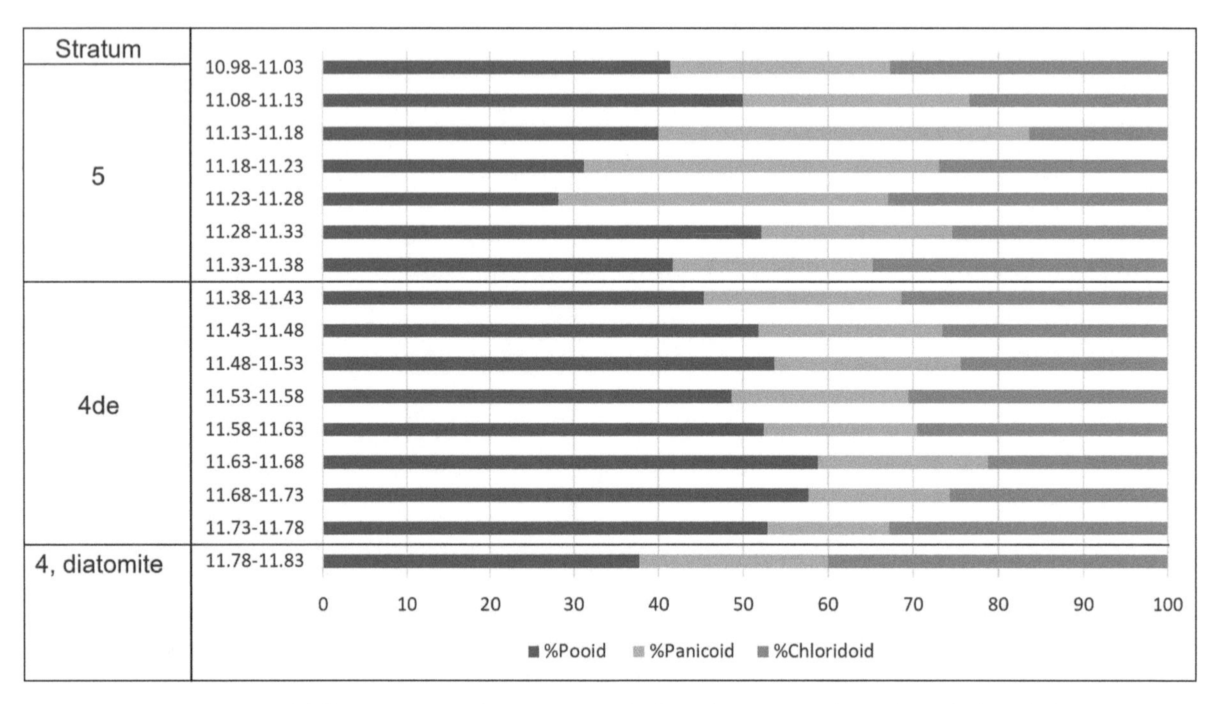

Figure 8.5. Percentages of different groups of grass short cells based on the total number of identified short cells. Pooid includes RONDELS and TRAPEZIFORMS, Panicoid includes the lobate types such as BILOBATES, POLYLOBATES, and CROSSES, and chloridoid includes the SADDLES.

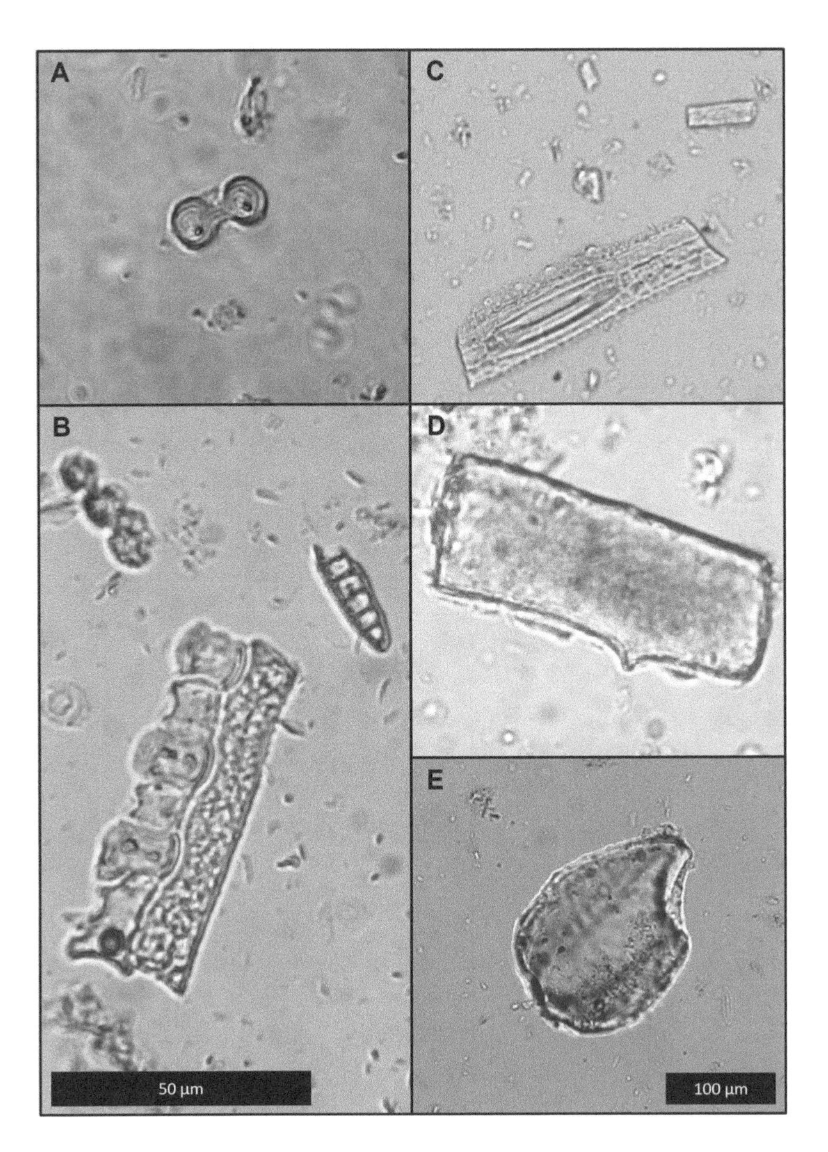

Figure 8.6. Examples of phytolith morphotypes mentioned in the text. Phytoliths extracted from sediment samples associated with micromorphological samples FDM 4.1 (A, B) from upper 4de, FDM 1-5 (D, E) from lower 4de, and a sample within Stratum 5 (C): (A) BILOBATE grass short cell; (B) POLYLOBATE short cell in upper left corner, diatom in upper left corner, and phytoliths in anatomical connection including saddle short cells, and an PARALLELEPIPED ELONGATE RUGULATE; (C) stomata in anatomical connection with short and long cells; (D) PARALLELIPEDAL BULLIFORM; and (E) CUNEIFORM BULLIFORM.

there may be multiple populations of phytoliths in the uppermost samples and these populations had different taphonomic histories.

The morphotype analysis of the samples from El Fin del Mundo clearly supports other evidence for a wetland setting of the site. Previous research in other environmental systems has indicated that wetland adapted plants tend to produce certain suites of phytoliths that can in turn sometimes be related to specific plant families (Guo and others 2012; Novello and others 2012). For example, conical or hat-shaped phytoliths are produced in large numbers by plants in *Cyperaceae* (sedges), whereas *Phragmites australis* (the common reed) and other types of Arundinoid plants (grasses, including reeds that follow the C3 photosynthetic pathway) are associated with abundant saddle short cells (specifically the plateau type), CUNEIFORM BULLIFORMS, and silicified stomata (Guo and others 2012; Lu and Liu 2003). Other examples are less precise as, for example, RONDELS are found in large quantities in most grasses, and many different types of BILOBATES are produced by Panicoid grasses leading to difficulty in achieving further specificity. The Fin del Mundo phytolith assemblage includes all of the types of phytoliths associated with wetland or aquatic plants in large numbers in all of the analyzed samples.

One interesting aspect to note is the large number of extremely well silicified BULLIFORM cells (both PARALLELIPEDAL and CUNEIFORM) found throughout the sequence but peaking in the upper samples. There is some debate in the literature regarding how to interpret BULLIFORM cells, as some researchers have correlated the silicification of these cells with increasing aridity as these parts of the

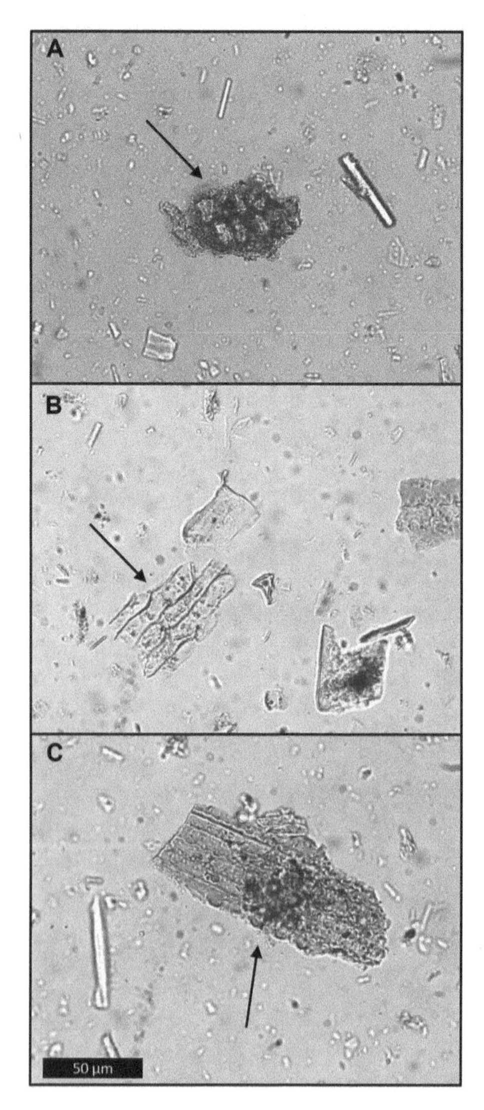

Figure 8.7. Various types of phytolith preservation in the upper layers (Stratum 5): (A) robust, well-silicified, stacked CUNEIFORM BULLIFORM phytoliths with a thin coating of calcium carbonate rich sediment; (B) well-preserved phytoliths in anatomical connection; and (C) phytoliths in anatomical connection with evidence for some surface damage and remnants of sediment adherence.

plant help to reduce evapotranspiration (Bremond and others 2004). However, another study that focused on modern phytolith samples collected around Lake Chad (Africa) found high numbers of specifically CUNEIFORM BULLIFORMS in swampy areas where plant roots were continually submerged (Novello and others 2012). Novello and colleagues (2012) interpret these high concentrations of cuneiform BULLIFORMS as suggesting that the continuous access to water in conditions of relatively low air humidity may have increased transpiration, and thus increased the silicification of the BULLIFORMS. It seems that the samples from El Fin del Mundo present a similar picture. The upper samples exhibit a combination of very high numbers of BULLIFORMS, very well-silicified phytoliths in general, and the lowest number of diatoms in the profile. This pattern could indicate that at the time of deposition of these plant remains, the area was no longer completely submerged, but that the plants still had consistent enough access to water to not suffer the effects of water stress. The appearance of the carbonates in this part of the profile may, however, indicate a trend towards aridification of the overall environment, increasing the rate of transpiration within the plant community.

Although the makeup of the phytolith assemblage remains largely similar throughout the profile, the slight differences noted above may indicate subtle shifts in the environment over time. Uppermost Stratum 3B and Stratum 4d are clearly related to somewhat deeper standing water, likely with stands of wetland grasses in and around this area as indicated by the mix of high concentrations of both diatoms and phytoliths. Upper Stratum 4de seems to show a slight decrease in the amount or depth of the standing water based on the slight decrease in the number of diatoms, consistent number of phytoliths, and beginning of the shift in the type of grass short cells. Finally, a clear change took place in the upper part of the sequence (Stratum 5) with a clear decrease in the number of diatoms, which seems to indicate that the amount of standing water perhaps shifted to a more swampy or muddy environment. This change towards a somewhat drier system appears to continue for some time. Despite general wetland conditions, lower humidity may have caused the combination of increased silicification in the BULLIFORMS and an increase in phytoliths in anatomical connection.

Acknowledgments

This work was carried out in the laboratory of the Geoarchaeology Working Group at the University of Tübingen. Many thanks to Dr. Susan Mentzer for collaborating on several aspects of the analysis and for her comments on earlier drafts of this manuscript, and to Dr. Vance Holliday for allowing me access to such interesting samples. Thanks are also due to Dr. Erin Ellefsen for her work on statistical analyses that were discussed for these datasets.

CHAPTER NINE

Diatom Paleoecology

Manuel R. Palacios-Fest

The diatomite and diatomaceous earth preserved and exposed at El Fin del Mundo provide an opportunity to reconstruct the environment at the site during and following the Clovis occupation. This is a long-standing issue in the U.S. Southwest and an important component of broader interest in terminal Pleistocene to early Holocene environments in the region (e.g., Antevs 1955; Martin 1963; Betancourt, Van Devender, and Martin 1990; Haynes and Huckell 2007; Haynes 2008; Ballenger and others 2011). This chapter reports the results of diatom analyses for sediments recovered from El Fin del Mundo and the environmental implications of those results.

Diatoms are microscopic siliceous algae of the division Bacillariophyta. The silica cell wall, called frustules, consists of two highly ornamented valves of diverse forms (Centric and Pennate; Round, Crawford, Mann and others 1990). The group evolved to form elaborate silica walls that reflect the types of habitats to which the particular species is adapted. Nearly all diatoms are microscopic, ranging in size from 2 mm to 500 mm (0.5 mm). They occur anywhere water and light are sufficient for forming an important part of the food web in the ecosystem (Starrat 2011). Sensitive to water chemistry, diatoms are valuable tools to identify recent and fossil environments. Many species have distinct ranges of pH and salinity tolerance, as well as other parameters such as nutrient concentration, suspended sediment, flow regime, water depth and chemistry, elevation, and different type of anthropogenic disturbance (Stoermer and Smol 1999). In the geologic record the fossilized silica frustules are the only remains of diatoms to survive and may be used to reconstruct past environments. Low energy environments favor preservation (Mannion 1987). High diversity at El Fin del Mundo includes more than 60 species and varieties, which permits the identification of specific environmental conditions and allows for the recognition of biofacies throughout the record (Starrat 2011). This same factor imposes a limitation, however, because identifying such a diverse assemblage is time-consuming and the ecological specificity of the diatoms also constrain spatial-temporal correlations. Some other restrictions include the ease of transportation of the organisms (reworking) and bioturbation; that is, animals and plants can mix the groups, mainly stratigraphically (Starrat 2011).

"Diatomite" is a term used to describe pure beds of diatoms, whereas "diatomaceous earth" refers to sediment with both diatoms and clastic material (Chapter 2). At El Fin del Mundo, upper Stratum 3B contains a thin lens of diatomaceous sediment within the alluvium. Above that, Stratum 4 consists of a thin bed of diatomite (4d), overlain by weakly bedded diatomaceous earth (4de) and by Stratum 5, which also includes diatomaceous earth (Chapter 2).

METHODS

Between 2007 and 2012, 39 sediment samples were collected from two profiles (10-4 and 08-1) in Locus 1 (Figure 4.3). Twenty-three samples were obtained from an 11-cm diatomite layer in lower Stratum 4 and uppermost Stratum 3 at profile 10-1 (Table 2.1). Sixteen samples were selected from the 75-cm thick Strata 5 and 4de sequence plus one from Stratum 3B in 08-1 (Table 2.1). Each set of samples was collected from their respective stratigraphic column based on the site datum and recorded in "meters below site datum" (mbd). Dr. Victoria Chraïbi from Tarleton State University prepared the slides and counted the diatom valves. The author completed the analyses.

Approximately, 0.3 g of sediment was first treated with 10 percent HCl to remove carbonates and then 30 percent H_2O_2 to digest organic material. Samples were rinsed five times with DI water to remove HCl once digestion was completed. Approximately 15 mL of 35 percent H_2O_2 was then added to digest the organic matter at room temperature. Samples were standardized to volume at 20 mL with DI water following five rinses with DI water to remove supernatant fluids after treatment. One mL of 5μ plastic microspheres were added to the samples before plating on glass slides. Rinsed samples were dried onto coverslips and mounted on slides with Naphrax, a permanent mounting medium with a high-refractive-index (Battarbee 1986).

When available, counting of ~300 diatom valves establish community assemblage (Smol 2002). Community assemblage percentages show abundance relative to the total assemblage counted. The data are plotted against depth. Significant changes in the community structure are identified using broken stick cluster analysis in the Rioja package of R statistical software (version 3.3.2; Juggins 2015) and using the software C2 (Figure E.1; Juggins 2014). A few "ghost-like" valves were recorded (where taxonomic features of valves were obscured). The analysis revealed two groups (Figure E.1A) based on the clustering (Figure E.1B).

Diatom concentration as a proxy of primary productivity is based on a ratio to the presence of a known concentration of plastic microspheres (~5 μm in diameter), corrected to the dry weight of the sediment in the subsample. Chrysophyte stomatocysts are included as a secondary proxy. They are also siliceous algae that under some circumstances become predators, feeding on bacteria or diatoms. They occur in cold, freshwater environments; a few forms are marine. The group is not part of this investigation but may shed some light on the patterns of environmental change at El Fin del Mundo.

DEPOSITION RATES

Age control for determining deposition rates follows the geochronology presented in Chapter 2 (Table 2.1). Charred plant fragments in upper Stratum 3B (i.e., just below the diatomite) dated to ≤13,415 cal yr B.P. Charcoal on top of Stratum 4d is ~11,125 cal yr B.P. Deposition of upper 3B and the diatomite therefore straddles the Pleistocene–Holocene boundary. The calculated sedimentation rate for the diatomite suggests a very low accumulation of about 0.005 cm/yr (or 204 yr/cm), whereas the diatomaceous sediments of Strata 4de and 5 accumulated more rapidly at about 0.07cm/yr (or 15 yr/cm). This rate provides an age estimate of ~10,440 cal yr for the Strata 4-5 contact.

RESULTS

The samples yielded 69 species and genera. When possible, valves were identified to the species level (Table 9.1). Some diatoms, especially those of very large valve size, were fragmented so consistently that species-level identification was uncertain. In that case, identification was restricted to the genus level. Most species prefer benthic or tychoplanktic habitats. Preservation was relatively good; fragmentation was moderate to high (>45%). The moderate to high fragmentation rate may have resulted from differential dissolution or to a lesser extent re-working (Abrantes and others 2005). Pennates fragment more readily than centric types, but a fragmentation index could not be determined. Even though many diatoms were discernible for identification, only fragments larger than 50 percent were counted to secure the most accurate interpretation possible.

Despite the moderate diversity encountered in the profiles, only 20 species are present in greater than 5 percent relative abundance in at least one sediment sample. These species are *Fragilaria tenera*, *Pseudostaurosira brevistriata*, *Punctastriata mimetica*, *Staurosira construens*, *Staurosirella pinnata*, *Ulnaria* spp. 1, *U. contracta*, *Achnanthidium minutissimum*, *A. minutissimum* v. *scotica*, *Diadesmis confervacea*, *Denticula* sp. cf. *D. tenuis*, *Epithemia* sp. cf. *E. adnata*, *Nitzschia amphibia*, *N. palea*, *Eunotia arcus/bilunaris*, *Hantzschia* sp., *Rhopalodia gibba*, *R. musculus*, *Aulacoseira italica*, and *A. islandica*.

Pennate forms comprised other, less abundant species (34), ranging from 1 percent to less than 5 percent in at least one sediment sample: *Achnanthidium exiguum*, *Sellaphora saugarresii* (aka *Navicula minima*), *Pinnularia borealis*, *P.* sp. cf. *P. saprophila*, *P.* spp. 1, *P.* spp. 2, *Amphora copulata-ovalis*, *A. pediculus*, *Anomoeoneis* spp., *Cocconeis placentula*, *Craticula* spp., *Cymbella mexicana*, *C. microcephala*, *Diploneis elliptica*, *Luticola* spp. 1, *Gomphonema johnsonii*, *G. consector*, *G.* spp. 1 (cf. *G. mexicanum*), *G. turgidum*, *Halamphora elongata*, *H. veneta*, *Karayevia* spp., *Navicula* spp. 1 (cf. *N. cryptocephala*), *Neidium* sp. cf. *N. dubium*, *Stauroneis smithii*, *Planothidium lanceolatum*, *Nitzschia* spp. 1, *N.* spp. 2, *Rhopalodia gibba*, *R. musculus*, *Surirella* sp. cf. *S. ovalis*, *Hantzschia* spp.; and the Centric forms *Cyclotella atomus*, and *Orthoseira* spp. Regardless of the high rate of fragmentation, these species are included in

Table 9.1. Diatom Assemblages and Ecological Preferences

Assemblage	*Species*	*Ecological Preferences**
I (Cosmopolitan)	*Achnanthidium levanderii*, ***A. minutissimum*, *A. minutissimum* var. *scotica***, *Achnanthidium* spp. 1, *Amphora copulata-ovalis, Cocconeis placentula, Diploneis elliptica, Halamphora elongata, H. veneta*, ***Navicula* spp. 1 (cf. *cryptocephala*)**, ***Nitzschia amphibia*, *N. palea***, *Pinnularia borealis*, ***Pinnularia* sp. cf. *P. saprophila*, *Pinnularia* spp. 2, *Planothidium lanceolatum*, *Staurosirella pinnata*, *Ulnaria contracta***	Habitat: from shallow lake to ephemeral pond, epilithic pH range: 5-9.1 Conductivity: Low to moderate (between 10 μS cm^{-1} and >1750 μS cm^{-1}) Nutrient tolerance: Oligotrophic to eutrophic (PO_4-P and NO_3+NO_2-N mainly range between 0.01->7 mg L^{-1} and 0.05->9 mg L^{-1}, respectively) Temperature range: ≥5->20°C
II (Deeper Pond)	*Achnanthidium exiguum, Amphora pediculus*, ***Aulacoseira islandica, Aulacoseira italica***, *Cyclostephanos* spp., *Cyclotella atomus, Cyclotella comensis, Cymbopleura, lapponica, Diadesmis confervacaea*, ***Eunotia arcus/bilunaris***, *Eunotia* spp., ***Fragilaria tenera***, *Gomphonema acuminatum*, , ***G. consector***, *Gomphonema* spp. 1 (cf. *mexicanum ?*), ***Pseudostaurosira brevistriata***, *Psammothidium didyum*, ***Punctastriata mimetica***, *Stauroneis smithii*, ***Staurosira construens, Ulnaria* spp. 1 (short, pinched waist)**	Habitat: Littoral zone of lake, benthic to tychoplanktic, epiphytic or epilithic pH range: 6-8.5 Conductivity: Low (between 10 μS cm^{-1} and 1000 μS cm^{-1}) Nutrient tolerance: Oligotrophic to mesotrophic (PO_4-P and NO_3+NO_2-N mainly range between 0.01-1 mg L^{-1} and 0.1-1 mg L^{-1}, respectively) Temperature range: 15-20°C *Eunotia arcus/bilunaris* may tolerate low temperatures (~10°C), and low pH (<6)
III (Shallow Pond)	*Anomoeneis* spp., *Craticula* spp., *Cymbella mexicana, C. microcephala, Cymbella* spp. 1, *Cymbella* spp. 2, *Diadesmis* spp. 1 (oval), *Gomphonema turgidum, G. kobayasii, Gyrosigma* spp. *Karayevia* spp., *Luticola* spp. 1, *Pinnularia* spp. 1 (glassy striae), ***Neidium* sp. cf. *N. dubium***	Habitat: Littoral zone of lake, shallow pond pH range: >7.5 Conductivity: Low to moderate (10 μS cm^{-1} to 2,000 μS cm^{-1}) Nutrient tolerance: Oligotrophic to mesotrophic (PO_4-P and NO_3+NO_2-N mainly range between 0.01-1 mg L^{-1} and 0.1^{-1} mg L^{-1}, respectively) Temperature range: 15-20°C
IV (Seasonal marsh)	***Denticula* sp. cf. *D. tenuis*, *Epithemia* sp. cf. *E. adnata***, *Epithemia* spp. 2, ***Gomphonema johnsonii, Hantzschia* spp., *Nitzschia* spp. 1 (large, prominent struts), *Nitzschia* spp. 2 (small, capitate,no striae)**, *Orthoseira* spp., *Planothidium abbreviatum*, ***Rhopalodia gibba, R. musculus***, *Sellaphora saugarresii* (aka *Navicula minima*), *Sellaphora* spp. 1 (cf. *pupula*), *Sellaphora* spp. 2, ***Surirella* sp. cf. *S. ovalis***	Habitat: Fresh to brackish shallow water, benthic, epiphytic or epipelic pH range: 6-9.5, carbonate-rich waters. Conductivity: Moderate (between 10 μS cm^{-1} and 4,000 μS cm^{-1}) Nutrient tolerance: Oligotrophic to eutrophic (PO_4-P and NO_3+NO_2-N mainly range between 0.05-1 mg L^{-1} and 0.1-2 mg L^{-1}, respectively) Temperature range: 15-20°C *Sellaphora saugrresii* may tolerate low temperatures (10-20°C). *Pinnularia borealis, Hantzschia* spp. and *Orthoseira* spp. are semiterrestrial.

Key: *The ecological preferences are generalized based on the relevant species (in bold).

Source: https://users.ugent.be/~pchaerle/strains/strains5.php?type=diatoms; accessed August 20, 2021; Potapova and Hamilton 2007; Battarbee and others 2011; Kim and Lee 2017; Cantonati and others 2020).

this report to refine El Fin del Mundo paleoecology. The remaining species occurring at concentrations less than 1 percent in any sample are not discussed (Table 9.1).

Based upon these dominant species, two assemblage shifts were identified by cluster analysis, at elevations of 11.81 and 11.835 meters below datum in Profile 10-4 (Figure 9.1; see also Appendix E). Both shifts depict an overall trend of early dominance of small benthic araphid diatoms such as *P. brevistriata* and *S. construens* (Figure 9.1). *P. brevistriata* and *S. construens* are morphologically diverse, common benthic-tychoplanktic diatoms that perform well in a wide range of pH (6-9.3), low to moderate turbidity, and carbonate-rich conditions (Potatova 2009; Morales 2010).

Around the 11.81 mbd, the assemblage shifted to the dominance of raphid benthic species including

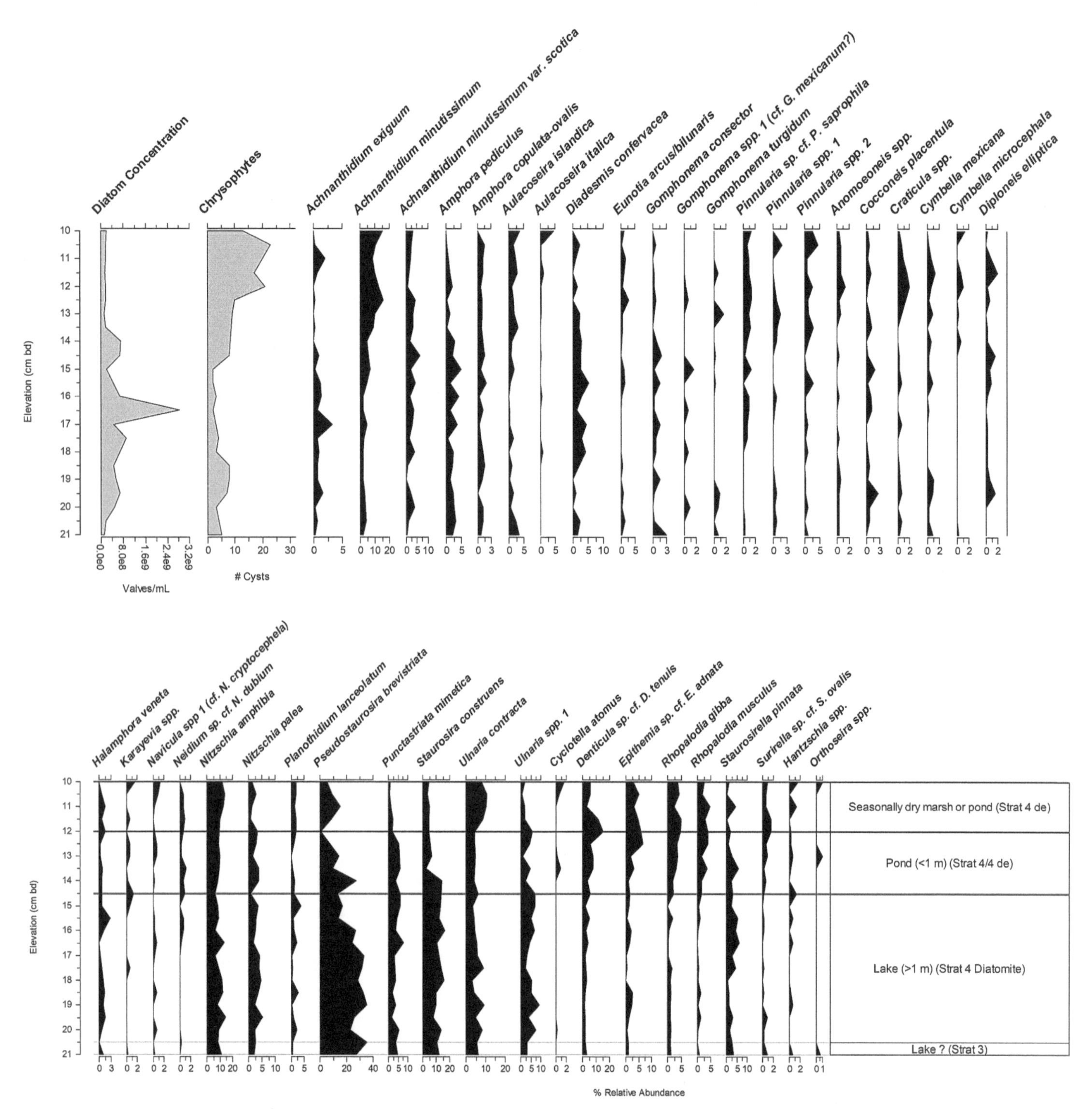

Figure 9.1. Relative abundance (%) trends of selected diatom species at El Fin del Mundo, Profile 10-4. Thirteen species are dominant (>5%); 29 occur between 1 and 5 percent; the shift (black) lines (12–14.5 cm bd) from the broken stick cluster analysis indicate the intervals where floral changes were more significant.

A. minutissimum, *Denticula* sp. cf. *D. tenuis*, and *Epithemia* sp. cf. *E. adnata* (Figure 9.1). *A. minutissimum* is a very common benthic (episammic/epilithic) diatom that prefers freshwater conditions (Potatova 2009). Temperature does not seem to be a factor for *A. minutissimum*; however, this species prefers streams over standing waters (Cantonati and Lange-Bertalot 2010; Leira, Meijide-Failde, and Torres 2017). *Denticula* sp. cf. *D. tenuis* (epilithic/epiphytic) occurs in diverse environments; however, it also thrives better in freshwater environments where carbonate-rich and moderate conductivity dominate (Spaulding and Edlund 2008). *Epithemia* sp. cf. *E. adnata* is an epiphytic or epipelic species occurring in fresh- to brackish water conditions. It is commonly found in carbonate-rich, alkaline waters, and high conductivity; it often possesses endosymbiotic nitrogen-fixing bacteria (Spaulding 2010).

In contrast, in Profile 08-1 the species richness of the overall diatom community is much lower than that from Profile 10-4. No new species were identified from Profile 08-1 with respect to Profile 10-4 (Figure 9.2). Low species evenness is shown by the few dominant species. Benthic forms such as *Rhopalodia gibba* and *R. musculus* along with the genus *Epithemia* (frequently fragmented and unsuitable for identification) are the most common species (>5%; up to 25%) in Profile 08-1. All three may be epiphytic and tolerate alkaline waters with low nitrate conditions because they either fix nitrogen or contain endosymbiotic cyanobacteria that fix nitrogen. Also dominant is *Denticula* sp. cf. *D. tenuis*, a species that prefers shallow, alkaline waters. Overall, diatom density in many samples is low. Rather, the samples are dense with siliciclastic debris that is likely phytoliths from grasses and sedges (Chapter 8). Chrysophyte cysts are also rarely present, but no chrysophyte scales were identified.

The biostratigraphy of taxa present in greater than 5 percent relative abundance remains largely stable through time, with no significant shifts in the dominance of species over time. The most notable increases are periodic peaks of *Nitzschia amphibia*, a benthic species that is common in low abundances in much of western United States and prefers neutral to alkaline pH. *Achnanthidium minutissimum* is common in rocky, fluvial, or near shore lake microhabitats. A relatively diverse assemblage of the benthic *Nitzschia* species is present in Profile 08-1 in low abundances. In a benthic dominated assemblage *Fragilaria tenera* (planktic or tychoplanktic) is common throughout the stratigraphic column associated with two species of the genus *Aulacoseira* (*A. islandica* and *A. italica*). *Hantzschia* sp. and several species of *Pinnularia* (*P. borealis* and others, Table 9.1), even though not as abundant may be associated with moist soils.

To improve the environmental signature, the relevant species present at El Fin del Mundo were grouped by preferred environments based on their ecological characteristics. Table 9.1 considers the controlling parameters affecting a shallow environment such as the pond at El Fin del Mundo. Following Cantonati and others (2009), pH, conductivity, and nutrient tolerance appear to be more important factors than temperature or depth. The experimental work of da Silva, Carvalho-Torgan, and Schneck (2019) demonstrates, however, that variations in temperature and surface runoff have interactive and complex effects on species richness and community composition of periphytic diatoms. For example, *A. minutissimum* is a eurythermic species not affected by changes in water temperature; *Navicula cryptocephala*, responded to increasing temperatures and runoff, while *Staurosira construens* was affected only by increasing temperature. By contrast, *Eunotia bilunaris* was reactive to runoff but not temperature. Using da Silva, Carvalho-Torgan, and Schneck's model, the assemblages identified in this study reflect the environmental conditions.

The cosmopolitan Assemblage I is common throughout the record at El Fin del Mundo. This assemblage is characterized by eurytopic species that tolerate a wide range of physical-chemical parameters documented in Table 9.1. The relevant species in this group include *Achnanthidium minutissimum*, *A. minutissimum* var. *scotica*, *Navicula* spp. 1 (cf. *cryptocephala*), *Nitzschia amphibia*, *N. palea*, *Pinnularia* sp. cf. *P. saprophila*, *Pinnularia* spp. 2, *Planothidium lanceolatum*, *Staurosirella pinnata*, and *Ulnaria contracta*. These species are abundant elsewhere (e.g., Winsborough 1995, 2016; Cantonati and others 2013).

Assemblage II, the freshwater deep pond, dominates most of the record of Profile 10-4 and starts to decline at the base of Profile 08-1. The cluster analysis of Profile 10-4 shows a break at 11.81 mbd and 11.835 mbd in Profile 10-4 (black lines in Figure 9.1) that extends up into the diatomaceous earth (Profile 08-1; Figure 9.2). This, in turn, constitutes the split of the four biofacies that characterizes the El Fin del Mundo assemblage. The two groups may co-exist, but the latter becomes dominant over the former over time. *Pseudostaurosira brevistriata*, *Staurosira construens*, and *Punctastriata mimetica*, occur in greater concentrations below 11.835 mbd and decline up through the section. At a minimum, the first two may adapt to tychoplanktic conditions, suggesting changes in the hydrologic system.

The scarce Assemblage III composed by a few species dominated by *Neidium* sp. cf. *N. dubium* indicates some

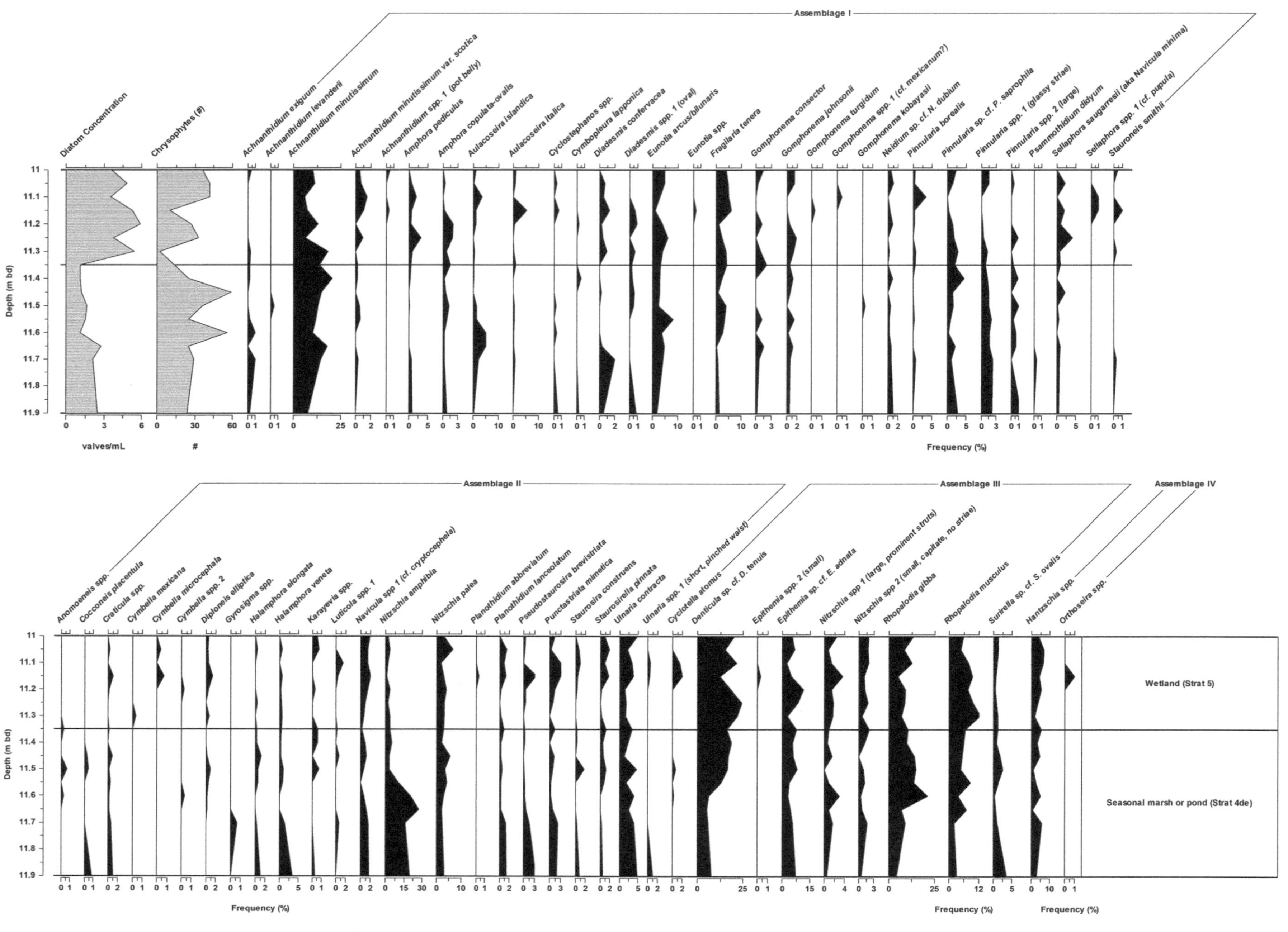

Figure 9.2. Relative abundance (%) trends of diatom species, Profile 08-1. Twelve species are dominant (>5%), 18 occur between 1 and 5 percent. The associated species are included to identify potential environmental assemblages.

Table 9.2. El Fin del Mundo Diatom Biofacies

Biofacies	*Interval (m bd)*	*Dominating Assemblage*	*Description*
1	11.905–11.835	Assemblage II: *Aulacoseira islandica, A. italica, Eunotia arcus/bilunaris, Fragilaria tenera, Gomphonema consector, Pseudostaurosira brevistriata, Punctastriata mimetica, Staurosira construens,* and *Ulnaria* spp. 1.	Lacustrine phase: Dilute, cold (?) freshwater as suggested by the occurrence of *E. arcus/bilunaris* (see Table 1)
2	11.835–11.825	Assemblage I: *Achnanthidium minutissimum, A. minutissimum* var. *scotica, Navicula* spp. 1 (cf. *cryptocephala), Nitzschia amphibia, N. palea, Pinnularia* sp. cf. *P. saprophila, Pinnularia* spp. 2, *Planothidium lanceolatum, Staurosirella pinnata,* and *Ulnaria contracta.* *Decline of Assemblage II and rise of Assemblage IV.* Assemblage IV: *Denticula* sp. cf. *D. tenuis, Epithemia* sp. cf. *E. adnata, Epithemia* spp. 2, *Gomphonema johnsonii, Hantzschia* spp., *Nitzschia* spp. 1 (large, prominent struts), *Nitzschia* spp. 2 (small, capitate, no striae), *Orthoseira* spp., *Planothidium abbreviatum, Rhopalodia gibba, R. musculus, Sellaphora saugarresii* (aka Navicula minima), *Sellaphora* spp. 1 (cf. *pupula), Sellaphora* spp. 2, *Surirella* sp. cf. *S. ovalis.*	Rheocrene phase: Increasing alkalinity and carbonate, ending of lacustrine phase
3	11.825–11.28	Assemblage IV: blooming of *Epithemia* sp. cf. *E. adnata, Nitzschia amphibia, Achnanthidium minutissimum, Rhopalodia gibba* and *R. musculus. Hantzschia* spp. is a common element of this assemblage.	Seasonal marsh phase
4	11.28–10.93	Assemblage IV: *Denticula* sp. cf. D. *tenuis, Rhopalodia musculus.* Re-appearance of *Amphora pediculus, A. copulata-ovalis, Diadesmis confervacea, Fragilaria tenera,* and *Sellaphora saugarresii.* Marked by the decline of *A. minutissimum, Pinnularia* spp. 1 and spp. 2, *Nitzschia amphibia,* and *Surirella* sp. cf. *S. ovalis.*	Wetland (steady water body)

variations between the deeper pond and the seasonal marsh environments at El Fin del Mundo. This assemblage appears to have little impact on the evolution of the pond.

Assemblage IV starts to dominate the environment above 11.835 mbd through the rest of the record (into Profile 08-1). *Denticula* sp. cf. *D. tenuis*, and *Epithemia* sp. cf. *E. adnata* were very common (Figure 9.2). They perform well in a wide range of pH (6-9.5), low nutrient, carbonate-rich waters of moderate conductivity, and wide temperature range (Spaulding and Edlund, 2008; Table 9.1). These two species are common in temporary ponds (seasonal marshes) such as the one that formed in El Fin del Mundo in the later portion of the stratigraphic package associated with *U. contracta*, *R. gibba* and *R. musculus*, and *Surirella* sp. cf. *S. ovalis* occur in greater relative abundance in the upper portion of Profile 10-4, especially above 11.81 mbd consistent with increasing salinization of the environment (Figure 9.1). Two subaerial species are lumped in Assemblage IV in support of the seasonal marsh conditions interpreted for this group, *Hantzschia* spp. and *Orthoseira* spp. (Table 9.1).

By contrast, no significant change is recorded throughout the Profile 08-1 biostratigraphy (Figure 9.2). Combining the uppermost part of Profile 10-4 (11.81-11.835 m bd) with the entire Profile 08-1, the presence and dominance of *Denticula* sp. cf. *D. tenuis*, *Rhopalodia gibba*, *R. musculus*, and *Achnanthidium minutissimum* describe the presence of a seasonal marsh.

To understand the evolution of the diatomite from Profile 10-4 into the diatomaceous earth of Profile 08-1 it is necessary to analyze the ecological patterns left by the diatom assemblages throughout the composite stratigraphic record. The diatoms also show some trends in concentration and diversity from base to top in the profiles herein recognized as Biofacies 1, 2, 3, and 4 (Table 9.2; Figure 9.3). Diatom concentration in Profile 10-4 is greater below 11.835 m bd with a significant peak between 11.85 and 11.855 m bd, which is Biofacies 1 dominated by Assemblage II (*Aulacoseira islandica*, *A. italica*, *Eunotia arcus/bilunaris*, *Fragilaria tenera*, *Gomphonema consector*, *Pseudostaurosira brevistriata*, *Punctastriata mimetica*, *Staurosira construens*, and *Ulnaria* spp. 1), the "lacustrine" phase. *Fragilaria*

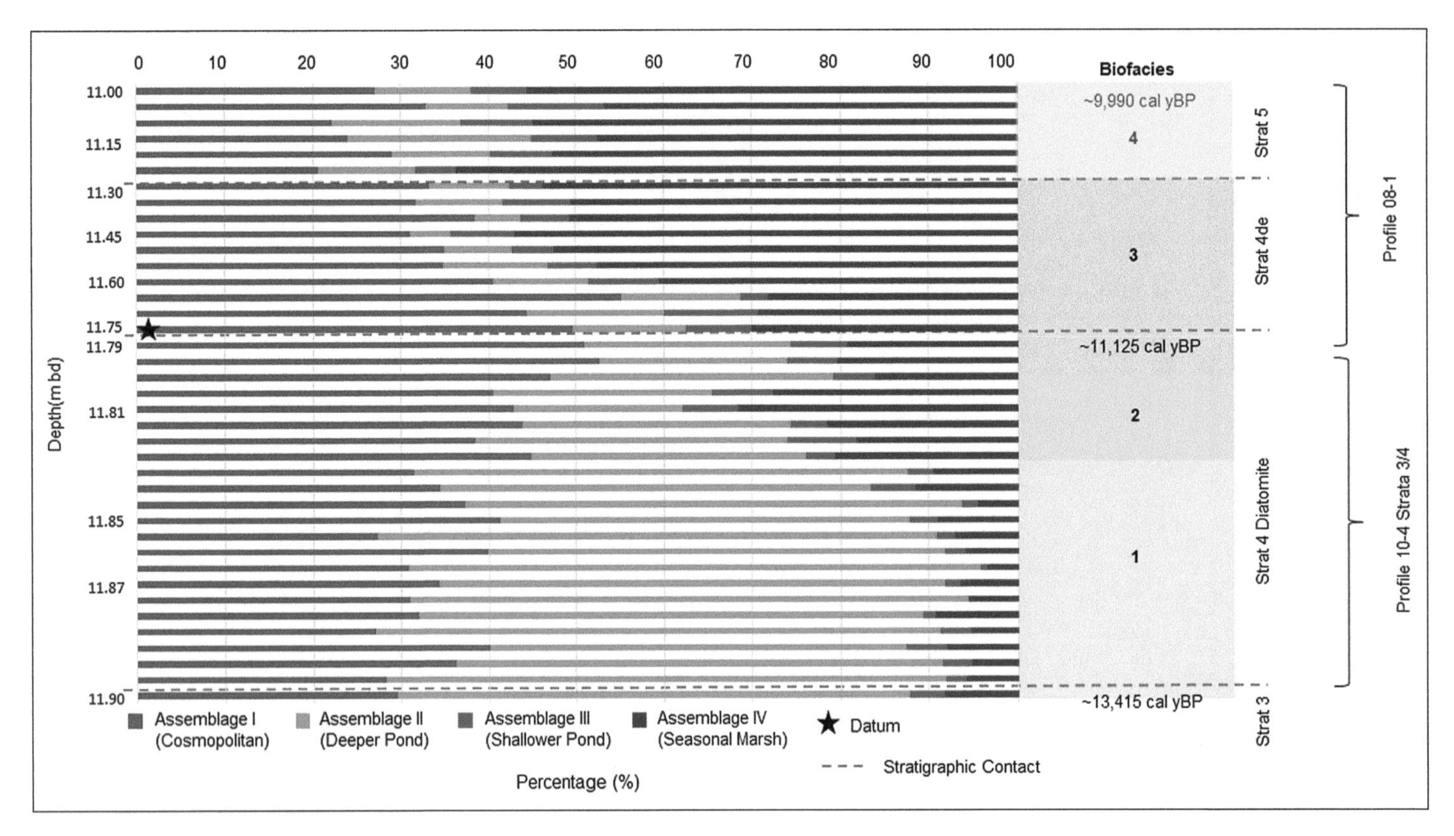

Figure 9.3 Schematic reconstruction of the paleoenvironmental evolution of El Fin del Mundo between ~13,415 cal yr B.P. and ~9,990 cal yr B.P. This diagram is a composite of Profiles 10-4 and 08-1 showing the four biofacies proposed in this investigation. The dashed lines indicate the strata breaks. Note the significant change in scale below and above the black star (center left). The samples below are from 10-4, collected in continuous 0.5 cm intervals, whereas samples above are from 08-1, collected in continuous 5.0-cm intervals. The former represents 11 cm of the stratigraphic column and the latter more than 70 cm. For visual purposes and to highlight the evolution of the pond, this composite representation is not shown at scale.

tenera and both species of *Aulacoseira* are considered planktic or tychoplanktic. Assemblage I is limited to a few species around or above 5 percent (in bold in Table 9.1) and discussed elsewhere in this investigation, whereas the other two assemblages represent less than 15 percent of the total diatom concentration. Even though there is not a significant change in floral composition or species concentration at the time of formation of the diatomite, the sedimentology (i.e., shift from stratum 3B to stratum 4) indicates that as water flow decreased only diatom frustules were accumulated in the pond. In this scenario, we interpret that Biofacies 1 formed throughout uppermost Stratum 3B, the mixed Strata 3/4, and the lower Stratum 4 diatomite (Figure 9.3). Biofacies 1 represents a deep pond stage, the "lacustrine" phase that includes the Younger Dryas Chronozone, interpreted by some researchers as a cold phase in the American Southwest (Hall and others 2012; Palacios-Fest and others 2021).

In Biofacies 2, the rheocrene phase, species diversity (e.g., *Achnanthidium, Gomphonema, Nitzschia*, and *Planothidium*), though relatively high, is lower than that below 11.835 m bd through about 11.81 mbd. Assemblage I continued to dominate the scenario, whereas Assemblage II decreased at the expense of the steady growth of Assemblage IV. The gradual increase of Assemblage IV implies alkalinization/salinization of the basin at the time of deposition of Biofacies 2. The co-occurrence of the freshwater and seasonal marsh groups suggests periods of dilute water input that may correspond to water table fluctuation (Chapter 2) and in good agreement with Cantonati and others (2012). Towards the end of Biofacies 2, Assemblage II significantly declined, whereas Assemblage IV became dominant consistent with this interpretation.

The uppermost 2 cm (11.79–11.81 mbd) in Profile 10-4 (lower Stratum 4de) is characterized by an increase of *Ulnaria contracta* indicating intervals of freshwater discharge into the system. *Achanthidium minutissimum*, *Nitzschia amphibia*, and *Denticula* sp. cf. *D. tenuis* are still relevant at the base of Biofacies 3. *Epithemia* sp. cf. *E. adnata* increase and *Rhopalodia gibba* and *R. musculus* start to bloom in this environment, suggesting formation of a seasonal marsh environment that characterizes the history of El Fin del Mundo through Profile 08-1.

At Profile 08-1, however, diatom concentration is low throughout Stratum 4de. Abruptly, diatom concentration increases at the contact of Strata 4de and 5 and remains high to the end of the record. The same species diversity found in Profile 10-4 occurs through Profile 08-1. Occurrence of *Denticula* sp. cf. *D. tenuis*, *Rhopalodia gibba*, *R. musculus*, and *Achnanthidium minutissimum* indicate an environmental change relative to the lower part of Profile 10-4 (below 11.81 m bd), suggesting that the marsh reached the alkaline/saline conditions characteristic of Biofacies 3. Diatomite deposition then transitioned to diatomaceous earth, suggested by the increase of clay and organic matter in the sediments. Biofacies 2 changed to Biofacies 3 within the diatomite where a brief zone of mixed diatomite and diatomaceous earth species co-exist. Biofacies 3 dominates the lower part of the diatomaceous earth (Stratum 4de). An abrupt change in the diatom concentration and species substitutions characterizes Biofacies 4 in Stratum 5 (Table 9.3). The floristic composition does not change but the species relative abundance does. This important change coincides with the sudden increase in the percentage of $CaCO_3$ in the system (Table 2.2; Figure 2.10), confirming that a geochemical change differentiates Stratum 4de from Stratum 5.

The transition from Strata 3 to 4 to 5 documents a shift from flowing water (a stream) to standing water (lake or pond) to surface or subsurface water (wetland). Streamflow from local springs or seeps decreased between Biofacies 1 and 2 creating a mixed zone between upper Stratum 3B and lower Stratum 4 diatomite (Figure 9.3; Chapters 2, 3). At 11.835 (Strata 3B/4d contact) Assemblage II is no longer dominant and starts to decline (Figure 9.3), whereas Assemblages I and IV show an increase. This hypothesis implies that a gentle stream washed fine sediments into a pond that was periodically subaerially exposed as indicated by the terrestrial flora (*Hantzschia* spp., *Orthoseira* spp.). The diatomite represents standing-water conditions (Assemblage II) and the diatomaceous earth represents more marshy conditions characteristic of a wetland environment with seasonal standing water (Assemblage IV) rich in organic matter. Stratum 5 (Assemblage IV) represents an abrupt increase in diatom concentration (Chapter 8) in a wetland (Figure 9.3).

Deposition of El Fin del Mundo diatomite started at the top of Stratum 3B, representing the transition between it and Stratum 4, dated between ~≤13,145 cal yr B.P. and ~11,125 cal yr B.P. (Chapter 2; Table 2.1). The transition from Stratum 3B alluvium to Stratum 4 diatomite was rapid. Some of the gomphothere bone was exposed above the top of Stratum 3B and encased by diatomite, but the weathering characteristics of the exposed bone is no different than the bone buried within Stratum 3B (Chapter 6). This is indicative of limited exposure of bone following the termination of Stratum 3B deposition but prior to the onset of Stratum 4 diatomite deposition.

The four assemblages discussed illustrate a more refined environmental signature (Figure 9.3). Diatom distribution displays a steady Assemblage II dominance throughout the diatomite facies of Stratum 4 with an upwards increasing influence of Assemblage IV. Assemblages I and III remain relatively constant throughout the record. The latter assemblage is important because it implies that some species thrived in meadow-like environments. *Orthoseira* sp., a benthic, acidophilous and oligotrophic form, is also common in subaerial systems (Bere 2014). This rare species at El Fin del Mundo may indicate episodes of freshwater input or subaerial conditions. *Hantzschia* spp. and *Orthoseira* spp. were identified at the base of the profile and above 11.82 mbd in Profile 10-4 and throughout Profile 08-4 (Figures 9.1 and 9.2). They suggest an environmental change towards aridification at El Fin del Mundo.

The four biofacies identified indicate the degree of influence of each of the four diatom assemblages throughout the history of the site. Biofacies 1, at the base of the diatomite profile, is heavily dominated by Assemblage II with a constant and stable presence of Assemblage I, and the introduction of Assemblages III and IV. Two possibilities may explain this biofacies. The first hypothesis is that diatoms bloomed as freshwater suddenly entered the system, either through flooding or water table rise, favoring the arrival but limited settling of *Orthoseira* spp. A rising water table is a preferred alternative to understand the formation of diatomite. The site favored the development of a slightly alkaline environment that permitted mainly the Assemblage II species to thrive in the system. The second hypothesis is that at this time the environment was desiccating, permitting the settling of *Orthoseira* spp. in a damp meadow. Assemblage II would likewise thrive under alkaline conditions. The problem with the latter hypothesis is that *Orthoseira* spp. are acidophilous to circumneutral, freshwater (subaerial) species that would not survive in alkaline waters or soils.

Biofacies 2 began to form as the result of the decline of Assemblage II and rise of Assemblage IV while Assemblage I remained relatively dominant until about 11.81 mbd in Profile 10-4 where Assemblage IV starts to dominate. A seasonal marsh is suggested by Biofacies 3 from this point to the end of the record in Profile 10-4 and to a depth of 11.28 mbd in Profile 08-1, where within Assemblage IV there is a change in microfloral dominance indicating a transition to a wetland environment (Biofacies 4). Even though the same species characterize this interval, *Denticula* sp. cf. *D. tenuis* and *Rhopalodia musculus. Amphora pediculus, A. copulata-ovalis, Diadesmis confervacea, Fragilaria tenera*, and *Sellaphora saugarresii* re-appeared. Other species such as *Epithemia* sp. cf. *E. adnata* and *Hantzschia* spp. remained steady throughout Strata 4 and 5, whereas *Nitzschia amphibia, Achnanthidium minutissimum, Pinnularia* spp. 1 and spp. 2, and *Surirella* sp. cf. *S. ovalis* declined, suggesting a more stable (less fluctuating) aquatic system. Assemblage III is negligible in the record. The alternating transition from Assemblage II to either Assemblages I or IV is a good indication of water table fluctuations over time. According to Cantonati and others (2012), variations in the water table favor changes in the floral composition of spring-fed ecosystems.

Between ~13,410 and ~9990 cal yr B.P., El Fin del Mundo experienced episodes of groundwater input and desiccation. The base of Biofacies 3, however, is the best indicator of decreasing water flow that ended the deep pond or shallow lake and gave rise to a pond or shallow wetland to the end of the record. The sharp increase in diatoms concentration at 11.28 to 10.93 mbd and the switch in dominant species are consistent with the idea of the establishment of a wetland at Fin del Mundo, Biofacies 4. An increase of Chrysophyte stomatocysts and siliciclastic debris (phytoliths) during this interval is in good agreement with periods of alternating subaquatic and subaerial episodes. The diatomite in Profile 10-4 and its transition to the diatomaceous earth of Profile 08-1 reflects the continuity of the spring-pond-marsh system that characterized El Fin del Mundo, but with declining water depth.

DISCUSSION

The diatom flora identified from El Fin del Mundo permitted recognition of three environments based on four assemblages. The trends show how fresh- to brackish-water diatoms interacted in this setting before and during the interval between ≤13,415 cal yr B.P. and ~9990 cal yr B.P., (late Bølling-Allerød, Younger Dryas, and earliest Holocene chronozones). These patterns broadly correspond to the ecological preferences of diatom species that, in turn, are good indicators of the environmental changes recorded in northwestern Sonora, Mexico. The geochronological resolution of the stratigraphy does not, however, allow differentiation of late Bølling-Allerød, Younger Dryas, and earliest Holocene time intervals.

The four assemblages identified in this study seem to respond to lithological and geochemical changes in the stratigraphic column between Strata 3 and 4. The transition from Stratum 3 into Stratum 4 is characterized by a mixed zone of diatomite and diatomaceous earth that may be associated with the deposition of aeolian dust. This is consistent with the interpretation that the water that fed Strata 3 and 4 originated from seeps or springs in the immediate surroundings of Locus 1. The Stratum 3B (alluvium) to Stratum 4 (diatomite) transition may represent a response to decreased runoff and rise in the water table producing the diatomite. Increasing alkalinity indicated by *Denticula* and *Epithemia*, between 11.895 mbd and 11.825 mbd in Strata 3/4, Profile 10-4 continued to increase during deposition of the Stratum 4 diatomaceous earth. The composite stratigraphic sequence (Figure 9.3) illustrates the transition of the lacustrine (standing water) environment into a palustrine environment of diatomaceous earth, a seasonal marsh or pond (Figure 9.2). Terrestrial sediment input may be due to increased aeolian dust infall as the water table declined, as indicated by dominance of the alkaline and subaerial microflora (Assemblage IV). The sudden increase in diatoms concentration did not modify the composition of Assemblage IV but it did change the role of several species that indicate the ponding of the site to create a wetland or seasonal marsh in Stratum 5.

The biological associations recognized in the previous section express a balance between inputs (precipitation and groundwater inflow) and outputs (evaporation, groundwater recharge) in the pond and wetland (Mason, Guzkowska, and Rapley 1994). Chemical changes were responsible for the physiological responses and species composition of El Fin del Mundo's diatom flora. In general, wetlands react to changes in hydrology and evapotranspiration by changes in water table and water chemistry. In this case, the initial accumulation of diatomite in uppermost Stratum 3B appears to occur in response to the water table rise, as suggested by the occurrence of two benthic-tychoplanktic species (*P. brevistriata* and *S. construens*) supported by the increasing groundwater discharge. But the initial accumulation of diatoms also coincides with a change from flowing to standing water (i.e., Stratum 3B alluvium to Stratum 4d diatomite). This change is best expressed towards the end of the diatomite deposition of lower Stratum 4. The seasonal marsh deposits (Biofacies 3), capping the diatomite and represented by the weakly bedded diatomaceous earth evolved from ~11,125 cal yr B.P. to ~10,440 cal yr B.P. when they shifted to a more stable wetland (Biofacies 4) that continued to aggrade until at least ~9990 cal yr B.P.

In semi-arid and arid regions such as El Fin del Mundo in the northwestern portion of Sonora, Mexico, where moisture balance (P-E) is negative, ponds such as that at Locus 1 usually tend to be closed basins. Diatoms show distributional patterns based both on salinity and brine composition (Fritz and others 1999). The hydrochemical evolution of any lake at any time plays a major role on its biological composition, diatoms included. For example, at El Fin del Mundo, freshwater pennate species such as *Achnanthidium minutissimum*, *A. minutissimum* var. *scotica*, *Eunotia* sp. cf. *E. adnata*, *Fragilaria tenera*, *Pinnularia borealis*, and *Pinnularia* sp. cf. *P. saprophila*, along with the Centric forms *Aulacoseira islandica* and *A. italica* increased in abundance at intervals of increasing freshwater input throughout the record. Even though abundant, these forms were relegated to a second level as the more salinity tolerant species of Assemblage II and later those of Assemblage III characterized the wetland. Assemblage IV remained reduced during the history of the wetland; however, it increases in Biofacies 3, especially towards the end of the record, consistent with episodes of subaerial exposure leading to wet meadow conditions (Chapter 2).

CONCLUSIONS

The paleoenvironmental investigations in El Fin del Mundo indicate that from ≤13,415 to ~9990 cal yr B.P., a local spring-fed depression formed a pond and, eventually, a wetland that, during Biofacies 3, was exposed periodically to form a wet meadow. The four assemblages identified at El Fin del Mundo show that an alkaline system formed at the transition from alluvial (Stratum 3B) to lacustrine (lower Stratum 4) deposition dominated by Assemblage II. This transition indicates a decline in water flow (stratum 3B) to standing water (lower Stratum 4 diatomite) to seasonal ponding (upper Stratum 4 diatomaceous earth) to a wetland (Stratum 5). Mainly fresh- to brackish water species thrived in the system. Freshwater benthic and a few tychoplanktic species co-existed or replaced each other, however, as environmental conditions shifted in response to less water flow.

Based on the autecologies of the most dominant diatom species in the site, deposition of Stratum 4 diatomaceous earth was likely characterized by a shallow alkaline marsh with low nutrient conditions, in particular, low nitrogen. Considering the high dominance of benthic diatoms, the low overall density of diatoms, the presence of chrysophyte cysts, and the high amount of phytoliths, the conditions of the site, at this time, may have been characterized as a shallow marsh periodically subjected to desiccation or to the formation of a wet meadow.

Ostracodes

Jordon Bright

In addition to diatoms, samples from El Fin del Mundo were analyzed for ostracodes. Ostracodes are crustaceans with calcite carapaces. In the geologic record only the carapaces of ostracodes remain but they are key to identifying the original living forms. These remains are important indicators for environmental reconstructions in paleolimnology (Cohen 2003). The samples were collected from Stratum 4, section 08-1 (Locus 1; Figure 4.3), the same column sampled for pollen (Chapter 7), phytoliths (Chapter 8), and diatoms (Chapter 9). The content and types of ostracodes were determined in the lab based on a sample size of roughly 5 gm of sediment per sample.

Ostracodes are present but not common and found only in Stratum 5 (Tables 2.1, 10.1). This distribution of ostracodes is due to the calcium carbonate make-up of the carapaces. They are unlikely to preserve in the noncalcareous depositional environment of Strata 4d or 4de.

Cypridopsis vidua and *C. okeechoebi* (not imaged) occurred as both adults and juvenile molts. They are very common ostracodes, preferring shallow, vegetated water. They live in the littoral zone of lakes but also are very common in and around springs with *C. okeechoebi* being more abundant with increasing proximity to the spring orifice. They are exceptionally desiccation resistant. One encrusted valve of *Strandesia meadensis* was present. They also prefer zones of spring discharge. Two Darwinulid ostracodes (*Microdarwinula* and possibly *Vestalenula*) could not be speciated. Nevertheless, they are typically subterranean/semi-terrestrial/spring genera.

A few gastropods were preserved in Stratum 5: a Pupillid (possibly *Pupilla* proper or *Vertigo*), probably *Discus*, and a possible Succinid (Table 10.1). They are all terrestrial to semi-terrestrial snails.

A variety of indicators suggest that the depositional environment of Stratum 5 was not particularly wet. This interpretation is based on the rarity of ostracodes, the absence of normal (and very common) surficial and standing water ostracodes (other than the two Cypridopsids), the groundwater/spring/semi-terrestrial nature of the small Darwinulids, and the terrestrial snails. This interpretation is somewhat at odds with the field observations of the lithology (Chapter 2) and the data from micromorphology (Chapter 3), phytoliths (Chapter 8), and diatoms (Chapter 9).

The out-of-sequence dating of the Succinids from Stratum 4, along with the broken character of the shell suggest that they were redeposited (Chapter 2). The intact ostracode valves, including juveniles, however, argue against redeposition of those specimens.

Table 10.1. Ostracodes from Profile 08-1

Sample, mbd	*Stratum*	*Gastropods*	*Ostracodes*
10.98–11.03	5	3 Pupillid 3 squat, coiled 2 disks	2 *Cypridopsis vidua*
11.03–11.08	5	7 Pupillid 7 squat, coiled 1 frag of a left-coiling snail	3 *Microdarwinula* sp. 1 *Cypridopsis vidua* juvenile fragment of *Cypridopsis vidua* valve
11.08–11.13	5	2 Pupillid 2 squat, coiled	3 *Vestalenula*-like Darwinulids 2 *Cypridopsis vidua* 1 encrusted valve of *Strandesia meadensis* Juveniles of *Cypridopsis vidua*
11.13–11.18	5	4 squat, coiled 3 disks 1 Succinid-type snail 1 Pupillid	6 *Microdarwinula* sp. 1 *Cypridopsis vidua*
11.18–11.23	5	6 disks 1 squat coiled	5 *Cypridopsis okeechobei*
11.23–11.28	5	2 Pupillid 1 squat coiled	2 *Cypridopsis vidua* 2 *Cypridopsis okeechobei* 1 small Candonid valve
11.28–11.33	5	3 disks 1 squat, coiled 1 Pupillid	1 articulated small Candonid 1 juvenile valve of *Cypridopsis vidua*
11.33–11.38	5	2 disks 1 squat coiled Fragments of Pupillid	1 articulated small Candonid 1 articulated juvenile of *Cypridopsis vidua*
11.38–11.43	4de	2 fragments of snail shell	None
11.43–11.48	4de	None	None
11.48–11.53	4de	None	None
11.53–11.58	4de	None	None
11.58–11.63	4de	3 fragments of snail shell	None
11.63–11.68	4de	1 fragment of snail shell	None
11.68–11.73	4de	None	None
11.73–11.78	4de	None	None
11.78–11.83	4d	None	None
11.83–11.85	4d	None	None

Note: Most of the gastropods are represented by fragments.

Key: mbd = meters below datum.

CHAPTER ELEVEN

Discussion and Conclusions

Vance T. Holliday, Guadalupe Sánchez, Ismael Sánchez-Morales,
and Joaquín Arroyo-Cabrales

THE SITE

El Fin del Mundo is a Clovis site represented by Clovis artifacts directly associated with the remains of two gomphotheres (*Cuvieronius* sp.) in a buried context along with an extensive surface component of Clovis artifacts on uplands in an arc 200 to 1,000 meters to the southeast, south, and southwest of the buried remains. The upland artifact assemblage also contains an Archaic component along with one possible Dalton/Golondrina point. The Clovis bonebed is the upper of two bonebeds discovered at the site. The lower one is paleontological and includes redeposited, fragmented remains of gomphothere (*Cuvieronius* sp.), mastodon (*Mammut americanum*), and mammoth (*Mammuthus* sp.). Both bonebeds were associated with alluvium from a small, perennial, spring-fed stream that flowed down a channel incised into an older alluvial fan. The lower bonebed is in coarse channel alluvium. The Clovis bonebed was preserved in fine-grained (i.e., very low energy) deposits representing the final stages of alluviation. Most of the stratigraphic record with the two bonebeds and younger deposits was removed by arroyo cutting, but the remaining sequence indicates that Clovis hunters encountered two younger gomphotheres (male and female aged ~2 to 8 and ~8 to 19 years old in African elephant equivalent years, respectively) on a flat alluvial plain.

El Fin del Mundo (FDM) is unusual among Paleoindian sites in North America. It provides the first evidence for the association of Clovis hunters with gomphotheres (*Cuvieronius* sp.), the first archaeological context for gomphothere in North America, and the northernmost post-LGM gomphothere in North America. It contains a rare association between a Clovis-megafauna feature and a campsite; it is the first excavated in situ Clovis site in northwestern Mexico; and it is the first excavated Clovis site outside of the United States. Further, the association of Terminal Pleistocene mastodon, mammoth, and gomphothere in the lower (paleontological) bonebed is rare in the greater Southwest and across North America. The site also yielded a stratigraphic record spanning the terminal Pleistocene-Holocene boundary.

Clovis is classically associated with late Pleistocene megafauna, particularly the proboscideans mammoth and mastodon, along with bison and camel (Grayson and Meltzer 2015). The research results presented here add a new species of proboscidean to the array of animals utilized by Clovis foragers. Evidence for hunting of *Cuvieronius* (as opposed to scavenging) is based on the presence of four presumed projectiles (Clovis points) among the bone concentrations, including one that likely snapped while hafted (#62942; see Chapters 4 and 5). Lithic debitage (N=23) consisting largely of microflakes from retouch made of high-quality materials were also found in and around the gomphothere remains (Chapter 4). Some argue (Eren and others 2021) that Clovis bifaces were not suited to be ballistic weapons but were more likely to be hafted knives. That proposal generated some debate (Kilby and others 2022; Eren and others 2022), but the debate seems to be one of degrees. Were Clovis bifaces designed and used primarily as ballistic projectiles or as knives? The incomplete bone and stone assemblage and the weathering of the remaining bone from FDM offer little evidence beyond the association of Clovis bifaces and gomphotheres. Nevertheless, the likelihood of two juvenile gomphotheres dying together at essentially the same time less than 4 meters apart on the same surface and then scavenged by Clovis foragers whose tools ended up in or near both bone piles seems remote.

Prolonged or repeated use of the site is indicated by the high numbers of diagnostic Clovis lithic materials indicative of aggregations of camps on the stable uplands southeast, south, and southwest of the gomphothere kill in Locus 1 (Chapter 5). Eleven whole and fragmentary Clovis points, 10 Clovis point preforms, 24 fragmentary secondary bifaces, 33 end scrapers, 52 blades, and 12 byproducts of blade manufacture were recovered from among an extensive surface lithic scatter. The diagnostic Clovis lithic artifacts are clearly differentiated from the Archaic lithic component and the possible Dalton point at the site based on technological and typological features diagnostic of Clovis lithic technology (Chapter 5). Close contemporaneity of the kill and the upland Clovis occupation could not be clearly demonstrated (e.g., via artifact refits). However, the same unusual chert varieties among the lithic assemblages in both settings, and presence of Clovis artifacts in both demonstrate at least a general contemporaneity.

The Clovis lithic assemblage from FDM includes bifaces, unifacial implements, and blades, similar to those seen in other Clovis sites in the region (e.g., El Bajío and Murray Springs; Chapter 5). The surface nature of most of the collection hinders the identification of additional lithic classes. The identified tool types and their contexts indicate that Clovis foragers at FDM manufactured bifaces and blades and utilized and discarded exhausted and damaged implements related to diverse tasks at the campsites located on the uplands.

The Sonoran sites of El Fin del Mundo and El Bajío, located ~110 km apart, are also distinguished from other Paleoindian sites in the region by the presence of good to high quality lithic raw material sources that were extensively exploited by Clovis foragers. This characteristic of the two Sonoran sites resulted in local tool-stones dominating the lithic collections, in particular at El Bajío. Nevertheless, high reliance on exogenous cherts is well documented at FDM, where most Clovis points and end scrapers were made on these raw materials.

Recovery of one possible Dalton/Golondrina point from the upland surface assemblage is significant for several reasons. Similar artifacts are reported from other sites in the region (Gaines, Sánchez, and Holliday 2009; Sánchez 2010, 2016; Sánchez and Carpenter 2012). For example, El Gramal (SON N:11:20-21) includes both Clovis and unfluted lanceolate styles similar to Dalton/Golondrina (Sánchez 2016). The Dalton style is common across the southeastern United States, from eastern Texas eastward, while Golondrina is common in south Texas and northeastern Mexico, with some overlap of the styles in eastern and southeastern Texas (e.g., Johnson 1989; Bousman, Baker, and Kerr 2004; Bousman and Oksanen 2012; Jennings, Smallwood, and Greer 2016; Hester 2015, 2017; papers in Miller, Smallwood, and Tune 2022). Dalton in its "heartland" (southern Missouri and Arkansas) appeared by at least ~12,500 cal yr B.P. and persisted into the early Holocene <11,500 cal yr B.P.) while its eastern variants overlapped and persisted later than "heartland Dalton" (Thulman 2022). Golondrina is an artifact style originally linked to Plainview (Johnson 1964), but subsequently differentiated as a younger and distinct artifact style not tied to Plainview (e.g., Kelly 1982; Holliday 2000a; Bousman, Baker, and Kerr 2004; Bousman and Oksanen 2012; Hester 2017; Jennings, Smallwood, and Waters 2015). Statistical analyses of artifact metrics indicate that Dalton and Golondrina are stylistically linked (Jennings, Smallwood, and Waters 2015). Golondrina artifacts are not well dated but appear to be early Holocene (Holliday 2000a; Bousman, Baker, and Kerr 2004; Jennings, Smallwood, and Greer 2016). Some argue for Golondrina representing a late expansion of the general Dalton style across southern Texas into neighboring states of Mexico (Johnson 1989; Hester 2015).

Presence of a possible Dalton/Golondrina component would suggest that the site area was attractive long after the Clovis occupation but for similar reasons. The pond and wetland system (Strata 4 and 5) was present and evolved from >11,130 to ~10,000 cal yr B.P. and the seeps represented by upper Stratum C were active into the early Holocene. The diatomaceous earth discovered in Locus 19 suggests local presence of a wetland in the middle Holocene.

Our investigations at FDM also resulted in recovery of more than 120 projectile points diagnostic of the Archaic and the Early Agricultural periods. They were collected from surface contexts during the systematic surveys of the upland loci of the site (Allaun D'Lopez 2019). The typology of this assemblage indicates that groups of hunter-gatherers continued to occupy the site through the middle and later Holocene. An environmental record is not preserved for that time, but the archaeological record suggests that water continued to be available. However, the striking contrast in raw materials observed between the Clovis lithics and the Archaic points, which are highly dominated by milky white quartz, suggest different patterns of landscape use (Sánchez-Morales 2012).

DATING CLOVIS

The age of the bonebed was initially proposed to be ~13,415 cal yr B.P. (~11,550 ^{14}C yr B.P.; AA-100181A), based on dating of fragmented and charred plant remains recovered from Stratum 3B in association with the feature in Locus 1 (Sánchez and others 2014). Subsequent dating of charred plant fragments and Succinids from the alluvium along with reanalysis of other dated samples (e.g., Table 2.4) indicates the likelihood that both shell and plant material were redeposited. The date of ~13,415 cal yr B.P. is the youngest of the group but is not necessarily the same age as the Clovis-gomphothere bonebed. It is a maximum limiting date for the feature; that is, the bonebed is ≤~13,415 cal yr B.P. The date is not unique for a Clovis occupation. The Aubrey Clovis site in north Texas yielded two similar dates: ~13,400 and ~13,450 cal yr B.P. (11,540 +/- 110 and 11,590 +/- 90 ^{14}C yr B.P., respectively; Ferring 2001, 2012). The Clovis occupation at the Gault/Friedkin complex in central Texas may also be about the same age, but the dating has far less precision because it is based on OSL. The Clovis level at Gault dates to 12,900 ± 700 years (Rodrigues and others 2016). The "Folsom-Clovis Zone" (10cm thick) at Friedkin is 13,590 ± 720 to 11,980 ± 490 years (Waters and others 2018:6). Because of the low precision of the dating, the age ranges for Clovis and Folsom at Gault/Friedkin span later pre-Clovis, all of Clovis, and most or all of Folsom time (following Surovell and others 2016; Waters, Stafford, and Carlson 2020; and Buchanan and others 2022).

In contrast, Waters, Stafford, and Carlson (2020) use radiocarbon dating on samples (most are purified amino acids from bone) from 10 Clovis sites.[7] They propose an age range of 11,110 ± 40 to 10,820 ± 10 ^{14}C yr B.P. providing a maximum calibrated (cal) age range for Clovis of ~13,050 to ~12,750 cal yr B.P. across North America. They provide a reasonable age estimate, but likely not the full range of Clovis. Surovell and others (2016: 85) noted that a small sample of dates is unlikely to represent the entirety of the age range of a Paleoindian techno-complex, with reference to using 12 dates to establish the age range of Folsom sites over a smaller (subcontinental scale) region. Similarly, Prasciunas and Surovell (2015:33) examined the duration of Clovis and the error margin introduced by sample bias through statistical analyses and argue that "although Waters and Stafford (2007) [referring to their first iteration of the age range of Clovis] provide very precise dates for some Clovis sites, this does not mean they have accurately defined the age range of Clovis. With a sample of only 11 sites, an observed age range between 200 and 450 calendar years can be expected even when the duration of the colonization event took as long as 1,500 years."

The issue of sampling is apparent in the geographic distribution of the sites that produced the dates reported by Waters, Stafford, and Carlson (2020). Six sites cluster in the Northern Plains/Rockies, three sites are on the Southern Plains, and three sites are in in the Midwest/Middle Atlantic region. This is not a large nor geographically representative sample given the purported coast-to-coast distribution of Clovis across North America south of the retreating ice sheets (though perhaps exaggerated as such). The dating highlights the age of a small number of sites in three limited areas with a high degree of preservation. Moreover, the highest density of Clovis finds, essentially east of the Mississippi River (based on Waters, Stafford, and Carlson 2020:Figure 1; following the Paleoindian Database of the Americas, http://pidba.utk.edu/), produced no reliable Clovis dates according to the authors. The dated sites used by Waters, Stafford, and Carlson (2020) essentially fall around the fringe of this high-Clovis-density area. A testable hypothesis can be proposed that the dated sites represent late movement of Clovis foragers out of a "Clovis core." For example, the Clovis-like Northeast fluted points in New England and along the St Lawrence Valley are younger than "classic" Clovis, suggesting a later movement of the Clovis style out of the "core" (Miller, Holliday, and Bright 2013; Smith, Smallwood, and DeWitt 2015). The limited number of Clovis dates from a limited geographic area strongly suggests that the age range proposed by Waters, Stafford, and Carlson (2020) is a minimum for the duration of Clovis in North America. Unfortunately, the chronology at FDM does not have the accuracy to inform on this debate.

CLOVIS AND LATER PALEOINDIAN LAND USE

El Fin del Mundo is unusual among Paleoindian sites in the greater Southwest and arid western North America. It is a Clovis site near the toe of an alluvial fan with an upland surface component on the fan and a buried component within fill of a local channel incised in the fan. The buried component was preserved in a small remnant of channel fill exposed by extensive arroyo incision. It is a

[7] Anzick (human and elk bone); Cactus Hill (charcoal); Colby (mammoth bone); Dent (mammoth bone); Domebo (mammoth bone); Jake Bluff (bison tooth and bone); Lange-Ferguson (mammoth bone); La Prele (mammoth bone); Shawnee-Minisink (carbonized seeds); and Sheridan Cave (bone).

small remnant of what was likely a much larger site. There are only a few comparable sites in the greater Southwest.

Clovis land-use patterns in north-central Sonora are consistent with those documented for the upper San Pedro River Valley of southeastern Arizona, ~250km to the northeast of El Fin del Mundo. In the San Pedro Valley, wide-ranging and highly mobile foragers positioned themselves on well-drained uplands near wetland hunting locations where large game were available (Haynes and Huckell 2007). The result was the preservation and discovery of six mammoth sites, at least four of which are confirmed Clovis sites (Haynes and Huckell 2007; Ballenger 2015; Grayson and Meltzer, 2015; Holliday, Haynes, and Huckell in press). Further, the Murray Springs site exhibits an upland Clovis camp adjacent to megafauna kills (Haynes and Huckell, 2007). Ferring (2001) compared the spatial dimension of occupations at the Aubrey site, Texas, with Murray Springs and several other fluted point sites, including surface sites such as the Debert site in Nova Scotia, the Fisher site in Ontario, and the Vail and Michaud sites in Maine. In most of them, the spatial scale was on the order of no more than a few hundred meters. The extensive dispersion of artifacts over several localities at FDM suggests that human activities were scattered over a larger area. Much of the site is missing due to extensive erosion, however, which undoubtedly affects the Clovis occupation record of the upland surface as well as that buried in Locus 1. Furthermore, between the bonebed in Locus 1 and the main area of the Clovis camp to the southwest, much of the surface has a veneer of carbonate that could be part of the early Holocene facies of Stratum C. Archaeological evidence closer to Locus 1 may be buried in this area, but our investigations have not yielded buried archaeological contexts so far.

Alluvial fans, both active and inactive, are obvious settings for foragers in the Basin and Range settings of the U.S. Southwest and northwest Mexico. During the Terminal Pleistocene, the basins were crossed by seasonal and perennial drainages. The alluvial fans provided viewsheds of the valleys and access to lakes, wetlands, and water courses on the valley floors. Depending on the bedrock lithologies in adjacent mountain ranges, the fans contained raw material suitable for tool manufacture. The fans also allowed access to suitable bedrock exposures in the mountains along with access to game and fresh water.

A few other Clovis sites are reported from the surface of alluvial fans or along drainages in the U.S. Southwest (e.g., Wessel and others, 1997; Holliday, Harvey, Cuba, and Weber 2019; Holliday, Condon, Cuba, Fenerty, and Bustos, in press). The most extensive and best documented is the Mockingbird Gap site in the Jornada del Muerto of central New Mexico (Holliday and others 2009; Hamilton and others 2013). It is a large Clovis surface site adjacent to what was once a deep (>25m) drainage with Clovis-age wetlands or flowing water. Depth of burial precludes discovery of a buried kill or other occupation in the drainage, but the possibility of such a feature seems high considering the extent of the surface occupation.

Folsom and later Paleoindian sites are very common on alluvial fans and associated with adjacent drainages (e.g., Wessel and others 1997; Holliday, 2015; Holliday, Condon, Cuba, Fenerty, and Bustos, in press). In terms of spatial scale and similar geomorphic and stratigraphic context, the Water Canyon site in west-central New Mexico has striking similarities to El Fin del Mundo. It is on and inset into a large alluvial fan/bajada system at least 25,000 years old based on degree of soil development in the fan deposits (Machette 1988). The site contains multiple bison-kill/processing features buried in wetland deposits inset into the toe of the fan (Holliday Dello-Russo, and Mentzer 2020). Dark gray muds with a Cody bonebed and younger bison bone processing features buried a small remnant of a Clovis-age wetland that does not contain archaeological remains. The wetland deposits rest on alluvium, but whether fan or channel alluvium was not determined. The origins of the ancient wetlands are also unclear. There is no obvious evidence for spring activity. A higher water table could have been a driver. The site area drains into Water Canyon Creek, less than 500 m from the site. A perennial stream could have raised the water table, or alluvium could have dammed the small site drainage. The wetland deposits were buried by younger, localized fan sediments derived from the channel that fed the wetland. Two fragments of late Paleoindian Cody points were found on the modern surface nearby, including a small, low terrace with a lithic scatter immediately adjacent to the buried wetland (Dello-Russo 2010). A Clovis point was found on the surface of the older, larger alluvial fan that forms the regional landscape. The Water Canyon site was largely preserved and discovered only due to a small arroyo exposure.

Both Water Canyon and El Fin del Mundo are unusual and significant in that they are the only documented Paleoindian sites buried in deposits inset into larger alluvial fans. Settings with flowing, impounded, or emergent water or some combination will be an attraction to people and animals and have some potential for burial. These settings provide additional insights into how early foragers used the landscape and where their sites may be preserved.

AMERICAN GOMPHOTHERES AND *CUVIERONIUS*

Remains of gomphotheres (Proboscidea, Gomphotheiidae) are commonly found in Upper Cenozoic deposits throughout the Americas, with *Cuvieronius* the only genus found in North, South, and Central America (Smith and DeSantis 2020:42). However, their abundance, diversity, and geographic range in North America rapidly dropped off after the appearance of mammoths. In fact, gomphotheres "are rarely represented in Rancholabrean faunal assemblages" (Smith and DeSantis 2020:41; see also Pérez-Crespo and others 2020).

Cuvieronius is known from the southern United States (Lucas and others 1999; Pasenko and Lucas, 2011; Smith and DeSantis 2020), throughout Mexico, and across Central and South America (Dudley 1996; Lambert 1996; Sanders 2002; Montellano-Ballesteros 2002; Corona-M. and Alberdi 2006; Pérez-Crespo, Arroyo-Cabrales, and Corona-M, and others 2015; Pérez-Crespo, Carbot-Chanona, Morales-Puente, and others 2015; Alberdi and Corona-M 2005; Alberdi and Prado 2022; Arroyo-Cabrales and others 2007; Mead, Arroyo-Cabrales, and Swift 2019). Besides FDM, numerical age control for this genus in North America is available only from a site in northeast Sonora (~43-40k cal yr B.P.; Mead and others 2006; Mead, Arroyo-Cabrales, and Swift 2019; Bright, and others 2010) and in southeast Texas (probably ~25k cal yr B.P.; Lundelius and others 2019). A tooth from a possible Late Pleistocene *Cuvieronius* is reported from the West Palm Beach site, Florida (Converse 1973); it has been radiocarbon dated to ~25,000 cal yr B.P. (Koch, Hoppe, and Webb 1998). The tooth is gomphothere but unequivocal identification as *Cuvieronius* is not possible (R. Hulbert, personal communication, August 2021). In addition, the method of dating the tooth is not specified (Buckley and Willis 1972) and thus cannot be accepted. In western Mexico a gomphothere *Stegomastodon* is dated to ~27k (Alberdi and others 2009). An association of stone tools with gomphotheres is reported from Valsequillo (Puebla, Mexico), but the association is not confirmed (Ochoa-Castillo and others 2003). In South America, however, gomphotheres in archaeological contexts are well-documented (Prado and others 2005; Prado and others 2015), though only a few are *Cuvieronius* (Jackson, Méndez, and de Souza 2004; Prado and others 2015).

The association of Clovis artifacts with *Cuvieronius* at FDM provides the youngest reliable age estimate for the genus and for gomphotheres in general in North America, indicating that they, too, were part of both the Rancholabrean Land Mammal assemblage and the late Pleistocene fauna that became extinct across the continent around or before the beginning of the Younger Dryas Chronozone. Grayson and Meltzer (2015:Table 6) further recognize "[A]rchaeological sites with evidence suggesting human predation on now-extinct Pleistocene genera" at 12 mammoth (*Mammuthus*) sites, 2 mastodon (*Mammut*) sites, 1 camel (*Camelops*) site, 1 horse (*Equus*) site, and 1 gomphothere (*Cuvieronius*) site. The *Cuvieronius* from FDM thus joins the small set of extinct Late Pleistocene fauna that are well documented as being associated with human activity in North America.

PALEOENVIRONMENTAL SETTING

A variety of indicators provide clues to the paleoenvironmental evolution of El Fin del Mundo and vicinity beginning in the LGM. Stratigraphy (Chapter 2) and micromorphology (Chapter 3) inform on depositional environments, site formation processes, and landscape evolution across the site. The megafauna at the base of Stratum 3A and in upper 3B (Chapter 6) provide snapshots for their environments. Several microbiological indicators derived from samples collected continuously through the Locus 1 stratigraphic sequence provide a higher resolution record of environmental change. Pollen is available from Strata 2 through 4de (Chapter 7); phytoliths were derived from Strata 4 and 5 (Chapter 8); diatoms are from upper 3B through 5 (Chapter 9); and ostracodes were recovered from Stratum 5 (Chapter 10).

The proxy indicators of past environments preserved and exposed at FDM provide an opportunity to reconstruct the local and possibly the regional environment during and following the Clovis ocupation of the site. This is a critical issue because of the debates surrounding the character of Clovis and later Paleoindian environments in the greater Southwest and, more broadly, the environmental characteristics before, during, and after the Younger Dryas Chronozone (e.g., Eren 2008; Haynes 1991, 2008; Meltzer and Holliday 2010; Strauss and Goebel 2011; Palacios-Fest and Holliday 2017). El Fin del Mundo also initially appeared to offer an opportunity to sample across the Terminal Pleistocene-early Holocene boundary, but the geochronological resolution of the stratigraphy did not allow differentiation of late Bølling-Allerød, Younger Dryas, and earliest Holocene time intervals.

Initial phases of landscape evolution in the immediate area of FDM are represented by deposition of Strata 1, 2, and C. The sedimentology and microstratigraphy of Strata 1 and 2 indicate deposition by stream flow. Both are part of the alluvial fan that comprises the regional landscape. Stratum C is a carbonate (marl) resting on top of the fan uplands of either Strata 1 or 2 near Locus 1, deposited by seeps or springs.

Dates for fan activity range from ~41,095 cal yr B.P. in Stratum 1 to ~25,850 cal yr B.P. in Stratum 2, with carbonate lenses forming during Stratum 1 deposition ~30,000 cal yr B.P. These ages place the final stages of fan formation in Marine Isotope Stage (MIS) 3 and the beginning of MIS 2. Carbonate of Stratum C began forming on the fan surface near Locus 1 by ~19,315 cal yr B.P. Gastropods and micromorphology suggest deposition of the carbonate via seeps or springs. Deposition of Lower Stratum C on top of a well-expressed Big Red soil (Locus 3) that formed Stratum 1 (dated ~41,095 cal yr B.P. in Locus 4) is indicative of prolonged fan stability. Stratum 2 was inset into Stratum 1 around Loci 1 and 2, probably in cycles ~31.9k to ~25.9k cal yr B.P. The well-expressed facies of Big Red in Locus 2 suggests prolonged stability followed by burial under Stratum C, probably the early Holocene facies.

Following deposition of Stratum 2, erosion incised a small channel in the toe of the fan through the area now identified as Locus 1. The incision was followed by a phase of highly localized deposition represented by Strata 3A, 3B, 4, and 5. Stratum 3A is somewhat coarser with more obvious cut-and-fill sequences, indicative of somewhat more energetic flow compared to the overlying 3B. Stratum 3B is indicative of slow aggradation and therefore less energetic flow. Upper Stratum 3B with the Clovis bonebed may reflect a perennial stream with locally common standing water.

Pollen from upper Stratum 2 (~25,850 cal yr B.P.; i.e., roughly at the onset of the Last Glacial Maximum, LGM) indicates an open *Pinus* forest in the site area along with high levels of nonarboreal pollen. Vegetation diversity was high at this time, and likely characterized by colder and more humid conditions relative to the modern climate. Stratum 3A (~15,410->13,415 cal yr B.P.) is devoid of arboreal pollen. *Pinus* and other arboreal pollen along with some grasses are common in some intervals of Stratum 3B, including *Pinus* (at 11.93-11.98 mbd) with the Clovis occupation, but not above that zone in 3B nor in 4d. Deposition of 3B (≤13,415 cal yr B.P.) on and above the Clovis occupation zone and onset of 4d deposition represented a brief time interval, however, raising the likelihood that the *Pinus* at 11.93 to 11.98 mbd was from below the occupation and redeposited into it. The pollen in 3B is indicative of a steppe-like vegetation community with a mix of trees and grassland. This plant community apparently changed to a grassland at about the time of the Clovis occupation or earlier. This vegetation change was also at least roughly coincident with the declining energy of stream flow. Overall, the pollen from upper 3B indicates a mixed steppe-grassland evolving to warmer, drier grasslands ≤13,415 cal yr B.P. Hydrologically, diatoms from uppermost Stratum 3B suggest that the deposits were incorporated into the initial lake phase that came to be dominated by formation of the diatomite.

The vertebrate fauna from the Lower Bonebed in Stratum 3A provides an interpretive conundrum, however. The recovery of mastodon and gomphothere remains along with mammoth from other areas of the site is anomalous (Chapter 6). Few other sites in Mexico contain this association. The bone is redeposited, but from unknown stratigraphic contexts. The Lower Bonebed could represent a mix of bone of a variety of ages, but the bone is similar in degree of preservation, suggesting an origin in a similar stratigraphic context if not the same stratigraphic context and same age. As noted in Chapter 6, the Lower Bonebed fauna includes animals such as horse (*Equus* sp.), mammoth (*Mammuthus* sp.), and giant ground sloth (*Paramylodon harlani*) that are frequently associated with open habitats, as well as those found in closed habitats such as mastodon (*Mammut americanum*), and tapir(*Tapirus* sp.). The assemblage also includes fauna associated with a mix of the two habitat types including gomphothere (*Cuvieronius* sp.) and camel (Camelidae). Stable isotope analyses of teeth from the Lower Bonebed similarly indicate the past presence of both C_3 plants and C_4 grasses on the landscape, with gomphotheres consuming a mixed diet and mastodon and tapir consuming a diet strongly dominated by C_3 plants (Pérez-Crespo and others 2016; Pérez-Crespo end others 2020).

The diversity in isotopic composition reflected in the fauna appears to reflect the plant diversity indicated by the pollen assemblage (Chapter 7). Those data, moreover, may be indicative of plant and animal assemblages not previously recognized in the region. In any case, as noted in Chapter 6, paleoecological data from El Fin del Mundo indicate that local conditions allowed three varieties of proboscideans to co-exist for some time prior to the Clovis occupation but only *Cuvieronius* sp. persisted until Clovis times.

Strata 4 and 5 (≤13,415 to ~9900 cal yr B.P.), which span the final millennia of the Terminal Pleistocene and early Holocene, represent a dramatic shift in local environmental conditions from the flowing water of Stratum 3. A relatively brief period of standing water with no clastic input persisted in a small lake or pond, as indicated by the nearly pure layer of diatomite in Stratum 4. At ~11,130 cal yr B.P., the pond gave way to a seasonal, freshwater marsh (Stratum 4de), which evolved into a more alkaline wetland (Stratum 5) that persisted until at least ~10,000 cal yr B.P. The initial wetland phase produced diatomaceous earth (Stratum 4de), dominated by diatoms and phytoliths with organic matter from plants. The depositional environment changed ~10,440 cal yr B.P. to a wetland with little surface water, dominated by marl genesis with additional accumulation of diatomaceous earth and phytoliths (Stratum 5). The final phases of this wetland might have persisted as a stable (non-aggrading) wetland after ~10,000 cal yr B.P. with development of a dense carbonate cap across the surface of Locus 1. The seep-derived carbonate of Stratum C continued to accumulate on the uplands into the early Holocene (at least until ~8870 cal yr B.P.), likely a facies of Stratum 5.

Paleobiological indicators support and amplify the interpretations from the sedimentology and stratigraphy of Strata 3B and 4. Pollen indicates that from -≤13,415 to ~11,130 cal yr B.P. (i.e., upper 3B to top of 4d), a grassland persisted with increases of Asteraceae pollen, probably reflecting change to warmer conditions than previous periods. Diatoms (Chapter 9) also suggest a gradual increase in water alkalinity through this time. They also suggest that the source of water shifted from runoff supporting alluviation to groundwater (via springs and seeps) supporting the diatomite pond of lower Stratum 4. Phytoliths (Chapter 8) from upper Stratum 3B through lower Stratum 4 are indicative of standing water with stands of wetlands grasses. Diatoms, phytoliths, and ostracodes (Chapter 10) show that the standing water conditions of Stratum 4de evolved into a seasonal marsh or pond (Stratum 4d). These palustrine conditions supported by groundwater persisted in the early Holocene from <11,130 to ~10,000 cal yr B.P. and likely beyond. But during that time there was a change in water geochemistry from conditions favoring formation of noncalcareous diatomaceous earth (in Stratum 4de) to those favoring precipitation of calcium carbonate in a diatomaceous context (Stratum 5). This is further evidence for progressive drying of drainages and wetlands in the early Holocene. Exactly how wet the depositional environment was as Stratum 5 aggraded is unclear, based on somewhat contrasting interpretations from phytoliths and diatoms versus those from ostracodes. But the different interpretations may be more of degree (i.e., very wet vs somewhat wet).

Absence of evidence for weathering as Stratum C accumulated (~20,000 to ~8900 cal yr B.P.) suggests continued flow of water, almost certainly feeding the Stratum 3 stream and Strata 4 and 5 pond and wetlands. The end of deposition of the carbonate in both Strata 5 and C in the early Holocene and post-depositional weathering in Stratum C across the site are further evidence of increasing Holocene aridity. Also, on the uplands on top of the Pleistocene fan alluvium is Stratum S, a thin, localized sandy sheetwash deposit likely derived from erosion of the fan surface. The absence of evidence for weathering in Stratum S indicates that it is a late Holocene deposit.

LGM and Terminal Pleistocene paleo-hydrological and paleovegetation records for northern Sonora and neighboring Chihuahua, New Mexico, and Arizona are from a variety of contexts, none directly comparable with El Fin del Mundo, particularly for the LGM and earliest post-LGM. Paleolake Babícora in western Chihuahua ~370km east of El Fin del Mundo records a lake deeper than today but declining during the LGM, with increased runoff 28 to 29k, 25k, and 21 to 19k cal yr B.P. This contrasts with increased lake salinity and eolian input 24k, 18k, 15k and 12k cal yr B.P. (Roy and others 2013). Pollen evidence shows *Pinus* and *Picea* dominating, indicative of overall colder and drier conditions (Melcalfe and others. 2002), but perhaps increased effective moisture. That record roughly coincides with indicators of drier conditions reflected in Strata 2 and 3A. However, that lake basin is on the eastern flank of the Sierra Madre Occidental at 2138m, a significantly different landscape position from FDM. Records of runoff from steeper mountain drainages at the paleolake may not be appropriate for comparisons.

Other records in the region are largely from packrat middens and thus from discontinuous records at higher elevations. *Pinus*, *Juniperus*, and *Quercus* dominate LGM and early post-LGM paleovegetation records at elevations between 600 and 1,675 m above sea level in the region (Van Devender 1990a, 1990b; Thompson and others 1993; Holmgren and others 2003; Holmgren, Betancourt, and Rylander 2006; Holmgren, Norris, and Betancourt 2007). In contrast, *Pinus* is discontinuously present during that time at FDM. It is absent in Stratum 3A, which is likely early post-LGM. The site is at a lower elevation (~630m) than most of the other studies, however (at 600 to 1675m above sea level), which could account for an earlier shift to more xeric plant communities. Further, only one pollen sample dating to the earliest LGM was available. It was in upper Stratum 2, just below a significant unconformity that separates Stratum 2 from Stratum 3. The amount of time missing from the record is unknown but likely represents some thousands of years.

A drying trend persisted across the region through the Terminal Pleistocene (Van Devender 1990a, 1990b), similar to the FDM record. In the Peloncillo Mountains of southeastern Arizona, Holmgren, Betancourt, and Rylander (2006) note a decline in *Pinus* after ~15,410 cal yr B.P., a brief disappearance ~13,925 cal yr B.P. as more xeric species expanded, a brief rebound of *Pinus* ~12,405 cal yr B.P. (during the YDC), and a final disappearance along with other mesic woodland species sometime after ~12,100 cal yr B.P. To the southeast, in the Playas Valley of southwestern New Mexico, pinyon-juniper was extirpated by the early YDC (Holmgren and others 2003). At the Lehner Clovis site in the lower elevations of the San Pedro Valley, Arizona, Mehringer and Haynes (1965) likewise saw a reduction in pine pollen and a spread of desert grassland through the YDC. Paleolake Babícora records high amplitude fluctuations in runoff through the post-LGM (Metcalfe and others 2002; Roy, Jonathan, Pérez-Cruz, and others 2012; Roy, Rivero-Navarette, and Lopez-Balbiaux, and others 2013), including cycles of reduced precipitation after ~14k cal yr B.P. and continuing through much of the Younger Dryas Chronozone and into the earliest Holocene.

Two lingering questions about the environmental evolution of FDM persist, but given the extensive erosion at the site, remain unanswerable. One question is what drove the incision into the fan sometime between ≤25,850 and ~15,410 cal yr B.P.? This age range includes the onset of Stratum C deposition via seeps or springs. Stratum C is unique to the toe of the fan and likely reflects water stored within the fan that subsequently emerged along the distal fan surface. LGM paleoenvironmental records noted previously suggest that there were some wetter cycles in the region. Overland flow of emergent water across the toe of the fan could have initiated erosion. Incision could also be initiated by internal geomorphic controls independent of regional environmental changes (Bull 1991). This is illustrated by extensive modern arroyo incision.

Another perplexing question is why the local depositional environment abruptly changed from alluvial (Stratum 3) to lacustrine (Stratum 4 diatomite). Some clues might be available from the remarkably similar geoarchaeological situations at the well-known Clovis (Blackwater Draw Locality 1) and Lubbock Lake archaeological sites (Cotter 1937, 1938; Sellards 1952; Haynes and Agogino 1966; Haynes 1975, 1995; Hester, Lundelius, and Fryxell 1972; Stafford 1981; Johnson 1987; Holliday 1985, 1997) along drainages on the Southern High Plains of northwest Texas and eastern New Mexico. Further, both sites are or were much more extensive and their broader stratigraphic contexts were traceable along hundreds of kilometers of their respective drainages (Holliday 1995), thus providing insights into the environmental drivers of the geologic records. The drainages with the two sites and other drainages across the region (Holliday 1995) all exhibit a change from alluvial conditions to lacustrine or palustrine environments from the terminal (post-LGM) Pleistocene into the early Holocene; identical to FDM. The evolution of this record is attributed to declining spring discharge or declining runoff or both. Stratigraphic and other environmental indicators support this interpretation of progressive drying from the terminal Pleistocene into the early Holocene on the Southern High Plains (Holliday 1995, 2000b; Holliday, Gustavson, and Hovorka 1996; Holliday, Mayer, and Fredlund 2008) and in similar settings in basins of the Southwest (Holliday and others 2006; Holliday and others 2009; Ballenger and others 2011; Palacios-Fest and Holliday 2017).

Drawing a direct comparison between the sites on the Southern High Plains and El Fin del Mundo is difficult given the erosion and very limited exposure at the latter, but a progressive decline in the water table is a logical conclusion to draw. That interpretation contrasts with some diatom evidence supporting a rising water table coincident with the transition from Stratum 3B to the diatomite of Stratum 4. A rising water table or increased runoff should maintain stream flow, yet the flowing water conditions of Stratum 3 gave way to the standing water conditions of Stratum 4. Some sort of blockage (local landslide or beaver activity) could have stopped or slowed stream flow, creating a lake, but continued flow into such a lake should have resulted in the eventual overtopping of any dam. Perhaps such an event initiated the widespread erosion that dominates the site area. So far, we are unable to find evidence that will answer this question.

CONCLUSIONS

The interdisciplinary archaeological investigations at El Fin del Mundo are notable for several reasons. They began with a local landowner contacting specialists and inviting them to start a research project (and materially supporting it). It brought together researchers, including students and other volunteers, along with both private and government funding from two countries and multiple public and private organizations to conduct and conclude an interdisciplinary archaeological endeavor that resulted in a series of remarkable discoveries in archaeology, vertebrate paleontology, and paleobiology. It also resulted in subsequent research projects, cited throughout this volume and including the sections of this volume written by graduate students (Sánchez-Morales and Worthey).

Together, these investigations provided new and invaluable information about a period of human and environmental history that was previously minimally investigated and poorly understood in the region and have paved the path for new investigations focused on the first peoples of what would become the Sonoran Desert of Northwest Mexico. On a larger scale, FDM and the rest of the Pleistocene archaeological record of Sonora offer a remarkable contribution to the understanding of the earliest settlers of the continent. This research has expanded our knowledge of Clovis subsistence strategies and their interactions with the North American Pleistocene landscapes.

Dating Bone and Teeth from the Upper and Lower Bonebeds

Gregory Hodgins

The gomphothere and mastodon bones were poorly preserved. Samples of bone obtained from the Upper Bonebed in Locus 1 did not yield collagen, so attempts at direct radiocarbon dating of bone were abandoned. Further efforts focused on gomphothere teeth from the Upper Bonebed and mastodon teeth from the Lower Bonebed because they appeared better preserved.

A 3-cm long fragment of M1-right from the gomphothere mandible found in Bone Concentration 2 was split open along existing fracture lines that ran from a cusp in the molar crown down to the enamel-root margin. The fragment included adhering dentin. An attempt was made to extract collagen from the exposed tooth dentin, but this also failed to yield any protein.

A strategy was adopted to try to date the gomphothere using the enamel radiocarbon content, despite the caution expressed by others (Hedges, Lee-Thorp, and Tuross 1995). Dentin removal exposed the inner enamel surface of the tooth. Metcalfe (2011) indicates that the inner enamel surface can provide higher resolution, time-series stable-isotope data, suggesting better enamel preservation there. The inner enamel surface became the target for enamel radiocarbon dating.

Enamel was drilled from the inner surface down to a depth of approximately 0.5 mm. The drillings were divided into two portions, one from the cusp end of the fragment (distal) and the other from the enamel-root margin end of the fragment (proximal). Both samples were soaked in 2.5% NaOCl for 24 hours, followed by 0.1M acetic acid for 4 hours, to remove exogenous carbonates. The resulting powder was rinsed extensively with DI water and dried. Hydrolysis was carried out using anhydrous phosphoric acid at 70 C overnight. This protocol was developed for the treatment of tooth enamel prior to stable isotope measurement (Koch, Tuross, and Fogel 1997). Graphitization, measurement, and calibrations were conducted by the AMS laboratory using established protocols.

The distal inner enamel surface generated 194.8 mgs of powdered enamel and 0.95 mg of carbon for dating (0.8%C). The proximal enamel surface yielded 223.1 mg of powdered enamel and 1.28 mg of carbon (0.87%C), of which 0.74 mgs was used for dating (Table A.1).

Both of the resulting dates are similar, suggesting that the enamel radiocarbon content was homogeneous throughout the tooth enamel at this crude spatial resolution. Both dates appear to be younger than expected based upon the association of the bone and teeth with Clovis points and the charcoal dates previously discussed. The results suggest that the inner enamel surface is not exempt from diagenetic uptake of younger carbon. A systematic pattern of younger enamel radiocarbon dates versus collagen radiocarbon dates from the same tooth has been documented by others (Hedges, Lee-Thorp, and Tuross 1995; Surovell 2000).

The sample preparation and dating of tooth enamel from a mastodon M1 molar found in the Lower Bonebed tooth used methods identical to those outlined for the gomphothere tooth. The sample preparation and dating of the gomphothere and mastodon teeth were carried out in parallel.

A 3.7-cm long fragment of the mastodon inner enamel surface was exposed and sampled for carbon and oxygen stable isotope analyses. (Figure A.1). Larger samples of enamel were milled from the distal and proximal end of the tooth cusp. The distal inner enamel sample weighed 49.3 mg; this yielded 0.39 mg of carbon (0.8%C). The

Table A.1. Radiocarbon Dating of Proboscidean Teeth from Locus 1

Lab Number	*Sample Context*	*Stratum*	*C (mg)*	*δ13C (‰)*	*^{14}C yr B.P.*	±
AA95550	***G*** M1 right proximal	3B	0.74	-5.6	9,804	93
AA95550	***G*** M1 right distal	3B	0.95	-5.3	9,755	92
AA95551	***M*** tooth	3A	0.37	-8.3	10,960	220
AA95551	***M*** tooth	3A	0.39	-9.1	11,440	220

Key: ***G*** = gomphothere; ***M*** = mastodon. Stratum 3B is the Upper Bonebed; Stratum 3A is the Lower Bonebed.

proximal inner enamel sample weighed 41.2 mg and produced 0.37 mg of carbon (0.9%C). The masses of distal and proximal inner enamel powder obtained from the mastodon tooth fragment were significantly smaller than those obtained from the gomphothere. The percent carbon yields from the mastodon enamel samples were the same as yields from the gomphothere enamel.

The mastodon enamel radiocarbon dates are older than the gomphothere dates, consistent with their stratigraphic relationship (Table A.1). The radiocarbon dates from the mastodon distal and proximal enamel samples differed from each other more than the gomphothere distal and proximal samples. They just missed overlap at one standard deviation but were within two standard deviations, so in formal statistical terms, they are identical. The smaller size of the mastodon graphites resulted in higher radiocarbon measurement uncertainties that obtained from the gomphothere enamel. Like the gomphothere dates, the mastodon dates are younger than expected based upon comparisons with other radiocarbon dates from within stratum 3A.

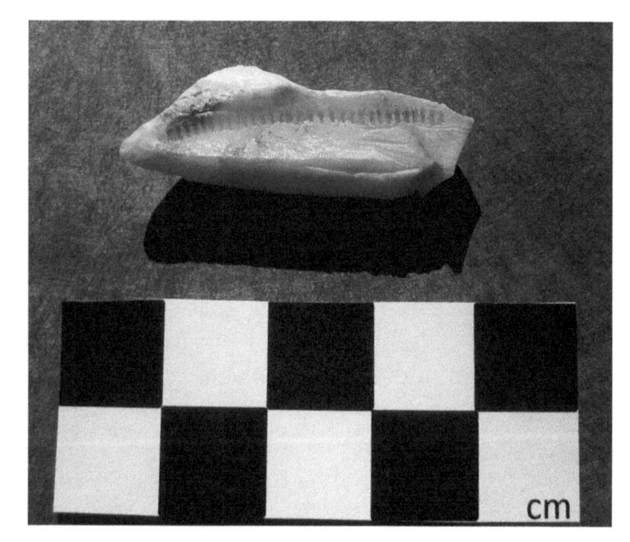

Figure A.1. A fragment of mastodon inner enamel surface (3.7 cm long), sampled for stable isotope analysis. Photograph by Gregory Hodgins.

Luminescence Dating of Sediments from El Fin Del Mundo

James K. Feathers

Six sediment samples from El Fin del Mundo, Locus 1, were submitted for luminescence analysis to the University of Washington Luminescence Dating Laboratory. Samples were processed for the sandy clay alluvium (Stratum 3B) that largely contained the bonebed, the overlying diatomite (Stratum 4d), which encased some bone protruding from upper Stratum 3B, and the lower diatomaceous earth (Stratum 4de). The basic idea was to date the surface of Stratum 3B prior to its burial by the diatomite.

The luminescence method dates sediments to when they were last exposed to sunlight. Certain minerals such as quartz and feldspars build up a latent luminescence signal as a function of natural radioactivity. Radiation ionizes atoms in the crystal structure creating charged electrons and electron vacancies. Some of this charge can be trapped at crystalline defect sites, where it may be held for lengthy periods of time. The charge is released upon exposure to sufficient sunlight, producing light called luminescence. The intensity of the luminescence is proportional to the duration of time since the material was last exposed to sunlight, usually at the time of burial. An age is the quotient of the equivalent dose (D_e), which is the laboratory estimate of the accumulated trapped charge, and the dose rate, which is the rate at which ionizing radiation impinges upon the sample (See Duller 2008 for an introduction to luminescence dating written for archaeologists).

Samples were collected by coring from the unexcavated surface of Locus 1 just west of the excavation area (as of March 2010). Stratum 5 crust and Stratum 4 diatomaceous earth were augured out to just above the diatomite, and then samples were collected in a standard opaque core barrel with a transparent plastic sleeve that was 20 cm long inside the barrel. Each sample straddled Stratum 4 diatomite and upper Stratum 3B, which contained the bonebed. Seven cores were recovered (Table B.1). One was exposed to light to serve as a comparative sample of the stratigraphy that could be viewed in both white light and the laboratory red light. After retrieval, the corer was placed in a large photographic film bag, where the plastic sleeve was removed without exposure to light and placed inside an opaque plastic PVC pipe of slightly larger diameter. The ends of the pipe were taped shut with opaque duct tape. Radiocarbon dating of charred plant material indicates that the bonebed in upper Stratum 3B dates to <13,415 cal yr B.P. and charcoal from the eroded surface of the diatomite in lower Stratum 4 dates to >11,125 cal yr B.P. (Chapter 2).

Table B.1. OSL Sample Provenience

UW Lab Number	*Sample Identification*	*Depth, meters below datum*
UW 2386	FDM OSL 10-1	11.80–11.95
UW 2387	FDM OSL 10-2	11.83–11.88
UW 2388	FDM OSL 10-3	11.78–11.93
UW 2389	FDM OSL 10-4	11.78–11.93
UW 2390	FDM OSL 10-5	11.81–11.96
UW 2391	FDM OSL 10-6	11.79–11.94
UW 2392	FDM OSL 10-x^2	11.81–11.99

Note: Sample UW 2392/FDM OSL 10–x^2 was exposed to sunlight.

In the laboratory, two subsamples were extracted from each sample. The contact between the diatomite and the underlying Stratum 3 was easy to detect in the plastic tubes in red light by color change and also when they were opened by abrupt change in texture. One sample was collected from Stratum 4d in each tube; these were labeled Subsample A. The other was taken from the top 3 to 4 cm of Stratum 3B; these were labeled Subsample B. The expectation was that Subsample B would date the deposition of the bonebed. For dose rate measurements, four subsamples were also extracted from each core: one each from A and B, one from the top of the core (above A) and one from the bottom of the core (below B). UW2392, the exposed sample, was cut into five 3-cm segments for dose rate measurements. All these samples were taken for laboratory measurements of the dose rate. In the field $CaSO_4$ dosimeters, which measure gamma and cosmic radiation in situ, were also placed near the diatomite contact in nearby locations and left for nine months.

DOSE RATE

The dose rate is composed of alpha, beta, gamma, and cosmic radiation. The terrestrial radiation mainly stems from ^{238}U and daughters, ^{232}Th and daughters, and ^{40}K. It was measured in the laboratory by thick source alpha counting, beta counting, and flame photometry (see Feathers and Nami 2018 for a detailed description of the procedures followed). Cosmic radiation was determined following Prescott and Hutton (1994), based on geographic location and burial depth. Dose conversion followed Adamiec and Aitken (1998). Comparison of the gamma and cosmic dose rate determined by the laboratory measurements with that measured by the on-site dosimeters showed that the dosimeters failed to provide useful information. They yielded varied gamma and cosmic dose rates ranging from 0.57 to 2.04 Gy/ka. The reason for this variation is apparent when looking at the radioactivity measurements for the different segments of UW2392 (Table B.2). The concentrations of all three major radioactive nuclides (^{40}K is nearly a constant proportion of total K) increase significantly with depth. The dose rate measured by the dosimeter depends critically on where exactly it was in relation to the diatomite, and such locational control was not possible. Calculating the dose rate therefore depended on assaying the radioactivity of the four subsamples from each sample in the laboratory and prorating the dose rate according to the distance among them (following Aitken 1985:Appendix H).

Table B.3 provides the concentration of the three major radiation contributors for each subsample, the gamma and cosmic dose rate calculated for both A and B, and the total dose rate for A and B. The dose rates for the A sections are

Table B.2. Radioactivity of UW2392

Depth from Top of Sample	*^{238}U (ppm)*	*^{232}Th (ppm)*	*K (%)*
0–3	1.90±0.11	1.40±0.47	0.45±0.12
3–6	2.63±0.17	3.74±0.79	0.63±0.04
6–9 (diatomite)	2.62±0.21	12.20±1.25	1.61±0.15
9–12	2.65±0.23	13.58±1.57	2.08±0.14
12–15	3.89±0.32	18.31±1.87	2.28±0.14

less than those for the B sections in accordance with the downward increase in radioactivity. Beta dose rates were also calculated in two ways: one directly for beta counting and the other by derivation from the alpha counting and flame photometry results assuming secular equilibrium. The data are not shown but of 21 measurements representing all four subsamples, only four yielded significant differences between the two measures and none for subsamples A or B. The downward trend in concentrations suggests there could be downward movement of radionuclides through time, increasing the dose rate through time for the lower sections. This is possible for U and K, but less plausible for Th, which is insoluble. Perhaps the Th moved down with clay translocation, but micromorphological analysis by project personnel did not reveal any clay movement. The difference in radioactivity between the diatomaceous silts and the sandy clay therefore is suspected to be lithological in origin, not post-depositional.

Dating analysis involved potassium feldspars, so a contribution to the dose rate from internal ^{40}K requires calculation. This was done by an energy dispersive X-ray analysis on a scanning electron microscope on 50 grains used for luminescence analysis from UW2386, UW2387 and UW2389. The K content averaged 8 ± 4 percent, which is less than the 14 percent expected for pure orthoclase, but not atypical of K-feldspar extracts in luminescence dating (Smedley and others 2012). The internal K is responsible for only about 10 percent of the total dose rate. Moisture content was measured from field samples and averaged about 11 percent. The moisture content was likely higher during the Terminal Pleistocene and early Holocene, as discussed later in this appendix.

EQUIVALENT DOSE

Luminescence was measured on single-grains of 180-212μm potassium feldspar. Quartz at this site was too insensitive to make single-grain dating practical.

Table B.3. Dose Rate Data

Sample	^{238}*U (ppm)*	^{233}*Th (ppm)*	*K (%)*	*External Dose Rate (Gy/ka)*	*Total Dose Rate (Gy/ka)*
UW2386-top	9.60±1.00	14.06±2.09	1.42±0.08		
UW2386-A	4.18±0.30	14.31±1.67	2.29±0.14	1.92±0.13	4.74±0.33
UW2386-B	3.16±0.32	23.78±2.21	2.30±0.21	1.92±0.12	4.87±0.38
UW2386-bottom	3.70±0.28	15.46±1.59	2.46±0.21		
UW2387-top	2.89±0.21	8.72±1.13	1.19±0.04		
UW2387-A	2.97±0.16	15.85±1.59	1.96±0.14	1.54±0.11	4.03±0.32
UW2387-B	3.34±0.26	13.81±1.47	2.54±0.12	1.61±0.11	4.47±0.33
UW2387-bottom	2.46±0.27	20.20±1.94	2.20±0.09		
UW2388-top	2.70±0.17	3.60±0.71	0.67±0.06		
UW2388-A	2.75±0.22	11.08±1.40	2.18±0.11	1.35±0.09	3.82±0.30
UW2388-B	3.16±0.26	15.11±1.69	2.11±0.18	1.48±0.10	4.07±0.34
UW2388-bottom	3.19±0.27	15.97±1.75	2.66±0.24		
UW2389-top	2.00±0.12	1.21±0.46	0.50±0.03		
UW2389-A	3.37±0.23	9.43±1.19	1.60±0.32	1.25±0.08	3.39±0.37
UW2389-B	3.60±0.26	10.91±1.40	2.56±0.11	1.42±0.10	4.24±0.32
UW2389-bottom	2.20±0.26	20.40±1.99	2.58±0.19		
UW2390-top	2.54±0.16	4.06±0.75	0.54±0.03		
UW2390-A	2.98±0.23	11.51±1.43	2.12±0.14	1.38±0.09	3.86±0.30
UW2390-B	4.08±0.33	18.83±0.20	2.06±0.17	1.55±0.10	4.33±0.35
UW2390-bottom	3.70±0.31	19.89±1.84	2.55±0.27		
UW2391-top	2.92±0.21	7.60±1.15	1.47±0.07		
UW2391-A	2.22±0.17	7.55±1.02	1.35±0.07	1.41±0.09	3.18±0.26
UW2391-B	3.39±0.31	21.60±1.93	2.40±0.12	1.70±0.11	4.68±0.33
UW2391-bottom	2.24±0.29	25.23±2.27	2.30±0.10		

Sample preparation followed standard procedure. The 180-212 μm fraction was isolated by sieving and then treated with HCl and H_2O_2, followed by a density separation using lithium metatungstate set at 2.58 specific gravity, on the assumption that most material lighter than that will be K-feldspars. Single-grain measurements were made on a TL/OSL DA-20 reader, equipped with an IR single-grain attachment. Stimulation for single-grains used a 150mW 830nm IR laser, set at 30 percent power and passed through an RG 780 filter. Emission was collected by the photomultiplier through a blue-filter pack, allowing transmission in the 350-450nm range. IRSL measurements were made at 50°C, and a preheat of 250°C for 1 minute at 5°C/s proceeded each measurement. Exposure for single-grains was for 0.8s, using the first 0.06s for analysis and the last 0.15s for background. Doses were delivered by a ^{90}Sr beta source, which provides about 0.12 Gy/s to 180-212μm grains.

Equivalent dose (D_e) was determined using the single-aliquot regenerative dose (SAR) protocol (Murray and Wintle 2000; Wintle and Murray 2006), as applied to feldspars by Auclair, Lamothe, and Huot (2003). The SAR method measures the natural signal and the signal from a series of regeneration doses on a single aliquot or grain. The method uses a small test dose (6 Gy in this case) to monitor and correct for sensitivity changes brought about by preheating, irradiation or light stimulation. SAR consists of the following steps: (1) preheat, (2) measurement of natural signal (OSL or IRSL), L(1), (3) test dose, (4) preheat, (5) measurement of test dose signal, T(1), (6) regeneration

Table B.4. Measured and Accepted Grains

Sample Number	*Number of Grains Measured*	*Number of Rejected Grains*				*Number of Grains Accepted*	*Acceptance Rate (%)*
		Low signal	*Recycle failure*	*Too high*	*Recuperation*		
UW2386b	594	384	41	8	55	106	17.8
UW2387b	592	400	47	4	52	89	15.0
UW2388b	594	412	50	8	28	96	16.1
UW2389b	491	328	36	4	47	76	15.5
UW2390b	686	501	74	8	29	74	10.8
UW2391b	788	572	59	4	59	94	16.4
Totals	**3745**	**2597**	**307**	**36**	**270**	**535**	**14.3**
UW2386a	195	164	4	10	3	14	7.2
UW2387a	196	178	1	3	6	8	4.1
UW2388a	390	267	19	20	12	72	18.5
UW2389a	196	121	14	0	26	35	17.9
UW2390a	199	150	5	1	5	38	19.1
UW2391a	193	167	6	0	4	16	8.3
Totals	**1369**	**1047**	**49**	**34**	**56**	**183**	**13.3**

Note: See text for definition of terms.

dose, (7) preheat, (8) measurement of signal from regeneration, L(*i*), (9) test dose, (10) preheat, (11) measurement of test dose signal, T(*i*), and (12) repeat of steps 6 through 11 for *i* regeneration doses. A growth curve is constructed from the L(*i*)/T(*i*) ratios and the equivalent dose is found by interpolation of L(1)/T(1). A zero-regeneration dose and a repeated regeneration dose are employed to insure the procedure is working properly.

An advantage of single-grain dating is the opportunity to remove from analysis grains with unsuitable characteristics by establishing a set of criteria grains must meet. Grains were eliminated from analysis if they (1) had poor signals (as judged from net natural signals not at least three times above the background standard deviation), (2) did not produce, within 20 percent, the same signal ratio (often called recycle ratio) from identical regeneration doses given at the beginning and end of the SAR sequence, suggesting inaccurate sensitivity correction, (3) yielded natural signals that did not intersect saturating growth curves (too high), or (4) had a signal larger than 10 percent of the natural signal after a zero dose (recuperation).

A dose recovery test was performed on about 200 grains from two samples, UW2386B and UW2387B, from which 48 grains gave a usable signal. The luminescence was first removed by exposure to IR diodes for 400 s followed by 1 s exposure to the laser. A dose of known magnitude (about 20 Gy) was then administered. The SAR procedure was then applied to see if the known dose could be obtained. Successful recovery is an indication that the procedures are appropriate. The weighted average (central age model) of the ratio of derived/administered dose was 1.06 ± 0.5; 69 percent of the grains were within 1-sigma of the given dose and 88 percent were within 2-sigma, close to what would be expected randomly. The procedures therefore seem satisfactory. There was 18 ± 5 percent over-dispersion observed in the dose recovery distribution. This would suggest that variation due to intrinsic measurement factors (such as machine error and varied luminescence properties among grains) is at least this much.

Feldspars are affected by loss of signal through time, called anomalous fading. Anomalous fading rates were measured using the procedures of Auclair, Lamothe, and Huot (2003) for single grains. Age was corrected following Huntley and Lamothe (2001). Storage times after irradiation of up to three to five days were employed. A fading-corrected age is obtained for each suitable grain.

Table B.4 gives the number of grains measured for each sample, the number of grains rejected as unsuitable, the number of accepted grains, and the acceptance rate. Most grains were rejected due to low signals not distinguishable

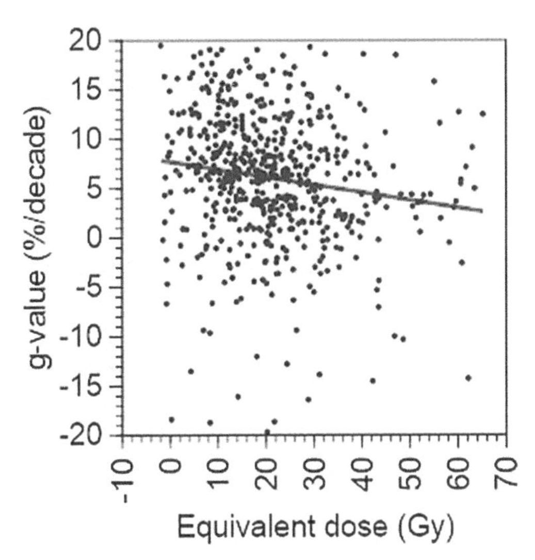

Figure B.1. Equivalent dose versus fading rate (or g-value, measured as % per decade, where a decade is a power of 10).

from the background. Recycle failures and recuperation were two other main causes of rejection. There were five negative D_e values and these were eliminated from further analysis.

Fading rates were highly variable. Of 561 measures, 76 were negative (presumably statistical artifacts of zero fading) but 174 had g-values greater than 10. The weighted average was 6.0 ± 1.1, which is relatively high for sediments. On 127 grains, the fading rate was sufficiently high that an infinite age correction resulted. This correspondingly reduced the number of grains for which an age could be calculated. Figure B.1 plots the fading rate (g-value) against equivalent dose for all samples. The solid line shows a regression. There is some tendency for the equivalent dose to get larger as the fading rate decreases, which is what is expected.

AGE DISTRIBUTIONS

The D_e for each grain was divided by the bulk dose rate for the sample to obtain an age for each grain, while at the same time correcting for fading following Huntley and Lamothe (2001). Some of the resulting variation may be attributed to differential dose rate at the single-grain level, the most likely cause being variation in internal K values. However, as discussed previously, the internal K contributed only 10 percent of the total dose rate, and a 50 percent error term was included on it, so it is not expected that this accounts for too much over-dispersion. Most variation must be due to different-aged grains.

Table B.5. Central Age and Over-Dispersion

Sample Number	*Number of Grains*	*Age (ka)*	*Over-Dispersion (%)*
UW2386b	78	5.91±0.48	36.3±7.7
UW2387b	64	6.38±0.52	31.1±7.9
UW2388b	85	8.40±0.54	25.0±7.0
UW2389b	63	7.67±0.63	32.4±8.7
UW2390b	59	8.00±0.58	21.8±8.2
UW2391b	72	6.47±0.49	33.2±7.3
Total all b grains	**421**	**7.09±0.23**	**33.0±3.2**
UW2386a	11	5.72±0.97	0
UW2387a	7	5.89±1.45	0
UW2388a	54	7.93±0.67	33.6±7.7
UW2389a	29	9.30±1.14	33.8±11.7
UW2390a	31	7.66±0.88	37.1±10.4
UW2391a	13	6.40±1.55	47.7±21.4
Total all a grains	**145**	**7.79±0.44**	**35.8±5.2**

Note: All ages are reported at 1-sigma. The base date for ka (1000 years ago) is 2010.

Table B.5 gives the fading corrected ages using the central age model and the over-dispersion for each sample. The central age is a weighted average using log values (Galbraith and Roberts 2012). The over-dispersion is a measure of spread beyond what can be accounted for by differential precision. The ages for the A segments are about in the same range as those for the B segments, if somewhat older, but both segments statistically are about the same age. The over-dispersion is about 30 percent for all samples, higher than the 15 to 20 percent one might expect for a single-aged sample, based on dose recovery. Some mixture of different-aged grains is thus apparent.

A finite mixture model was applied to look at the structure of the age distributions. The finite mixture model (Galbraith and Roberts 2012) uses maximum likelihood to separate the grains into single-aged components based on the input of a given σ_b value (15% in this case) and the assumption of a log normal distribution of each component. The model estimates the number of components, the weighted average of each component, and the proportion of grains assigned to each component. The model provides two statistics for estimating the most likely number of components, maximum log likelihood (llik) and Bayes Information Criterion (BIC). The finite mixture model is appropriate for samples that have discrete age populations due to post-depositional mixing.

Table B.6. Finite Mixture Model

Sample	*Component*	*Age (ka)*	*Proportion (%)*
B-segments	1	0.09±0.05	0.6
	2	4.01±0.50	22.6
	3	8.12±0.39	76.8
A-segments	1	3.52±0.49	10.6
	2	8.62±0.39	89.4

Because all six samples behaved similarly, grains from all of them were combined to maximize sample size. Table B.6 shows that the B segments can be divided into three single-age components, while the A segments divide into two. Nearly 90 percent of the grains in the A segments fall into a single component, suggesting that mixture is fairly minimal. For the B segments, 77 percent fall into one component, suggesting somewhat more mixing, mainly of younger grains, but not a lot. The youngest component can probably be attributed to contamination of the sample either during collection or preparation.

The ages for the main components of each set of segments are statistically indistinguishable. In fact, if one looks graphically at the distributions, they appear almost identical, save for differences in sample size. Figure B.2 shows radial graphs of the two distributions. Radial graphs plot standardized age values against precision. The age values are standardized by the number of standard errors each point is away from some reference point. The axis on the right side is drawn so that lines from the origin through any point will intersect the axis at the derived age. For both segments, the reference was chosen as the value of the main finite mixture component, 8.1 ka cal yr B.P., of segment B. A second reference was drawn at 13 ka (the approximate age of Clovis archaeology). The shaded areas around each reference encompass all points within two standard errors of the reference.

What the graphs also show is that there are very few grains that are statistically older than Clovis. The 13 ka line almost forms an age barrier beyond which no grains date. An attempt was made to calculate a maximum possible age from the B-segment distribution. Not surprisingly this yielded an age of 8.4 ± 0.4 cal yr B.P., not much different from the third component. This is not surprising because there is no group of grains statistically older than that component. It is clear from the graphs, however, that there are a few grains of Clovis age, just not enough to make a statistical difference.

DISCUSSION

One way to interpret these data is to assume a very rapid deposition of the clayey sand, the diatomite, and the diatomaceous silts so that the ages of all are indistinguishable. Some mixing in of younger grains occurs throughout. This might be the interpretation of someone with no prior information about the site.

However, another interpretation is that Stratum 3B was deposited at the time of Clovis, but it formed a stable surface for several thousand years. Bioturbation during this time brought many grains to the surface where they were exposed. This was followed by a rapid deposition of the diatomite and the diatomaceous silts, burying Stratum 3 and preventing any further exposure. The top layer of Stratum 3, which is what was dated, would then have very few grains that had not been exposed since the deposition of that unit (and the bonebed). The dates of this top portion would approximate the depositional date of the overlying sediment, which in fact it does. One way to test this would be to date a profile down the stratum (e.g., over the top 20 cm) to see if ages get progressively older up to a maximum as the reach of turbation gets less. This has been a way luminescence has detected stable surfaces (e.g., Feathers and others 2015). An argument against this interpretation is that no sedimentary evidence of prolonged weathering is observed. Plus, 8.0 cal yr B.P. also seems an underestimate for the diatomite or diatomaceous silts as well, judging from the radiocarbon date of >11.1 cal yr B.P. for the top of the diatomite.

Another possibility is that the water content has been underestimated, particularly if wetter conditions pertained during the early history of the deposit. The age for UW2388 was recalculated assuming an average water content of 35 percent, instead of the 11 percent used in the original calculation. The age for this sample, which from distribution analysis appears as a single-aged sample, increases to 10.0 ± 0.6 cal yr B.P., still short of a Clovis age. One could also assume there was clay translocation so that the dose rate of the B segments is overestimated. Using the dose rate of the A segment instead produces less of a change than the 35 percent water assumption. Possibly, both the water content is underestimated and the dose rate overestimated. Then perhaps ages in the range of Clovis could be obtained. But this relies on a lot of assumptions and certainly not the most parsimonious interpretation of the data. Also, one would have to have an independent method of estimating what the moisture content and the dose rate should be to arrive at an estimated age and avoid circular reasoning.

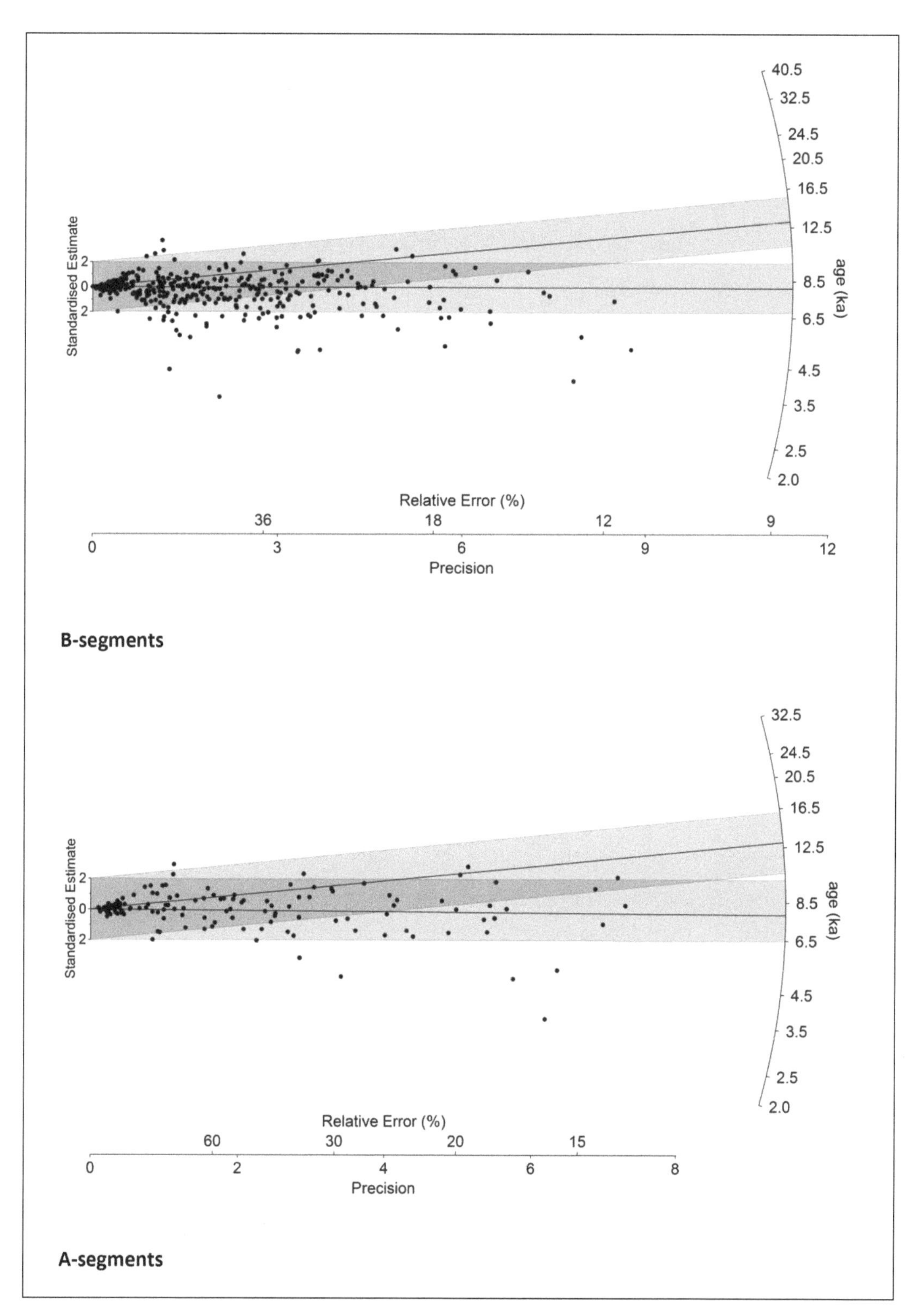

Figure B.2. Radial plots of age distributions for B-segments (top) and A-segments (bottom). See text for explanation of the plots. Age (ka) refers to thousand cal yr B.P.

Metric and Descriptive Data of the Clovis Lithics from El Fin el Mundo and the San Pedro Valley Sites

Table C.1. Secondary Bifaces from El Fin del Mundo

Bag Number	*Locus*	*Condition*	*Raw Material*	*Munsell Color*	*Weight (in g)*	*Max. Length (in mm)*	*Max. Width (in mm)*	*Max. Thickness (in mm)*	*Body Morphology*	*Base Morphology*	*Overshot Flaking*
46021	1	Distal	Rhyolite	Light reddish brown 5YR 6/2	—	91.2	43.4	16.4	Lanceolate		Yes
46022	1	Medial	Quartz crystal	Translucent	—	39.8	32.2	10.4			
59070	5	Medial	Rhyolite	Light reddish brown 2.5YR 6/4	20.5	37.8	46.5	11.1			
59076	5	Proximal	Rhyolite	Light reddish brown 2.5YR 5/3	20	48.2	36.1	12.8		Convex	
59090	5	Medial	Rhyolite	Reddish brown 5YR 5/4	33	37.9	49.4	17.1			
59096	5	Medial	Vitrified basalt	Dark greenish gray Gley2 4/10BG	25.8	53.1	37.6	11	Lanceolate		
59103	5	Proximal	Rhyolite	Light reddish brown 5YR 6/3	20	46.3	36	12.1		Straight	Yes
63233	10	Proximal	Rhyolite	Reddish brown 5YR 5/4	14.3	36.4	32	9.4		Slightly concave	Yes
63239	10	Distal	Not Identified Igneous	Very dark gray Gley1 3/N	30.9	51.9	40	17.8	Lanceolate		
63269	10	Distal	Rhyolite	Reddish brown 5YR 5/3	8.85	33.7	37.6	7.9	Lanceolate		
63271	10	Distal	Rhyolite	Light gray 5YR 7/1	16.2	42.8	44	10.3	Lanceolate		
63281	10	Distal	Rhyolite	Light reddish brown 5YR 6/2	7.6	39.8	38.2	5	Lanceolate		
63282	10	Proximal	Rhyolite	Reddish brown 5YR 5/3	13.9	38.7	33.3	11.6			
63288	10	Distal	Rhyolite	Light reddish brown 2.5YR 6/3	18.6	49.1	36.6	12.6	Lanceolate		
63296	10	Proximal	Rhyolite	Light reddish brown 5YR 6/3	49.6	49.6	50.3	16.2		Convex	
63305	10	Proximal	Rhyolite	Light reddish brown 5YR 6/3	47.7	62.5	48.8	15		Convex	Yes
63310	10	Proximal	Rhyolite	Pinkish gray 5YR 7/2	8.5	41.6	35.5	9.4			
63321	10	Proximal	Rhyolite	Light reddish brown 5YR 6/3	18.4	33.8	36.3	14.5		Convex	
63334	10	Indeterminate	Rhyolite	Reddish brown 5YR 5/4	25.1	55.8	40.8	14.1	Lanceolate		
63334	10	Proximal	Rhyolite	Reddish brown 5YR 5/4	13.1	29.3	37	10.9		Convex	
63334	10	Proximal	Rhyolite	Reddish brown 5YR 5/4	47.8	62.1	47.8	15.7	Lanceolate	Convex	Yes
63355	10	Medial	Rhyolite	Dark reddish gray 10R 4/1	28.3	52.3	34.6	12.5	Lanceolate		
63343	17	Proximal	Rhyolite	Reddish brown 5YR 5/3	10.1	27	42.3	9.2		Convex	
63354	17	Distal	Rhyolite	Reddish brown 5YR 5/3	16.9	45.9	38.4	10.1			
63368	22	Proximal	Rhyolite	Light reddish brown 5YR 6/3	27	47.1	38.4	14		Convex	
63360	22	Proximal	Rhyolite	Light reddish brown 5YR 6/3	11.3	28.4	36.2	9.1		Straight	

Table C.2. Clovis Point Preforms from El Fin del Mundo

Bag Number	Locus	Condition	Raw Material	Munsell Color	Weight (in g)	Max. Length (in mm)	Max. Width (in mm)	Max. Thickness (in mm)	Diagnostic Elements	Base Morphology	LF 1 (in mm)	LF 2 (in mm)	LMG 1 (in mm)	LMG 2 (in mm)
59287	2	Proximal/ Reworked	Rhyolite	Duskey red 10R 3/2	6.3	31.2	26.1	7.1	—	Concave	14.9	—	25.8	14.8
59083	5	Complete/ 2 fragments	Quartzite	Grayish brown 10YR 5/2	21.2	67.1	30.4	9.4	Fluting	Straight	18.6	—	—	—
59307	5	Proximal	Chert	Strong brown 7.5YR 5/8 mottled with reddish yellow 7.5YR 7/6	8.9	34.8*	26.7*	7.3*	Fluting on both faces	Concave	19.8	22.4	—	—
59324	5	Proximal	Quartz	Milky White	10.9	40.6*	29.8*	7.9*	Fluting	Convex	19.5	—	—	—
62756	5	Proximal	Rhyolite	Light reddish gray 2.5YR 7/1	22.9	37*	48*	12.4*	Fluting	Straight	25.1	—	—	—
63214	10	Proximal	Rhyolite	Light reddish gray 2.5YR 7/1	11.9	31.1*	39.6*	10*	Fluting	Convex	—	—	—	—
63224	10	Proximal	Not Identified Igneous	Black Gley1 2.5/N	10.7	24.9*	36.8*	9.9*	Fluting	Convex	—	—	—	—
63332	17	Proximal	Rhyolite	Reddish brown 5YR 5/4	11.7	27.6*	38.1*	11.4*	Fluting/ overshot flaking	Convex	—	—	—	—
63333	17	Proximal	Rhyolite	Light reddish brown 5YR 6/2	10.9	32.6*	30.2*	10.2*	Fluting	Convex	19.2	—	—	—
63092	21	Proximal	Jasper	Dark yellowish brown 10YR 4/4	16.9	35.7*	43.6*	11.5*	Fluting	Convex	20.1	—	—	—

Key: *= Incomplete measurement; LF1= Length of fluting on side 1; LF2= Length of fluting on side 2; LMG 1= Extent of lateral margin grinding 1; LMG 2= Extent of lateral margin grinding 2.

Table C.3. Clovis Points from El Fin del Mundo

Bag Number	Locus/Unit	Condition	Raw Material	Munsell Color	Weight (in g)	Max. Length (in mm)	Max Width (in mm)	Max Thickness (in mm)	Basal Width (in mm)	Basal Concavity (in mm)	LF 1 (in mm)	LF 2 (in mm)	LMG 1 (in mm)	LMG 2 (in mm)
46023	1/surface	Complete	Chert	Light gray 2.5Y 7/1 banded with grayish brown 2.5Y 5/1	24.4	88.5	28.5	7.9	24.7	1.6	25.7	28	31.4	36.5
59342	1/surface	Complete	Quartz crystal	Translucent	7.2	49.2	23.1	5.7	19.9	2.8	12.4	19.8	22.8	23.8
59569	1/disturbed	Complete	Chert	Brown 7.5YR 5/2	6.1	52.4	17.7	6.4	17.6	2.2	13.7	12.7	20.9	22.6
62942	1	Distal	Chert	White 10YR 8/1 speckled with dark reddish gray 10R 4/1 and reddish gray 6/1, and banded by red 2.5YR 4/4	7.1	50*	23.7*	7.3*						
62943	1	Complete/ repaired	Chert	Light gray 2.5Y 7/1 with dark yellowish brown 10YR 4/4, speckled with dusky red 10R 3/2	4	41.2	19.5	5.8	14.7	1.6	19.3	17.8	15.7	15.6

(continued)

Table C.3. (continued)

Bag Number	*Locus/Unit*	*Condition*	*Raw Material*	*Munsell Color*	*Weight (in g)*	*Max. Length (in mm)*	*Max Width (in mm)*	*Max Thickness (in mm)*	*Basal Width (in mm)*	*Basal Concavity (in mm)*	*LF 1 (in mm)*	*LF 2 (in mm)*	*LMG 1 (in mm)*	*LMG 2 (in mm)*
63008	1	Complete/ repaired	Chert	White 10R 8/1 mottled with weak red 10R 4/2 and light olive brown 2.5Y 5/4	4.4	46.1	18.3	6	15.5	2.6	13.6	25.8	20	20
63177	1	Complete	Rhyolite	Dark brown 7.5YR 3/2 with reddish brown 5YR 5/3	28.7	95.5	30.3	8.3	26.1	2.9	24.3	26.3	39.1	39.9
59082	5	Proximal	Chert	Gray 5YR 6/1 banded with Very pale brown 10YR 8/3	6.6	29*	26.9*	7.5*	21	3.3	23.3	23		
59332	5	Semicomplete/ one broken corner/highly reduced	Quartzite	Brown 10YR 4/3	9.5	43	27	7.1	23.6	1.6	17.9	22.1	19.3	19.3
59603	5	Proximal	Chert	Light gray 2.5Y 7/2, mottled with Dark reddish brown 2.5YR 2.5/3	5.8	22.3*	27.4*	8.1*	26.3	2.5				
59604	5	Semicomplete/ one broken corner/highly reduced beveled	Chert	Very dusky red 10R 2.5/2	5.5	38	24	6.5		5	14.7	18.7	17.6	-15.8
63983	5	Distal	Rhyolite	Brown 7.5YR 5/4	11	49.7*	30.8*	7.7*						
36516	5	Proximal	Rhyolite	Brown 7.5YR 5/4	1.6	15.1*	19.2*	5.5*	18.9	4				
33487	5	Proximal with catastrophic impact fracture	Quartz crystal	Translucent	2.6	26.5*	18*	5*	14.8	2				
33030	5	Proximal	Chert	Yellow 2.5Y 8/8	2.3	15.7*	24.4*	-	22.8	3				
59593	8	Proximal	Chert	Translucent Light yellowish brown 2.5Y 6/3	6	31.1*	25.5*	6.4*	22.6	2.9	21.1			
59727	9	Semicomplete/ one broken corner/highly reduced	Rhyolite	Light reddish brown 2.5YR 7/3	8.1	48.1	23	7.2		5.5	10.4	8.3	31.5	-29.4
63220	10	Proximal	Chert	Greenish gray Gley 1 5/10GY mottled with Very dark greenish gray Gley 1 3/5GY	2.6	14.4*	24.9*	6.2*	24.2	2.7				

Key:: *= Incomplete measurement; () = Estimated measurement; LF1= Length of fluting on side 1; LF2= Length of fluting on side 2; LMG 1= Extent of lateral margin grinding 1; LMG 2= Extent of lateral margin grinding 2.

Table C.4. Unifaces from El Fin del Mundo

Bag Number	*Uniface Type*	*Locus*	*Raw Material*	*Munsell Color*	*Condition*	*Made on*	*Weight (in g)*	*Max. Length (in mm)*	*Max. Width (in mm)*	*Max. Thickness (in mm)*	*WE IL (in mm)*	*WE 1A*	*Spur*
45980	SS	1	Chert	Pale yellow 2.5Y 7/3	Complete	Thick flake		90.5	75.2	20.7			
59054	ES	5	Chert	Very pale brown 10YR 8/2, with shades of olive yellow 2.5Y 6/6	Complete	Flake	9.7	35.3	24.4	9.8	21.6	60°	
59058	ES	5	Chert	Strong brown 7.5YR 5/6 banded with light gray 2.5Y 7/1 and red 2.5YR 4/6	Complete	Flake	8.7	29.8	23.6	11.5	25.6	65°	Yes
59060	ES	5	Chert	Light reddish brown 5YR 6/3	Complete	Blade	8.5	31.3	24.4	9	23.3	50°	
59065	ES	5	Chert	White 2.5YR 8/1	Distal	Blade	2.95	18.9*	22.9	5.9	22	70°	Yes
59071	ES	5	Chert	Brown 7.5YR 4/4	Complete	Blade	8	24.9	22.4	8.8	22.2	40°	
59072	ES	5	Chert	Gray Gley 1 6/N, with lines of yellow 2.5Y 7/6	Complete	Flake	10.4	37.3	30	8.9	30	60°	Yes
59073	ES	5	Chert	Dark reddish brown 2.5YR 3/4, mottled with very pale brown 10YR 7/3	Complete	Flake	5.1	24.9	24.3	5.7	25.3	50°	
59085	ES	5	Chert	Pale yellow 2.5Y 8/3, with shades of pale yellow 2.5Y 8/4 and 8/2	Nearly complete	Blade	3.3	24.3	19.1	6.6	19	65°	Yes
59092	ES	5	Jasper	Weak red 10R 4/4	Complete	Flake	15.75	40.8	31.4	11.8	34.4	70°	
59413	ES	5	Chert	Translucent white 7.5YR 8/1	Complete	Blade	4.65	23.9	24	6.1	23.9	50°	
59625	ES	5	Obsidian	Black	Complete	Blade	4.1	26.6*	20.4	6	20.7	60°	Yes
59655	ES	5	Chert	Pink 7.5YR 8/3	Complete	Flake	7.45	35.5	24.4	19.1	19	70°	
59771	ES	5	Chert	Translucent Brown 10YR 4/3	Complete	Flake	6.95	33.5	25	6.9	25	40°	
59793	ES	5	Chert	Very pale brown 10YR 8/3, banded with gray 10YR 6/1	Complete	Blade	5.25	33.1	21.5	7.6	21.7	60°	Yes
59794	ES	5	Not Identified Sedimentary	Greenish gray Gley 1 5/10GY	Complete	Flake	15	30.6	35.2	10.2	34.9	55°	Yes
59810	ES	5	Chert	Pale yellow 2.5Y 7/3	Complete	Blade	4.9	37.5	17.7	7	17.6	55°	
59925	ES	5	Jasper	Dusky red 10R 3/4	Nearly complete	Blade	5.45	30.1*	20.1	7.6	19.5	45°	
59927	ES	5	Chert	Very pale brown 10YR 8/3, banded with gray 10YR 6/1	Complete	Blade	14	44.7	27.3	12.4	36.5	40°	
59932	ES	5	Chert	Light olive brown 2.5Y 5/6, mottled with pale yellow 2.5Y 8/3	Complete	Flake	8.1	32.6	20.7	9.9	20.4	65°	Yes

(continued)

Table C.4. (continued)

Bag Number	*Uniface Type*	*Locus*	*Raw Material*	*Munsell Color*	*Condition*	*Made on*	*Weight (in g)*	*Max. Length (in mm)*	*Max. Width (in mm)*	*Max. Thickness (in mm)*	*WE IL (in mm)*	*WE 1A*	*Spur*
59934	ES	5	Chert	Reddish brown 2.5YR 4/3, mottled with weak red 2.5YR 5/2	Complete	Blade	3.65	26.4	19.5	6.8	19.5	65°	Yes
63082	ES	5	Jasper	Weak red 10R 4/3	Complete	Blade	8.8	39.23	23.4	9.3	23.5	80°	Yes
63085	ES	5	Chert	Translucent light brownish gray 2.5Y 6/2	Complete	Flake	7.9	35.5	24.4	9.8	24.08	50°	Yes
63240	ES	5	Chert	Yellow 10YR 7/6, banded with very pale brown 10YR 8/2 and yellowish red 5YR 4/6	Complete	Flake	6.85	29.6	23.1	9.8	23.3	45°	
63974	ES	5	Chert	Weak red 10R 4/2	Complete	Blade	3.5	34.3	17.5	7	17.7	70°	Yes
63984	ES	5	Chert	Yellow 10YR 8/6 and white 10YR 8/1 in some areas	Complete	Blade	6.8	37.1	20.8	9.9	20.3	55°	
64052	ES	5	Chert	Yellowish brown 10YR 5/4, and red 2.5YR 4/8 in the proximal end	Complete	Blade	11.4	39.7	25.7	11.9	22.9	75°	Yes
33031	ES	5	Chert	Dark reddish brown 2.5YR 3/4	Distal	Blade		17.98*	19.68	5.77	18.98	75°	
33268	ES	5	Chert	Gley 1 5/N	Complete	Blade		34.85	26.4	10.31	25.94	80°	Yes
33269	ES	5	Chert	Reddish brown 2.5YR 5/3	Complete	Blade		35.1	26.19	8.51	26.17	80°	Yes
33473	ES	8	Chert	White 2.5Y 8/1	Complete	Blade		38.24	25.77	8.12	25.95	65°	Yes
74584A	ES	5	Jasper	Dark reddish brown 2.5YR 2.5/4	Complete	Blade		30.18	21.88	7.01	20.45	65°	Yes
74588A	ES	5	Chert	Very pale brown 10YR 8/2	Complete	Blade		34.87	19.95	8.72	19.9	65°	Yes
74601A	ES	5	Chert	Light gray Gley 1 7/N	Complete	Blade		43.02	24.98	9.51	23.59	85°	Yes

Key: SS= side scraper; ES= end scraper; *= Incomplete measurement; () = Estimated measurement; WE 1L= Length of working edge 1; WE 1A= Angle of working edge 1.

Table C.5 Blade Cores from El Fin del Mundo

Bag Number	*Locus/ Area*	*Core Type*	*Condition*	*Weight (in g)*	*Max. Length (in mm)*	*Max. Width (in mm)*	*Max. Thickness (in mm)*	*Raw Material*	*Munsell Color*
59088	5	Wedge-shaped core	Complete/ exhausted	36.8	41.7	39.1	21.9	Chert	Very pale brown 10YR 8/2, speckled with dark reddish brown 2.5YR 2.5/44
33271	5	Conical core	Complete/ exhausted	75.83	52	42.9	34.2	Vitrified basalt	—

Table C.6. Blade-Core Rejuvenation Tablet Flakes from El Fin del Mundo

Bag Number	*Locus*	*Condition*	*Weight (in g)*	*Max. Length (in mm)*	*Max. Width (in mm)*	*Max. Thickness (in mm)*	*Raw Material*	*Munsell Color*
59335	5	Complete	76.3	64.93	44.63	19.38	Vitrified basalt	Yellowish brown 10YR 5/4 patina, natural color dark bluish gray Gley2 4/10B
59524	5	Complete	75.8	57.86	56.96	21.43	Quartzite	Light yellowish brown 2.5Y 6/4
59795	5	Complete	42.1	53.91	46.48	13.72	Quartzite	Light yellowish brown 2.5Y 6/4 with Light gray 2.5Y 7/1 on one edge
33270	5	Proximal	—	43.15*	48.5	10.94*	Vitrified basalt	Yellowish brown 10YR 5/4 patina, natural color olive gray 5Y 5/2

Table C.7. Crested Blades from El Fin del Mundo

Bag Number	*Locus*	*Condition*	*Weight (in g)*	*Max. Length (in mm)*	*Max. Width (in mm)*	*Max. Thickness (in mm)*	*Raw Material*	*Munsell Color*
63099	19	Complete	28.1	55.68	27.7	20.3	Quartzite	Gray Gley1 6/N
63313	10	Complete	11.2	56.08	18.85	14.82	Rhyolite	Light gray 10R 7/1
63253	10	Complete	68.5	90.4	45.1	22.49	Rhyolite	Gray 7.5YR 6/1
63370	22	Complete	146.4	57.33	28.7	249.51	Rhyolite	Light gray 10R 7/1

This table includes only artifacts found up to 2015 (two more crested blades were found during the 2018 field season).

Table C.8. Cortical Blades from El Fin del Mundo

Bag Number	*Locus*	*Condition*	*Weight (in g)*	*Max. Length (in mm)*	*Max. Width (in mm)*	*Max. Thickness (in mm)*	*Raw Material*	*Munsell Color*	*Cortex % of Dorsal Surface*	*Platform*	*Macro use-wear*
59250	5	Complete	58.9	99.44	37.95	17.42	Not Identified Sedimentary	Yellowish brown 10YR 5/4 patina	50	Cortical	
63208	10	Complete	37	77.62	36.5	13.9	Not Identified Sedimentary	Bluish gray Gley2 6/5PB, and Yellowish brown 10YR 6/4 patina	50	Plain	
63372	22	Complete	45.3	79.76	34	17.75	Quartzite	White 5Y 8/1, mottled with red 2.5YR 4/8	25	Cortical	Yes
63388	23	Complete	78.3	87.53	43.7	22.88	Quartzite	Pale yellow 2.5Y mottled with red 2.5YR 4/8, brown 7.5YR 5/4 in cortex	50	Plain	
63423	25	Complete	136	113.9	46.31	29.96	Rhyolite	Pinkish gray 5YR 6/2	50	Cortical	

Table C.9. Noncortical Blades from El Fin del Mundo

Bag Number	*Locus*	*Condition*	*Weight (in g)*	*Max. Length (in mm)*	*Max. Width (in mm)*	*Max. Thickness (in mm)*	*Raw Material*	*Munsell Color*	*Macro use-wear*
59079	5	Proximal	5	36.35*	20.63*	7.09*	Chert	Pale yellow 2.5Y 8/4	
59080	5	Distal	11.4	35.72*	33.46*	11.73*	Quartzite	Dark yellowish brown 10YR 4/6	
59087	5	Distal	5.8	35.12*	21.31*	6.98*	Chert	Pale yellow 2.5Y 8/2, clouded with light gray 2.5Y 7/1	Yes
59091	5	Complete	16.6	68.04	29.11	12.8	Chert	White 5Y 8/1	
59097	5	Proximal	4.6	40.23*	18.5*	18.45*	Chert	White Gley1 8/N, clouded with white 2.5Y 8/1	
59134	5	Complete	64.1	116.91	36.64	14.79	Vitrified basalt	Black Gley1 2.5/N, and brownish yellow 10YR 6/6 patina	
59139	5	Medial	4.8	43.28*	20.3*	6.29*	Rhyolite	Weak red 2.5YR 5/2	Yes
59249	5	Medial	5.1	40.4*	19.97*	5.8*	Jasper	Duskey red 10R 3/4	Yes
59255	5	Proximal	7.8	32.41*	23.98*	10.83*	Quartzite	Pale yellow 2.5Y 8/2	
59309	5	Proximal	10.8	46.75*	29.94*	8.86*	Chert	Brownish yellow 10YR 6/6 Patina	
59334	5	Medial	27.6	66.35*	34.38*	10.81*	Chert	White 2.5Y 8/1, with a band of very dark gray Gley1 3/N	Yes
59530	5	Complete	19.5	63.8	33.67	10.19	Not identified sedimentary	Gray Gley1 5/N, and very pale brown 10YR 7/4 patina	
59772	5	Complete	5.6	60.77	17.26	5.23	Rhyolite	Light reddish brown 5YR 6/4	Yes
59802	5	Complete	12	53.99	22.3	12.06	Rhyolite	Brown 7.5YR 5/3	
59909	5	Complete	10.3	41.84	20.1	8.62	Not identified sedimentary	Light gray 2.5Y 7/1	
59931	5	Proximal	5.9	39.39*	23.1*	8.78*	Chert	Pale yellow 5Y 8/2	
61558	5	Complete	99	115.99	45.82	22.99	Rhyolite	Pinkish white 5YR 8/2, banded with dark gray 5YR 4/1	Yes
62493	5	Complete	5	37.16	17.73	6.25	Chert	Reddish gray 10R 6/1, mottled with white 10R 8/1	Yes
62720	5	Nearly complete	11.2	43.2	23.9	9.4	Quartzite	Greenish gray Gley1 6/10Y, banded with reddish yellow 5YR 6/6	Yes
62751	5	Complete	22.6	58.79	33.2	9.08	Vitrified basalt	Light yellowish brown 10YR 6/4 patina	Yes
63401	5	Complete	29.5	73.88	30.79	10.73	Not identified sedimentary	Brownish yellow 10YR 6/6 Patina	
63402	5	Medial	20.4	63.06*	36.5*	10.8*	Quartzite	White 10YR 8/1, and brownish yellow 10YR 6/6 patina	
66065	5	Complete	16.1	61.43	28.06	10.66	Chert	Brownish yellow 10YR 6/6, and very pale brown 10YR 8/2	Yes

Table C.9. (continued)

Bag Number	*Locus*	*Condition*	*Weight (in g)*	*Max. Length (in mm)*	*Max. Width (in mm)*	*Max. Thickness (in mm)*	*Raw Material*	*Munsell Color*	*Macro use-wear*
59594	8	Proximal	12.1	40.39*	31.3*	7.9*	Chert	Pale yellow 2.5Y 8/2 speckled with bluish black Gley2 2.5/5PB	Yes
63201	10	Distal	22.3	35.18*	37.12*	13.7*	Quartzite	Light greenish gray Gley1 7/10Y	Yes
63205	10	Complete	9.2	51.9	19.36	10.38	Dacite	Very dark gray Gley1 3/N	
63249	10	Complete	18.7	61.38	28.9	11	Not identified sedimentary	Dark greenish gray Gley1 4/5GY	Yes
63298	10	Proximal	5.3	29.25*	29.6*	6.43*	Chert	Dusky red 10R 3/2, mixed with red 10R 4/6, and bluish gray Gley2 5/5PB	Yes
63307	10	Complete	38.3	79.94	35.12	11.72	Quartzite	Light greenish gray Gley1 7/10Y, cortex reddish yellow 7.5YR 6/6	
63312	10	Nearly complete	10	55.46	25.87	7.9	Rhyolite	Light reddish gray 2.5YR 7/1	Yes
63313	10	Complete	3.1	47.08	20.01	5.02	Rhyolite	Light gray 10R 7/1	
62711	14	Complete	36.4	63.75	40	14.85	Quartzite	Yellowish red 5YR 5/6 patina	Yes
63370	22	Complete	146.4	341.33	120.53	59.1	Rhyolite	Reddish gray 2.5YR 5/1	
63384	23	Proximal	52	71.1*	44.35*	14.02*	Dacite	Very dark gray Gley1 3/N	
63392	24	Nearly complete	22.6	65.85	30.55	14	Rhyolite	Reddish gray 2.5YR 6/1	
63423	25	Proximal	31.9	66.57*	34*	13.98*	Rhyolite	Reddish gray 2.5YR 5/1	
63426	25	Complete	31	73.58	31.4	16.4	Rhyolite	Gray 5YR 5/1	Yes
63426	25	Complete	42.9	92.45	38.86	14.37	Rhyolite	Weak red 2.5YR 5/2	Yes
63430	25	Complete	139	134.57	42.42	24.47	Rhyolite	Brown 7.5YR 5/3	Yes
63432	25	Complete	49.4	92.66	31.39	18.06	Quartzite	White 5Y 8/1 speckled with yellowish red 5YR 4/6	
63435	25	Proximal	62	88*	48.4*	19.9*	Rhyolite	Weak red 2.5YR 5/2	Yes
63436	25	Complete	178	122.16	60.6	31.31	Rhyolite	Reddish brown 5YR 5/3	
63440	25	Complete	53.1	84.81	41.5	18.9	Rhyolite	Pale red 2.5YR 6/2	Yes
63472	25	Complete	30.1	66.79	32.2	14.3	Rhyolite	Pale red 2.5YR 6/2	
63473	25	Nearly complete	29.9	74.71	34.22	15.38	Rhyolite	Gray 5YR 6/1	Yes
63474	25	Complete	43	106.26	32.85	11.66	Vitrified basalt	Bluish gray Gley2 2.5/5PB	Yes
74618	5	Proximal	—	33.95*	27.85*	8.26*	Chert	Very pale brown 10YR 8/2	

Keys: *= Incomplete measurement.

Table C.10. Clovis Points from the San Pedro Valley Sites

ASM Number	*Site*	*Condition*	*Raw Material*	*Munsell Color*	*Weight (in g)*	*Max. Length (in mm)*	*Max. Width (in mm)*	*Max. Thickness (in mm)*	*Basal Width (in mm)*	*Basal Concavity (in mm)*	*LF 1 (in mm)*	*LF 2 (in mm)*	*LMG 1 (in mm)*	*LMG 2 (in mm)*	*Remarks*
A-31231	Escapule	Nearly complete/ tip fracture	Chalcedony	Gray 5Y 5/1	31	97.4 (108)	30.2	8.9	26	4.7	34.8	32.4	28.3	29	
A-31232	Escapule	Nearly complete/ tip fracture	Chert	Olive gray 5YR 5/2	25	90.3 (95)	34	6.8	28.7	3.8	16	36.3	29	31	
A-12676	Lehner	Complete	Chalcedony	Translucent dark gray, mottled with light and dark shades of brown 10 YR 3/2	25	82.5	29.2	9.8	26.8	3	61.4	38	41.6	36.6	
A-12677	Lehner	Complete	Chert	Translucent dark gray 7.5R 4/0, with bands of darker gray 5YR 4/1	17	74.2	28.3	7.6	23.1	4.4	21.7	25.6	27.7	26.2	
A-12679	Lehner	Complete	Chert	Translucent gray 2.5Y 6/0, mottled with lighter and darker specks of gray	15	61.7	31	7.7	25.2	2.4	26.4	20.8	29.9	29.8	
A-12681	Lehner	Complete	Quartz crystal	—	6	46.8	20.6	6.7	20.5	3.3	Not fluted	Not fluted	18.6	15.2	
A-12682	Lehner	Complete	Chalcedony	Translucent gray 10YR 5/1, mottled with specks of pale red 7.5YR 5/4	10	55.7	24.6	7.2	23	1.6	17.2	15.9	24.5	24.7	
A-12685	Lehner	Complete	Chert	Opaque gray-brown 10YR 5/2, streaked light gray and red brown10R 3/6	25	96.3	30.3	8.2	26	3	33	25.7	34.7	35.4	
A-12686	Lehner	Complete	Jasper	Opaque pale red 7.5R 4/2	11	51.1	28.1	7.3	25.7	1.9	23.5	18.4	21.5	21.6	
A-12675	Lehner	Complete	Chert	Opaque dark gray-brown 2.5Y 4/2	14	78.66	19.2 (22)	7	19.9	2.5	24.2	20.6	24.1	27.7	
A-12674	Lehner	Distal	Calcedony	Very dark gray 10YR 3/1, banded with black 10YR 2/1	25	85.5* (87)	30.5	7.9	—	—	19.9*	32.4*	17.2*	16.2*	
A-12680	Lehner	Distal	Chert	Opaque light brownish-gray 10YR 6/2, with a band of light gray 10YR 7/1	17	80.5* (81)	28.7	7.3	25	—	—	—	25.1*	26.4*	

Table C.10. (continued)

ASM Number	*Site*	*Condition*	*Raw Material*	*Munsell Color*	*Weight (in g)*	*Max. Length (in mm)*	*Max. Width (in mm)*	*Max. Thickness (in mm)*	*Basal Width (in mm)*	*Basal Concavity (in mm)*	*LF 1 (in mm)*	*LF 2 (in mm)*	*LMG 1 (in mm)*	*LMG 2 (in mm)*	*Remarks*
A-12683	Lehner	Nearly complete/ one broken corner	Quartz crystal	—	2	30.8	16.6	4.7	16.4*	1.1	Not fluted	Not fluted	11.9	11.2	
A-12678	Lehner	Nearly complete/ tip fracture	Quartz crystal	—	4	35.6 (36)	16.8	7.4	15.2	1.5	Not fluted	Not fluted	14.5	12	
A-12684	Lehner	Nearly complete/ tip fracture and one broken corner	Chalcedony	Translucent gray 7.5R 5/1, mottled with dark gray brown spots 10YR 4/2	21	78.1 (78)	29.7	8	22.3	1.3	27.6	24.1	31.5	30.5	
A-24127	Leikem	Complete	Chert	Pale brown 10YR 8/3 with a band of light yellowish brown 10YR 6/4	33	94.6	31.7	10.2	28.8	5.6	47.5	40.9	38.7	38.4	
A-32718	Murray Springs/Area 3	Complete	St. David chalcedony	Translucent light gray 5Y 7/1 mottled with lighter shades of gray and white	14	56.8	28.3	9	23.7	2	16	22.3	Not ground	Not ground	
A-32716	Murray Springs/Area 3	Tip	Petrified wood	Dusky red 2.5YR 4/4 and reddish gray 10R 5/1, banded with yellowish brown 10YR 6/3 on one face	3	25.6*	26.3*	5.9*	—	—	—	—	—	—	
A-32719	Murray Springs/Area 3	Tip	St. David chalcedony	Translucent light gray 2.5Y 7/2	<1	11.7*	13.1*	2.5*	—	—	—	—	—	—	
A-33110	Murray Springs/Area 4	Complete	Chert	Light brownish gray 2.5Y 6/2	8	50.8	24.9	5.8	22.5	4.7	24.2	14.7	24	20	
A-33116	Murray Springs/Area 4	Complete	Obsidian	Translucent	6	41.8	22.3	6.4	16	2	18.8	30.4	20.9	22.5	
A-33114	Murray Springs/Area 4	Distal	Chert	Pale brown 10YR 6/3	18	54.8*	33.5	7.7	—	—	—	—	—	—	
A-33109	Murray Springs/Area 4	Nearly complete	Chert	Greenish gray Gley1 6/10Y	18	57	31	9.1	21	3	15	19	34	22.9	
A-33111	Murray Springs/Area 4	Nearly complete	Chert	Light brownish gray 2.5Y 6/2 mottled with lighter shades of gray	31	83.4 (89)	33.8	9.6	26.4	5.2	25.6	41.6	39.4	37.7	

(continued)

Table C.10. (continued)

ASM Number	Site	Condition	Raw Material	Munsell Color	Weight (in g)	Max. Length (in mm)	Max. Width (in mm)	Max. Thickness (in mm)	Basal Width (in mm)	Basal Concavity (in mm)	LF 1 (in mm)	LF 2 (in mm)	LMG 1 (in mm)	LMG 2 (in mm)	Remarks
A-33115	Murray Springs/Area 4	Nearly complete	Chert	Yellowish brown 10YR 5/4 banded with reddish brown 5YR 4/4	21	68.8 (81.1)	30.8	9.3	27	5	27.6	36.6	34	33.5	
A-33924	Murray Springs/Area 4	Nearly complete/ missing both corners	Chalcedony	Dark greenish gray Gley1 4/10GY	24	75.2	31.3	9.3	—	—	30.9	26.8	24*	25*	
A-33925	Murray Springs/Area 4	Tip	Chert	Reddish brown chert		5*	13*	4*	—	—	—	—	—	—	Not available at ASM. Metrics from Huckell (2007)
A-33917	Murray Springs/Area 5	Proximal	St. David chalcedony	Light gray 5Y 7/2	4	18.3*	28.3*	7*	27.1	3.6	—	—	—	—	
A-33922	Murray Springs/Area 5	Proximal	Chert	Very pale brown 10YR 7/4	4	23.9*	28*	6.5*	27.9	4.5	—	—	—	—	
A-32992	Murray Springs/Area 6	Complete	Chert	Brown 7.5YR 5/4 mottled with pale yellow 2.5Y 7/4	25	71.8	33.3	9.6	30.9	5	19	25.8	31.5	35.5	
A-32991	Murray Springs/Area 6	Midsection	Chert	Yellowish brown 10YR 5/4 mottled with light yellow and white	32	86.7*	32.3	10.2	—	—	—	—	—	—	
A-47139	Murray Springs/Area 7	Proximal	Obsidian	—		10.1*	19.7*	4.8*	19.5	2.5	—	—	—	—	Not available at ASM. Metrics from a cast at ASM
A-47140	Murray Springs/Area 7	Proximal	St. David chalcedony	Olive yellow 2.5Y 6/6	2	14.1*	29.5*	5.6*	27.4	1.5	—	—	—	—	
A-47162	Murray Springs/Area 7	Proximal	Chert	Yellowish brown 10YR 5/8 banded with yellowish brown 10YR 5/6	21	49.6*	33.2	11.2	27.3	4	31.8	45	28	30	
A-47283	Murray Springs/Area 7	Proximal	Chert	Yellow 7/6	5	22.2*	33.2*	6.7*	30.8	5	—	—	—	—	

Table C.10. (continued)

ASM Number	*Site*	*Condition*	*Raw Material*	*Munsell Color*	*Weight (in g)*	*Max. Length (in mm)*	*Max. Width (in mm)*	*Max. Thickness (in mm)*	*Basal Width (in mm)*	*Basal Concavity (in mm)*	*LF 1 (in mm)*	*LF 2 (in mm)*	*LMG 1 (in mm)*	*LMG 2 (in mm)*	*Remarks*
A-10899	Naco	Complete	Chert	Dark brown 7.5YR 5/2 to dark brown 10YR 5/3 banded	—	70.8	25.8	7.8	19.4	2.8	21.1	20.4	31	—	Cast
A-10902	Naco	Complete	Chert	Dark gray 2.5YR 4/0 grading to very dark gray 2.5YR 3/0. Banded, translucent at edges	—	67.7	31.3	7.5	26.8	4.7	19.6	21.4	30	—	Cast
A-10904	Naco	Complete	Chert	Dusky red 10R 3/2	—	57.8	27.8	7.3	21.6	2.3	29.7	23.4	32	—	Cast
A-11912	Naco	Complete	Chert	Dusky red 10R 3/2 to 10R 2/1; reddish black near tip and base	—	94.2	25.8	9.2	22.8	2	29.3	26.5	32	—	Cast
A-11913	Naco	Complete	Chert	Reddish gray 5YR 5/2 to dark gray 5YR 4/1. Some mottling	—	97	30.5	9.7	26.6	2.3	27.8	31.1	35	—	Cast
A-10900	Naco	Nearly complete/ one broken corner	Felsite	Very dark gray 2.5YR 3/0	—	67.8	27.1	7.8	27*	3.8*	31.3	34.6	35	—	Cast
A-10903	Naco	Nearly complete/ one broken corner	Chert	Gray 10YR 4/1 to light brownish gray 10YR 6/2 mottled	—	116.9	34	9.5	29.3*	5.7*	26.2	40.3	36	—	Cast
A-10901	Naco	Nearly complete/ tip fracture	Felsite	Very dark gray 2.5YR 3/0	—	80.8 (83)	29.8	7.9	27	4.8	24.1	27.2	35	—	Cast
A-11914	Naco	Nearly complete/ tip fracture	Chert	Gray 10YR 6/1	—	81.5	32	10	30	—	—	v	32	—	Cast not available. Metrics from Haury (1953)
A-42268	Navarrete	Midsection	St. David Chalcedony	Translucent Pale yellow 5Y 8/2	23	67.1*	30.7*	10.2*	—	—	—	—	—	—	

Key: *= Incomplete measurement; ( ) = Estimated measurement; ASM = Arizona State Museum; LF1= Length of fluting on side 1; LF2= Length of fluting on side 2; LMG 1= Extent of lateral margin grinding 1; LMG 2= Extent of lateral margin grinding 2.

Table C.11. Primary and Secondary Bifaces from Murray Springs

ASM No.	*Area*	*Raw Material*	*Munsell Color*	*Condition*	*Technological Stage*	*Weight (in g)*	*Max. Length (in mm)*	*Max Width (in mm)*	*Maximum Thickness (in mm)*
A-33113	4	St. David Chalcedony	5Y 7/1 Light gray	Proximal	Primary	47	36.4*	73.44*	22.12*
A-46365	6	Chert	10YR 5/6 Yellowish brown	Proximal	Primary	19	26.8*	66.48*	11.34*
A-47163/ A-47199	7	St. David Chalcedony	5Y 8/3 Pale yellow	Fragmentary/ 2 fragments refit	Primary	7	36*	33*	9*
A-32715	3	Chert	5Y 7/1 Light gray mottled with 2.5Y 7/4 Pale yellow	Distal	Secondary	18	48*	36*	11*
A-33923	5	Chalcedony	5Y 5/2 Olive gray, banded with 2.5 Y 5/4 Light olive brown	Complete	Secondary	20	63	36	9
A-47170	7	St. David Chalcedony	2.5Y 7/6 Yellow mottled with 10YR 6/8 Brownish yellow	Complete	Secondary	54	95.24	41.68	13.64
A-33918	5	Chalcedony	2.5Y 7/2 Light gray mottled with 2.5Y 8/3 Pale yellow	Complete/ 2 fragments refit	Secondary	50	89	42	13

Table C.12. Unifacial Tools from Murray Springs

ASM Number	*Area*	*Raw Material*	*Munsell Color*	*Condition*	*Blank*	*Tool Type*	*Weight (in g)*	*Max. Length (in mm)*	*Max. Width (in mm)*	*Max Thickness (in mm)*	*WE 1L (in mm)*	*WE 1A*	*WE 2L (in mm)*	*WE 2A*	*Spur*
A-33929	7	Chert	10YR 6/6 Brownish yellow banded with 10YR 7/4 very pale brown	Complete	Blade	End scraper	10	42	28	10	26	80°	—	—	
A-45615	7	Chert	10YR 7/6 yellow	Complete	Blade	End scraper	9	40	30	7	28	60°	—	—	
A-33928	7	Chert	7.5YR 6/8 Reddish yellow	Complete	Blade	End scraper	4	36.94	28	10	17	100°	—	—	Yes
A-33943	6	Chert	2.5Y 7/4 Pale yellow mottled with 7.5YR 5/8	Complete	Blade	End scraper	10	39	28	10	25		—	—	Yes
A-33112	1 (according to ASM label)/ 4(Huckell 2007)	Chalcedony	2.5Y 6/2 Light brownish gray banded with 5YR 5/4 Reddish brown	Complete	Flake	Composite tool	17	67	34	8	67	40°	34	40°	
A-45614	7	Chert	10YR 7/3 Very pale brown banded with 7.5YR 5/8 Strong brown	Complete	Flake	Lateral retouched flake	7	40	31	7	19*	90°	30*	70°	
A-47158	7	St. David chalcedony	2.5Y 5/4 Light olive brown	Complete	Flake	Lateral retouched flake	4	28	32	4	16*	65°	—	—	
A-47291	7	St. David chalcedony	2.5Y 6/8 Olive yellow	Fragmentary	Flake	Lateral retouched flake	10	45	35	5	47*	90°	29	*	
A-32717	1 (according to ASM label)/ 3(Huckell 2007)	Chert	7.5YR 5/3 Brown clouded with 10YR 7/2 Light gray	Semi-complete	Flake	Lateral retouched flake	2	28	32	3	31*	25°	—	—	

Table C.12. (continued)

ASM Number	*Area*	*Raw Material*	*Munsell Color*	*Condition*	*Blank*	*Tool Type*	*Weight (in g)*	*Max. Length (in mm)*	*Max. Width (in mm)*	*Max Thickness (in mm)*	*WE 1L (in mm)*	*WE 1A*	*WE 2L (in mm)*	*WE 2A*	*Spur*
74-52-1	7	Chert	10YR 6/6 Brownish yellow mottled with 7.5YR 5/6 Strong brown	Complete	Flake	Lateral retouched flake	52	78	53	13	73	35°	63	70°	
A-47264	7	Chert	10YR 6/6 Brownish yellow	Semi-complete/ refitted fragments	Flake	Lateral retouched flake	14	61	38	7	53	70°	46	60°	
A-46143	7	St. David chalcedony	2.5Y 6/6 Olive yellow clouded with 2.5Y 8/3 Pale yellow	Fragmentary	Flake	Side scraper	39	60	41	13	60	85°	—	—	
A-33919	5	Chalcedony	5Y 4/1 Dark gray	Complete	Flake	Lateral retouched flake	16	60	29	10	50	40°	—	—	
A-33935	3	St. David chalcedony	2.5Y 6/3 Light yellowish brown	Complete	Flake	Triple graver	1	29	14	3	3		3	—	
A-45703**	7	St. David chalcedony	—	—	Flake	Thin lateral retouched flake	—	53	52	9	39*	40°	—	—	
A-33347**	6	Chert	—	—	Flake	Lateral retouched flake	—	24	26	4	16*	70°	—	—	
A-32990**	6	Chert	—	—	Flake	Double graver	—	51	30	5	3		2	—	
A-46160**	3	St. David Chalcedony	—	—	Flake	Single graver	—	24	25	6	2*		—	—	
A-33936**	4	Chert	—	—	Flake	Lateral retouched flake	—	21	14	4	21*	30°	—	—	
A-47127**	7	Chert	—	—	Flake	Lateral retouched flake	—	18	13	5	10*	90°	—	—	
A-43053**	3	Chert	—	—	Flake	Lateral retouched flake	—	17	8	4	15*	70°	—	—	
A-45088**	3	Chert	—	—	Flake	Lateral retouched flake	—	28	9	4	12	50°	—	—	
A-43022**	7	Silicified siltstone	—	—	Flake	Lateral retouched flake	—	25	17	4	21*	30°	—	—	

Nine artifacts were not available for study at the Arizona State Museum and it is indicated in the "remarks" column. Data for these artifacts was obtained from Huckell (2007).

Key: *= Incomplete measurement; ** = not available at ASM; WE 1L= Length of working edge 1; WE 1A= Angle of working edge 1; WE 2L= Length of working edge 2; WE 2A= Angle of working edge 2.

Table C.13. Blades from Murray Springs

ASM Number	*Area*	*Condition*	*Weight (in g)*	*Max. Length (in mm)*	*Max. Width (in mm)*	*Max. Thickness (in mm)*	*Raw Material*	*Munsell Color*	*Macro Use-Wear*	*Remark*
A-46180	7	Complete	12	67.46	23.88	8.84	Chert	2.5Y 5/4 Light olive brown	Yes	
A-43024	7	Medial	6	30*	27*	8*	Chert	10YR 6/6 Brownish yellow	Yes	
A-33926	4	Complete	12	74	26	7	Chert	2.5Y 6/3 Light yellowish brown	Yes	
A-47266	7	Complete	16	61	27	12	Silicified siltstone	7.5YR 4/4 Brown	Yes	
A-46368	6	Distal	3	41*	18*	6*	Chalcedony	5Y 4/1 Dark gray		
A-32720	3	2 fragments refitted/proximal and medial	13	64*	27*	10*	Chert	2.5Y 7/4 Pale yellow	Yes	
A-43023	6	Distal	8	36*	19*	12*	Chert	2.5Y 7/4 Pale yellow		
A-43007	6	Complete	—	87	29	6	Chalcedony	—		Not available at ASM
A-46187	7	Complete	—	30	20	5	Silicified siltstone	—		Not available at ASM
A-46179	7	Complete	—	65	42	15	Silicified siltstone	—		Not available at ASM
A-33927	7	Distal	—	68*	42*	22*	Silicified siltstone	—		Not available at ASM
A-33930	7	Distal	—	113*	45*	20*	Silicified siltstone	—		Not available at ASM
A-45647	7	Indeterminate fragment	—	29*	15*	4*	Silicified siltstone	—		Not available at ASM

Note: Six artifacts were not available for study at the Arizona State Museum, as indicated in the "remarks" column. Data for these artifacts was obtained from Huckell (2007).

Key: ASM = Arizona State Museum; *= Incomplete measurement.

Table C.14. Unifacial and Expedient Tools from the Lehner Site

ASM Number	*Raw Material*	*Munsell Color*	*Condition*	*Blank*	*Tool Type***	*Weight (in g)*	*Max. Length (in mm)*	*Max. Width (in mm)*	*Max. Thickness (in mm)*	*WE 1L (in mm)*	*WE 1A (in °)*	*WE 2L (in mm)*	*WE 2A (in °)*
A-12688	Rhyolite	Grayish-brown 10YR, ranging from 5/2 through 7/2	Complete	Cobble	Cobble chopper	595	120.66	109.63	43.64	83.05	67°	—	—
A-12691	Chalcedony	Translucent light gray 7.5 R 7/10 mottled with streaks of white 10R 9/1, and weak red 10R 4/4	Fragmentary	Blade	Keeled scraper	26	64.04* (65)	31.38*	16.29*	48.0*	69°	47.08*	68°
A-12694	Rhyolite	Pale brown 10YR 6/3, speckled with dark brown 10YR 4/3	Fragmentary	Blade	Keeled scraper	23	63.47*	26.86*	15.3*	60.48*	52°	61.06*	55°
A-12693	Silicified limestone	Very dark gray 7.5R 3/0	Complete	Flake	Knife	55	70.71	53.92	13.36	38.2	20°	—	—
A-12692	Silicified limestone	Black 7.5R 2/0	Complete	Flake	Sidescraper	231	121.05	87.2	21.1	95.7	22°	—	—
A-12687	Chalcedony	Translucent grayish-brown 2.5Y 6/2, streaked with dark brown 10 YR 4/3	Complete	Flake	Sidescraper	46	65.36	46.11	15.46	64.11	74°	—	—
A-12689	Chalcedony	Opaque dark gray 5Y 4/1 with clouds of dusky red 10R 3/3	Complete	Flake	Sidescraper	96	111.02	57.79	19.54	108.12	50°	—	—
A-12690	Silicified limestone	Gray patina 5Y 6/1	Fragmentary	Flake	Sidescraper?	17	34.78* (35)	40.96*	7.34*	34.19*	58°	—	—

Key and abbreviations: *= Incomplete measurement; ** Tool type based on Haury, Sayles and Wasley (1959); () = Estimated measurement; WE 1L= Length of working edge 1; WE 1A= Angle of working edge 1; WE 2L= Length of working edge 2; WE 2A= Angle of working edge 2. ASM = Arizona State Museum

Upper and Lower Bonebeds, Raw Data

APPENDIX D

Table D.1. Upper Bonebed, *Cuvieronius* sp. and Proboscidean Bone and Teeth Consistent with *Cuvieronius* sp.

INAH Bag Number	*Bone Concen-tration*	*Taxon*	*Element*	*Portion of element*	*Side*	*State of fusion*	*Weather-ing stage*	*Orienta-tion to grid N*	*Unit*	*Level*	*Depth below datum (m)*	*E*	*N*	*Year Collec-ted*	*Comments*
58960	1	Proboscidean, consistent with *Cuvieronius* sp.	Calcaneus	Calcaneal tuberosity	Left	Nearly fused	4		10	5	11.96–12.01	153.5	544.42	2007	
46240	1	Proboscidean, consistent with *Cuvieronius* sp.	Caudal vertebra	Nearly complete		Unfused	3		9	6	11.84–11.87	151.5	544.39	2007	
46249	1	Proboscidean, consistent with *Cuvieronius* sp.	Caudal vertebra	Nearly complete		Unfused	3		9	6	11.85–11.89	151.48	544.25	2007	
58956	1	Proboscidean, consistent with *Cuvieronius* sp.	Caudal vertebra	Nearly complete		Unfused	3		9	6	11.89–11.93	151.58	544.12	2007	
58956	1	Proboscidean, consistent with *Cuvieronius* sp.	Caudal vertebra	Nearly complete		Unfused	3		9	6	11.85–11.89	151.78	544	2007	
58956	1	Proboscidean, consistent with *Cuvieronius* sp.	Caudal vertebra	Nearly complete		Unfused	3		9	6	11.85–11.89	151.78	544	2007	
58956	1	Proboscidean, consistent with *Cuvieronius* sp.	Caudal vertebra	Nearly complete		Unfused	3		9	6	11.83–11.86	151.8	544.15	2007	
58956	1	Proboscidean, consistent with *Cuvieronius* sp.	Caudal vertebra	Complete		Unfused	3		9	6	11.88–11.92	151.9	544.08	2007	
58983	1	Proboscidean, consistent with *Cuvieronius* sp.	Caudal vertebra	Body		Unfused	3		9	6	11.91–12.01	151.5–152	544–544.5	2007	
46250	1	*Cuvieronius* sp.	Fibula	Nearly complete	Left	Unfused proximal, fused distal	5	25	10	6	11.86–11.93	152.54–153.83	544.61–544.93	2007	
59132	1	*Cuvieronius* sp.	Fibula	Proximal >1/2	Right	Unfused	5		3	7	11.84	150.05	543.73	2007	
46249	1	Proboscidean, consistent with *Cuvieronius* sp.	First phalanx	Nearly complete		Fully fused	3		9	6	11.89–11.91	151.87	544.5	2007	
58984	1	Proboscidean, consistent with *Cuvieronius* sp.	First phalanx	Complete		Fully fused	3		10	5	11.91	152.34	544.5	2007	
59049	1	Proboscidean, consistent with *Cuvieronius* sp.	First phalanx	Complete	Right	Fully fused	3		6	7	11.94	153.52	543.6	2007	
58983	1	*Cuvieronius* sp.	Innominate	Nearly complete	Left	Nearly fused	5	5	9	6	11.81–12.01	151.5–152	544–544.5	2007	
59211	1	*Cuvieronius* sp.	Innominate	Acetabulum section, pubic body	Right		4	65	18	3	11.92	151.8	544.5	2007	
59053	1	Proboscidean, consistent with *Cuvieronius* sp.	Innominate	Iliac blade fragment			4		3	7	11.91	152.5-153	543–544	2007	Mild root etching

Table D.1. (continued)

INAH Bag Number	Bone Concentration	Taxon	Element	Portion of element	Side	State of fusion	Weathering stage	Orientation to grid N	Unit	Level	Depth below datum (m)	E	N	Year Collected	Comments
59050	1	Proboscidean, consistent with *Cuvieronius* sp.	Magnum	Complete	Left		3		6	7	11.92	153.75	543.75	2007	
46246	1	Proboscidean, consistent with *Cuvieronius* sp.	Metatarsal	Complete	Right	Fully fused	3		6	7	11.86–11.98	153–154	543–544	2007	
58996	1	Proboscidean, consistent with *Cuvieronius* sp.	Molar	Fragment					14	bloque	12.01	154.1	543.7	2007	
59121	1	Proboscidean, consistent with *Cuvieronius* sp.	Molar	Crown fragment					14	3	11.94	155.05	543.5	2007	
46249	1	Proboscidean, consistent with *Cuvieronius* sp.	First or second phalanx	Complete		Fully fused	3		9	6	11.87–11.9	151.94	544.44	2007	4cm gash-recent damage; root etching
46126	1	Proboscidean, consistent with *Cuvieronius* sp.	Rib	Short diaphysis			4	125	3	8	11.95	151.54	542.9	2007	
46126	1	Proboscidean, consistent with *Cuvieronius* sp.	Rib	Shaft fragment			4		3	8	11.95	151.54	542.9	2007	
46197	1	Proboscidean, consistent with *Cuvieronius* sp.	Rib	Shaft fragment			4			5	11.86	151.5	542.93	2007	
46239	1	Proboscidean, consistent with *Cuvieronius* sp.	Rib	Proximal shaft fragment			4	135	8	6	11.89–11.95	152.88	545	2007	
58966	1	Proboscidean, consistent with *Cuvieronius* sp.	Rib	Short diaphysis			3	55	10	5	11.9–11.95	152.7	544.55		
58972	1	Proboscidean, consistent with *Cuvieronius* sp.	Rib	Short diaphysis			4		3	8	11.92–11.94	151.55	542.98	2007	
58987	1	Proboscidean, consistent with *Cuvieronius* sp.	Rib	Mid-shaft fragment			4	145	6	7	11.91–11.97	152.45–152.8	543.47–543.7		
58988	1	Proboscidean, consistent with *Cuvieronius* sp.	Rib	Short diaphysis			3		3	8	11.96–11.98	151.8–152	543.2–544		
58989	1	Proboscidean, consistent with *Cuvieronius* sp.	Rib	Short diaphysis			4	55	3	8	11.93	151.85–152	543.27–543.38	2007	
58992	1	Proboscidean, consistent with *Cuvieronius* sp.	Rib	Shaft fragment					6	7	11.86–11.94	152.04–152.44	543.55–543.97	2007	
58995	1	Proboscidean, consistent with *Cuvieronius* sp.	Rib	Shaft fragment			5		6	7	11.99–12.02	152.55–152.68	543.59–543.71	2007	
59053	1	Proboscidean, consistent with *Cuvieronius* sp.	Sacral or caudal vertebra	Nearly complete		Unfused	3		3	7	11.91	152.5–153	543–544	2007	

(continued)

Table D.1. (continued)

INAH Bag Number	*Bone Concen-tration*	*Taxon*	*Element*	*Portion of element*	*Side*	*State of fusion*	*Weather-ing stage*	*Orienta-tion to grid N*	*Unit*	*Level*	*Depth below datum (m)*	*E*	*N*	*Year Collec-ted*	*Comments*
59053	1	Proboscidean, consistent with *Cuvieronius* sp.	Sacral vertebra	Nearly complete		Unfused	3		3	7	11.91	152.5–153	543–544	2007	
59053	1	Proboscidean, consistent with *Cuvieronius* sp.	Sacral vertebra	Nearly complete		Unfused	3		3	7	11.91	152.5–153	543–544	2007	
58957	1	Proboscidean, consistent with *Cuvieronius* sp.	Scaphoid	Nearly complete	Right		3		6	7	11.98–12.02	153.35	543.38	2007	5cm gash – recent damage
58969	1	Proboscidean, consistent with *Cuvieronius* sp.	Scapula	Proximal epiphysis					3	8	11.81–11.91	150.6–151.3	543.4–544	2007	
46246	1	Proboscidean, consistent with *Cuvieronius* sp.	Sesamoid	Nearly complete			3		6	7	11.86–11.98	153–154	543–544	2007	
58973	1	Proboscidean, consistent with *Cuvieronius* sp.	Sesamoid	Complete			3		3	8	11.93	151.95	543.42	2007	
58964	1	Proboscidean, consistent with *Cuvieronius* sp.	Thoracic or lumbar vert.	Incomplete			3		6	7	11.88–11.97	152.52–152.72	543.55–543.76	2007	
58965	1	*Cuvieronius* sp.	Tibia	Distal < 1/2	Left	Nearly fused	4	15	10	5	11.87–11.96	152.5–153	544–544.5		
59218	1	Proboscidean, consistent with *Cuvieronius* sp.	Tusk						18	3	11.94–12.04	154.5	543.7	2007	
46241	1	Proboscidean, consistent with *Cuvieronius* sp.	Vertebra	Neural arch			3		9	6	11.81–11.85	151.57	544.44	2007	
58969	1	Proboscidean, consistent with *Cuvieronius* sp.	Vertebra	Anterior epiphysis		Unfused	3		3	8	11.81–11.91	150.6–151.3	543.4–544	2007	
58981	1	Proboscidean, consistent with *Cuvieronius* sp.	Vertebra				4		8	6	11.89–11.93	152.5–153	545.5–546	2007	
59036	1	Proboscidean, consistent with *Cuvieronius* sp.	Vertebra						6	7	11.91–11.97	152–152.5	543–544	2007	
59046	1	Proboscidean, consistent with *Cuvieronius* sp.	Vertebra	Anterior epiphysis		Unfused	3		7	7	11.86–11.96	150–152	545–546	2007	
59053	1	Proboscidean, consistent with *Cuvieronius* sp.	Vertebra			Unfused	4		3	7	11.91	152.5–153	543–544	2007	
59053	1	Proboscidean, consistent with *Cuvieronius* sp.	Vertebra				4		3	7	11.91	152.5–153	543–544	2007	
59179	2	Proboscidean, consistent with *Cuvieronius* sp.	Caudal vertebra	Body		Unfused	3		19	4	11.89–11.99	154.5–155	546–546.5	2007	
59264	2	Proboscidean, consistent with *Cuvieronius* sp.	Caudal vertebra	Body		Unfused	3		16	3	11.99	155.95	544.72	2007	

Table D.1. (continued)

INAH Bag Number	Bone Concen-tration	Taxon	Element	Portion of element	Side	State of fusion	Weather-ing stage	Orienta-tion to grid N	Unit	Level	Depth below datum (m)	E	N	Year Collec-ted	Comments
59299	2	Proboscidean, consistent with *Cuvieronius* sp.	Caudal vertebra	Body		Unfused	3		19	3	11.92–11.96	155.04	546.03	2007	
59583	2	Proboscidean, consistent with *Cuvieronius* sp.	Caudal vertebra	Centrum			3		209	3	11.94–12.01	157.5–157.6	545.9–546	2008	
59218	2	Proboscidean, consistent with *Cuvieronius* sp.	Cranium	Fragment			3		18	3	11.97–12	155.9	545.3	2007	
59469	2	Proboscidean, consistent with *Cuvieronius* sp.	Cranium	Occipital condyle			4		18	4	11.99–12.06	155.5	545.5–546	2008	
59768	2	Proboscidean, consistent with *Cuvieronius* sp.	Cranium	Zygomatic	Left	Unfused	4	25	209	3	11.96–12.04	156.96–157.12	545.39–545.63	2008	
59183	2	Proboscidean, consistent with *Cuvieronius* sp.	Cuboid	Nearly complete	Right		3		19	4	11.89–11.99	155	546.76	2007	
59302	2	Proboscidean, consistent with *Cuvieronius* sp.	Cuboid	Complete	Left	Fully fused	3		19	3	11.94–11.97	155.55	546.49	2007	
59292	2	*Cuvieronius* sp.	Innominate	Nearly complete	Left	Nearly fused	5	75	19	3	11.94	155	546.25	2007	
59860	2	Proboscidean, consistent with *Cuvieronius* sp.	Long bone	Shaft fragment			3		209	4	12.07–12.08	156.14	545.5		Mild natural abrasion
59867	2	Proboscidean, consistent with *Cuvieronius* sp.	Long bone	Shaft fragment			3		209	4	12.07–12.08	156.05	545.54	2008	Mild natural abrasion
59469	2	Proboscidean, consistent with *Cuvieronius* sp.	Lumbar vertebra	Body		Unfused	4		18	4	11.99–12.06	155.7	545.75		Articulated with 4 other lumbar vertebrae
59469	2	Proboscidean, consistent with *Cuvieronius* sp.	Lumbar vertebra	Body		Partially fused	4		18	4	11.99–12.06	155.7	545.75		Articulated with 4 other lumbar vertebrae
59469	2	Proboscidean, consistent with *Cuvieronius* sp.	Lumbar vertebra	Body		Partially fused	4		18	4	11.99–12.06	155.7	545.75		Articulated with 4 other lumbar vertebrae

(continued)

Table D.1. (continued)

INAH Bag Number	*Bone Concen-tration*	*Taxon*	*Element*	*Portion of element*	*Side*	*State of fusion*	*Weather-ing stage*	*Orienta-tion to grid N*	*Unit*	*Level*	*Depth below datum (m)*	*E*	*N*	*Year Collec-ted*	*Comments*
59469	2	Proboscidean, consistent with *Cuvieronius* sp.	Lumbar vertebra	Body		Partially fused	4		18	4	11.99–12.06	155.7	545.75		Articulated with 4 other lumbar vertebrae
59469	2	Proboscidean, consistent with *Cuvieronius* sp.	Lumbar vertebra	Body		Partially fused	4		18	4	11.99–12.06	155.7	545.75		Articulated with 4 other lumbar vertebrae
62394	2	*Cuvieronius* sp.	Mandible with articulated dP3, dP4, M1	Complete	Left and right		4		209	3	11.82–11.97	156.3	545	2008	
59001	2	Proboscidean, consistent with *Cuvieronius* sp.	Molar	Fragment					14	2	11.86–11.93	155.5–156	543.5–544	2007	
59421	2	Proboscidean, consistent with *Cuvieronius* sp.	Molar	Fragment					207		11.94	156.8	546.9	2008	
59429	2	Proboscidean, consistent with *Cuvieronius* sp.	Molar	Fragment					207	3	11.96	156.64	546.86	2008	
59438	2	Proboscidean, consistent with *Cuvieronius* sp.	Molar	Fragment					207	3	11.84–12.0	156.5–157	546.5–547	2008	
59439	2	Proboscidean, consistent with *Cuvieronius* sp.	Molar	Root					207	3	11.95–11.97	156.57–156.61	546.42–546.46	2008	
59440	2	Proboscidean, consistent with *Cuvieronius* sp.	Molar	Root					207	3	11.93–11.96	156.7–157	546.45–546.55	2008	
59442	2	Proboscidean, consistent with *Cuvieronius* sp.	Molar	Fragment					206		11.91–11.92	156.56	544.56	2008	
59516	2	Proboscidean, consistent with *Cuvieronius* sp.	Molar	Fragment					206		11.905–11.89	156.25	544.68	2008	
59759	2	Proboscidean, consistent with *Cuvieronius* sp.	Molar	Fragment					209	3	11.84–11.92	156.5–157.6	545.5–546	2008	
59778	2	Proboscidean, consistent with *Cuvieronius* sp.	Molar	Root			4		206	3	11.88–11.96	156.5–157	544.5–545	2008	
62347	2	Proboscidean, consistent with *Cuvieronius* sp.	Molar	Fragment					209	4	11.95–12.00	156–156.5	545–545.5	2008	
59183	2	Proboscidean, consistent with *Cuvieronius* sp.	Navicular	Nearly complete			3		19	4	11.94	155.14	546.84	2007	
59183	2	Proboscidean, consistent with *Cuvieronius* sp.	Phalanx	Nearly complete			3		19	4	11.89–11.99	155.4	546.9	2007	

Table D.1. (continued)

INAH Bag Number	Bone Concentration	Taxon	Element	Portion of element	Side	State of fusion	Weathering stage	Orientation to grid N	Unit	Level	Depth below datum (m)	E	N	Year Collected	Comments
59861	2	Proboscidean, consistent with *Cuvieronius* sp.	Phalanx	Complete		Unfused	3		19	4	11.98–12.02	155	546.75	2008	
59233	2	Proboscidean, consistent with *Cuvieronius* sp.	Radius	Proximal <1/2	Right	Fully fused	5		18	3	11.94–12.02	154.58	545.5	2007	Mild root etching
46050	2	Proboscidean, consistent with *Cuvieronius* sp.	Rib	Shaft fragment			5		2	7	12.12	158.9	543.15	2007	
59123	2	Proboscidean, consistent with *Cuvieronius* sp.	Rib	Proximal shaft fragment			4		19	3	11.89	153–153.32	546.5–546.6	2007	
59206	2	Proboscidean, consistent with *Cuvieronius* sp.	Rib	Shaft fragment			4	55	18	3	11.95–11.99	155.24–155.36	545.41–545.49	2007	Moderate root etching
59217	2	Proboscidean, consistent with *Cuvieronius* sp.	Rib	Short diaphysis			4		19	3	11.96–11.99	155.15	546.63	2007	
59217	2	Proboscidean, consistent with *Cuvieronius* sp.	Rib	Short diaphysis			4	0	19	3	11.99	155.7	546.08	2007	
59217	2	Proboscidean, consistent with *Cuvieronius* sp.	Rib	Short diaphysis			4	145	19	3	12–12.02	155.7	546.04	2007	
59217	2	Proboscidean, consistent with *Cuvieronius* sp.	Rib	Shaft fragment			4	20	19	3	11.94–11.99	155.8	546.63	2007	
59217	2	Proboscidean, consistent with *Cuvieronius* sp.	Rib	Shaft fragment			4	100	19	3	11.94–11.97	155.9	546.26	2007	
59217	2	Proboscidean, consistent with *Cuvieronius* sp.	Rib	Short diaphysis			4	110	19	3	11.92–11.95	155.65–156	546.65	2007	Mild root etching
59218	2	Proboscidean, consistent with *Cuvieronius* sp.	Rib	Shaft fragment				125	18	3	11.96–12.04	154.55	545.53	2007	
59218	2	Proboscidean, consistent with *Cuvieronius* sp.	Rib	Shaft fragment				40	18	3	12.02–12.04	155.12	545.92	2007	
59218	2	Proboscidean, consistent with *Cuvieronius* sp.	Rib	Short diaphysis				25	18	3	12.02–12.04	155.3	545.95	2007	
59218	2	Proboscidean, consistent with *Cuvieronius* sp.	Rib	Proximal <1/2		Unfused	3	60	18	3	11.98	155.4	545.9	2007	
59218	2	Proboscidean, consistent with *Cuvieronius* sp.	Rib	Shaft fragment				85	18	3	12.07	155.45	545.62	2007	

(continued)

Table D.1. (continued)

INAH Bag Number	Bone Concentration	Taxon	Element	Portion of element	Side	State of fusion	Weathering stage	Orientation to grid N	Unit	Level	Depth below datum (m)	E	N	Year Collected	Comments
59218	2	Proboscidean, consistent with *Cuvieronius* sp.	Rib	Short diaphysis				95	18	3	11.97–12.04	155.5	545.69	2007	
59218	2	Proboscidean, consistent with *Cuvieronius* sp.	Rib	Short diaphysis				135	18	3	11.99–12.08	155.68	545.92	2007	
59218	2	Proboscidean, consistent with *Cuvieronius* sp.	Rib	Shaft fragment			3	80	18	3	11.96–11.98	155.88	545.44	2007	
59293	2	Proboscidean, consistent with *Cuvieronius* sp.	Rib	Short diaphysis			4	140	18	3	11.92–12.12	154–156	545–546	2007	
59294	2	Proboscidean, consistent with *Cuvieronius* sp.	Rib	Short diaphysis			4	80	19	3	11.94–11.95	155.26	546.23	2007	
59297	2	Proboscidean, consistent with *Cuvieronius* sp.	Rib	Distal <1/2			4		18	3	11.92–12.12	155–155.5	545.5–546	2007	
59303	2	Proboscidean, consistent with *Cuvieronius* sp.	Rib	Short diaphysis			4	110	19	3	11.94–11.96	155.7	546.43	2007	
59419	2	Proboscidean, consistent with *Cuvieronius* sp.	Rib	Proximal <1/2			4	95	207	3	11.92	156–156.18	546.2	2008	
59462	2	Proboscidean, consistent with *Cuvieronius* sp.	Rib	Long diaphysis			5		206	3	11.915–12.04	157.10–157.63	544.65–544.65	2008	
59466	2	Proboscidean, consistent with *Cuvieronius* sp.	Rib	Short diaphysis			4		18	4	11.96–12.04	155.5–156	545–545.5	2008	Mild root etching
59745	2	Proboscidean, consistent with *Cuvieronius* sp.	Rib	Proximal 1/2	Right		4		209	3	11.92–11.98	156–156.25	545.5–546	2008	
59745	2	Proboscidean, consistent with *Cuvieronius* sp.	Rib	Long diaphysis			4	5	209	3	11.92–11.98	156–156.25	545.5–546	2008	
59762	2	Proboscidean, consistent with *Cuvieronius* sp.	Rib	Proximal <1/2			5		209	3	11.91–11.96	156–156.20	545.43–545.63	2008	
59766	2	Proboscidean, consistent with *Cuvieronius* sp.	Rib	Proximal <1/2	Right		4	55	209	3	11.94–12.01	156.6	546		Three parallel marks– recent damage
59767	2	Proboscidean, consistent with *Cuvieronius* sp.	Rib	Proximal 1/2			5	10	209	3	11.93–11.99	156.38–156.48	545.68–546.08	2008	
—	2	Proboscidean, consistent with *Cuvieronius* sp.	Scapula	Nearly complete	Right	Unfused	5	50	209	3	11.89–12.01	156.2	545.3		
59441	2	*Cuvieronius* sp.	Stylohyoid	Half	Left	Unfused	3	175	207	3	11.93–11.98	156.72–156.79	546.63–546.79	2008	

Table D.1. (continued)

INAH Bag Number	Bone Concentration	Taxon	Element	Portion of element	Side	State of fusion	Weathering stage	Orientation to grid N	Unit	Level	Depth below datum (m)	E	N	Year Collected	Comments
59875	2	*Cuvieronius* sp.	Stylohyoid	Half	Right	Unfused	3	90	209	4	12	156.44	545	2008	
59183	2	Proboscidean, consistent with *Cuvieronius* sp.	Third phalanx	Nearly complete			3		19	4	11.95	154.6	546.8	2007	
59217	2	Proboscidean, consistent with *Cuvieronius* sp.	Thoracic vertebra	Incomplete			3		19	3	11.97–12.02	155.62	546.18	2007	
59469	2	Proboscidean, consistent with *Cuvieronius* sp.	Thoracic vertebra	Incomplete			4		18	4	11.99–12.06	155.5	545.5–546	2008	
59217	2	Proboscidean, consistent with *Cuvieronius* sp.	Thoracic vertebra	Incomplete			4		19	3	11.92–11.95	155.65–156	546.65	2007	
59584	2	Proboscidean, consistent with *Cuvieronius* sp.	Thoracic vertebra	Incomplete			3		209	3	11.95–12	157.52–157.6	545.92–546	2008	Mild root etching
59184	2	*Cuvieronius* sp.	Tibia	Nearly complete	Left	Unfused	4		19	4	11.91	154.7–155.86	546.7–546.9	2007	
—	2	*Cuvieronius* sp.	Tibia	proximal <1/2	Right	Unfused	4		18		11.98	154.5	545.75	2007	
58999	2	Proboscidean, consistent with *Cuvieronius* sp.	Tusk						14	2	11.86–11.93	154.5–155	543.5–544	2007	
59118	2	Proboscidean, consistent with *Cuvieronius* sp.	Tusk						19	3		154–154.5	546–546.5	2007	
46238	2	Proboscidean, consistent with *Cuvieronius* sp.	Vertebra	Body		Unfused	4		5	6	11.85–11.93	153.73	546.8	2007	
59179	2	Proboscidean, consistent with *Cuvieronius* sp.	Vertebra	Neural arch			3		19	4	11.89–11.99	154.5–155	546–546.5	2007	
59179	2	Proboscidean, consistent with *Cuvieronius* sp.	Vertebra	Anterior epiphysis		Unfused	4		19	4	11.89–11.99	154.5–155	546–546.5	2007	
59179	2	Proboscidean, consistent with *Cuvieronius* sp.	Vertebra	Anterior epiphysis		Unfused	4		19	4	11.89–11.99	154.5–155	546–546.5	2007	
59208	2	Proboscidean, consistent with *Cuvieronius* sp.	Vertebra				4		19	3	11.91–11.99	155.5–156	546.5–547	2007	
59217	2	Proboscidean, consistent with *Cuvieronius* sp.	Vertebra	Fragment			3		19	3	11.92–11.94	155.86	546.44	2007	Mild root etching
59288	2	Proboscidean, consistent with *Cuvieronius* sp.	Vertebra				3		18	3	12.01–12.02	155.66	545.44	2007	
59295	2	Proboscidean, consistent with *Cuvieronius* sp.	Vertebra				4	75	18	3	11.92–12.02	155.5–156	545–545.5	2007	

(continued)

Table D.1. (continued)

INAH Bag Number	*Bone Concen-tration*	*Taxon*	*Element*	*Portion of element*	*Side*	*State of fusion*	*Weather-ing stage*	*Orienta-tion to grid N*	*Unit*	*Level*	*Depth below datum (m)*	*E*	*N*	*Year Collec-ted*	*Comments*
59297	2	Proboscidean, consistent with *Cuvieronius* sp.	Vertebra				4		18	3	11.92–12.12	155–155.5	545.5–546	2007	
59298	2	Proboscidean, consistent with *Cuvieronius* sp.	Vertebra	Neural arch			4		18	3	11.92–12.12	155.5–156	545.5–546	2007	
59300	2	Proboscidean, consistent with *Cuvieronius* sp.	Vertebra	body		Unfused	4		19	3	11.92–12.12	155.5–156	546–546.5	2007	
59448	2	Proboscidean, consistent with *Cuvieronius* sp.	Vertebra				4		18	3	11.99–12.02	155.18–155.25	545.78–545.89	2008	
59469	2	Proboscidean, consistent with *Cuvieronius* sp.	Vertebra	Fragment			4		18	4	11.99–12.06	155.5	545.5–546	2008	
59757	2	Proboscidean, consistent with *Cuvieronius* sp.	Vertebra	Pre-zyga-pophysis			3		209	3	11.95–11.98	156.32–156.4	545.08–545.12	2008	
74856A		Proboscidean, consistent with *Cuvieronius* sp.	Carpal or tarsal		Left				404	10	12.00–12.03	147.5–148	546–546.5	2020	
74741A		Proboscidean, consistent with *Cuvieronius* sp.	Humerus	Proximal epiphysis		Fused			402	5	11.70–11.76	146.28–146.41	543.95–544.04	2020	
74800A		Proboscidean, consistent with *Cuvieronius* sp.	Ischium	Incomplete	Right	Unfused				5	11.72–11.885	146.83–147.09	545.2–545.7	2020	
74843A		Proboscidean, consistent with *Cuvieronius* sp.	Long bone	Epiphysis					404	9	11.95–12.0	147.09–147.14	546–546.05	2020	
74772A		Proboscidean, consistent with *Cuvieronius* sp.	Molar	Fragment					401	8	11.81–11.86	146.5–147	543.5–544	2020	
74787A		Proboscidean, consistent with *Cuvieronius* sp.	Molar	Fragment					401	9	11.86–11.91	147.5–478	543.5–544	2020	
74828A		Proboscidean, consistent with *Cuvieronius* sp.	Molar	Fragment					404	8	11.91–11.92	146.74–146.76E	546.19–546.22	2020	
74836A		Proboscidean, consistent with *Cuvieronius* sp.	Molar	Fragment					403	7	11.83–11.88	147–147.5	545–545.5	2020	
74834A		Proboscidean, consistent with *Cuvieronius* sp.	Phalanx						403	7	11.83–11.88	147.5–148	545–545.5	2020	
74860A		Proboscidean, consistent with *Cuvieronius* sp.	Rib	Proximal					402		11.90–11.97	146.94–147.03	544.4–544.47	2020	
62624		Proboscidean, consistent with *Cuvieronius* sp.	Thoracic vertebra	Incomplete			4		233	7	12.05–12.14	160	543.4	2008	

Table D.2. Upper Bonebed, *Paramylodon harlani*

INAH bag number	Taxon	Element	Count	Unit	Level	Depth below datum (m)	E	N	Year Collected
63057	*Paramylodon harlani*	Dermal ossicle	1	302A	15	12.07–12.12	158.0–158.5	546.0–546.5	2010
74965A	*Paramylodon harlani*	Dermal ossicle	1	403	11	12.03–.08	146–146.5	545–545.5	2020
74776A	*Paramylodon harlani*	Dermal ossicle	1	402	8	11.81–.86	147–147.5	544–544.5	2020
74917A	*Paramylodon harlani*	Dermal ossicle	1	403	9	11.93–.98	146.5–147	545.5–546	2020
74854A	*Paramylodon harlani*	Dermal ossicle	1	402	11	11.96–12.01	147–147.5	544.5–545	2020
59884	*Paramylodon harlani*	Dermal ossicle	1	16		11.99–12.07	154.51	544.99	2008
63043	*Paramylodon harlani*	Dermal ossicle	1	302A		12.09	158.41	545.36	2010
63032	*Paramylodon harlani*	Dermal ossicle	1	308A	2	11.956–12.036	151.24	544.33	2010

Table D.3. Upper Bonebed, Order Rodentia

INAH Bag Number	Taxon	Element	Portion of Element	Side	Weathering stage	Unit	Level	Depth Below Datum (m)	E	N	Year Collected
59881	Rodentia	Humerus	Distal epiphysis	Right	3	18	4	11.93–12.05	154–155	545–546	2008

Table D.4. Upper Bonebed, Order Testudines

INAH Bag Number	Taxon	Element	Portion of element	Unit	Level	Depth Below Datum (m)	E	N	Year Collected
74864A	Testudines	Plastron	Fragment	403	8	11.88–.93	147–147.5	545–545.5	2020

Table D.5. Lower Bonebed, *Cuvieronius* sp.

INAH Bag Number	Strati-graphic Context	Taxon	Element	Portion of Element	Side	Mastodont Tooth Wear Stage	Mastodont Tooth Wear Age Category	State of Fusion	Weather-ing Stage	Orienta-tion to Grid N	Unit	Level	Depth Below Datum (m)	E	N	Year Collected	Comments
74560A	3A above contact	*Cuvieronius* sp.	Femur	Nearly complete	Right					155	8	7	12.525–12.548	148.7–149.13	545.44–545.49	2018	
62626	Contact of 3A/2	*Cuvieronius* sp.	Cervical vertebra	Complete				Fully fused	3		231	8	13.105–13.14	161.05–161.25	545.57–545.76	2008	
74593A	3A above contact	*Cuvieronius* sp.	Fibula	Complete	Right			Distal fused, proximal unfused		55	17	1-3	12.32–12.41	148.74–148.91	544.71–544.82	2020	

(continued)

Table D.5. (continued)

INAH Bag Number	Strati-graphic Context	Taxon	Element	Portion of Element	Side	Mastodont Tooth Wear Stage	Mastodont Tooth Wear Age Category	State of Fusion	Weather-ing Stage	Orienta-tion to Grid N	Unit	Level	Depth Below Datum (m)	E	N	Year Collected	Comments
33492	Contact of 3A/2	*Cuvieronius* sp.	Fibula	Distal	Right					60	208	6	12.485–12.52	148.82–149.02	545.58–545.79	2018	
62631	Contact of 3A/2	*Cuvieronius* sp.	Fibula	Complete	Left			Fully fused	3	135	231	8	12.6–12.64	161.5	545.85	2008	
62625	Contact of 3A/2	*Cuvieronius* sp.	Innominate	Nearly complete	Right			Fully fused	4	140	227B	11	12.83–13.03	162.65–163.5	542.9–593.65	2008	
62628	Contact of 3A/2	*Cuvieronius* sp.	M2 (lower)	Nearly complete	Right	3	Young adult				227B	11	12.87–12.91	162.3–162.4	543.8–543.98	2008	
62542	Contact of 3A/2	*Cuvieronius* sp.	M2 (upper)	Nearly complete	Right	1	Youth				236	4	12.46–12.58	149.65–148.79	546.25–546.38	2008	
74566A	Contact of 3A/2	*Cuvieronius* sp.	M3 (upper or lower)	Incomplete		4	Adult late in life				8	6	12.461–12.535	153.7–153.21	545.18–545.28	2018	
33491	Contact of 3A/2	*Cuvieronius* sp.	Mandible with articulated M3	Incomplete	Left	2+	Adult				208	4-7	12.305–	149.06–149.2	545.69–545.95	2018	Paired with specimen 33493
33493	Contact of 3A/2	*Cuvieronius* sp.	Mandible with articulated M3	Nearly complete	Right	2+	Adult			150	7	4	12.33–12.47	150.37–150.68	545.64–548.15	2018	Paired with specimen 33491
33314	Contact of 3A/2	*Cuvieronius* sp.	Molar	Nearly complete		4					208	4	12.30–12.42	149.74–149.89	545.18–545.32	2018	
33374	Contact of 3A/2	*Cuvieronius* sp.	Molar	Nearly complete		4					7	5	12.49–12.545	150.78–150.91	545.96–546.03	2018	
74508	Contact of 3A/2	*Cuvieronius* sp.	Molar	Incomplete		4					8	7	12.557–12.635	152–152.14	545.15–545.3	2018	
33124	3A above contact	*Cuvieronius* sp.	Radius	Proximal	Left						8	3	12.34–12.36	152.13–152.23	545.53–545.65	2018	
33541	3A above contact	*Cuvieronius* sp.	Radius	Distal epiphysis	Right						5	9	12.39–12.459	153.13–153.26	546.26–546.38	2018	
62629	Contact of 3A/2	*Cuvieronius* sp.	Tibia	Nearly complete	Left			Fully fused	4	140	231	8	12.515–12.565	161.11–161.47	545.67–546.07	2008	
62630	Contact of 3A/2	*Cuvieronius* sp.	Tusk	Medial <1/2						165	231	8	12.61–12.75	162	546	2008	
33514	Contact of 3A/2	*Cuvieronius* sp.	Ulna	Proximal						50	7	8	12.61–12.63	151.29–151.37	545.4–545.46	2018	

Table D.6. Lower Bonebed, *Mammut americanum*

INAH Bag Number	Stratigraphic Context	Taxon	Element	Portion of Element	Side	Mastodont Tooth Wear Stage	Mastodont Tooth Wear Age Category	Orientation to Grid N	Unit	Level	Depth Below Datum (m)	E	N	Year Collected
74511A	Contact of 3A/2	*Mammut americanum*	M1 or M2 (upper)	Incomplete		4	Young adult/ adult		8	7	12.52–12.63	152.18–152.32	545.39–545.57	2018
62622	Contact of 3A/2	*Mammut americanum*	M2 (upper or lower)	Nearly complete		0-1	Youth		227B	11	12.89–12.92	162–164	542.1–544	2008
—	Contact of 3A/2	*Mammut americanum*	M2 (upper)	Nearly complete	right	3	Young adult		2	11	12.74	158–159	543–544	2007
59178	Contact of 3A/2	*Mammut americanum*	M3 (lower)	Nearly complete	right	4	Adult late in life		1	10	12.66–12.78	157.432–157.65	543.29–543.48	2008
62623	Contact of 3A/2	*Mammut americanum*	M3 (lower)	Fragment		4	Adult late in life		227B	11	12.9–13.01	162–164	542.1–544	2008
62546	Contact of 3A/2	*Mammut americanum*	molar	Fragment					231	9	12.97–12.98	161–162	543–544	2008
—	Contact of 3A/2	*Mammut americanum*	tusk	Distal <1/2				165	204	4	12.36–12.56	151	542.5–543	2008
74564A	Contact of 3A/2	*Mammut americanum*	tusk	Medial <1/2				50	5		12.326–12.49	152.65–153.76	545.94–546.66	2018

Table D.7. Lower Bonebed, *Mammuthus* sp.

Stratigraphic Context	Taxon	Element	Portion of Element	Year Collected	Comments
Eroded from Locus 1 landform	*Mammuthus* sp.	Molar	Fragments	2007	Concentration of 5 fragments
Surface 140m west of Locus 1	*Mammuthus* sp.	Molar	Fragments	2007	Concentration of 17 fragments

Table D.8. Lower Bonebed, *Equus* sp.

INAH Bag Number	Stratigraphic Context	Taxon	Element	Portion of Element	Side	State of Fusion	Weather-ing Stage	Orientation to Grid N	Unit	Level	Depth Below Datum (m)	E	N	Year Collected
46027	Bioturbated context	*Equus* sp.	Molar or premolar (lower)	Nearly complete	Left	—	—	—	—	—	12.234	164.33	540.46	2007
62382	Bioturbated context	*Equus* sp.	Molar or premolar (lower)	Fragment	—	—	—	—	227A	8	12.32–12.53	162–162.5	545–545.5	2008
46178	Contact of 3A/2	*Equus* sp.	Incisor (upper or lower)	Nearly complete	—	—	—	—	2	12	12.85	159.09	542.95	2007
62611	Contact of 3A/2	*Equus* sp.	Molar or premolar (upper)	Crown	Right	—	—	—	231	8	12.65–12.66	161.78–161.82	547.57–545.62	2008
62627	Contact of 3A/2	*Equus* sp.	Radioulna	Complete	Right	Fully fused	4	120	231	8	12.61	161.74–162.10	543.44–543.62	2008
—	Eroded from primary context near Locus 1	*Equus* sp.	Molar or premolar (lower)	Complete	Left	—	—	—	—	—	—	—	—	2007

Table D.9. Lower Bonebed, *Paramylodon harlani*

INAH bag number	*Stratigraphic context*	*Taxon*	*Element*	*Count*	*Unit*	*Level*	*Depth below datum (m)*	*E*	*N*	*Year Collected*
59892	3A above contact	*Paramylodon harlani*	Dermal ossicle	1	—	—	12.12–12.42	155.68–155.88	543.6–543.88	2008
74579A	3A above contact	*Paramylodon harlani*	Dermal ossicle	1	17	1	12.31–12.36	148.0–148.5	544.5–545.0	2020
74596A	3A above contact	*Paramylodon harlani*	Dermal ossicle	1	17	2	12.39	149.61	544.27	2020
74692A	3A above contact	*Paramylodon harlani*	Dermal ossicle	1	17	2	12.36–12.41	148.5–149.0	544.5–545.0	2020
74898A	3A above contact	*Paramylodon harlani*	Dermal ossicle	1	401	13	12.06–12.11	147.0–147.5	543.5–544.0	2020
74908A	3A above contact	*Paramylodon harlani*	Dermal ossicle	1	404	13	12.15–12.2	146.0–146.5	546.0–546.5	2020
74933A	3A above contact	*Paramylodon harlani*	Dermal ossicle	1	401	14	12.11–12.138	147.5–148.0	543.0–543.5	2020
74944A	3A above contact	*Paramylodon harlani*	Dermal ossicle	2	404	14	12.2–12.25	147.0–147.5	546.0–546.5	2020
74980A	3A above contact	*Paramylodon harlani*	Dermal ossicle	2	402	15	12.16–12.21	146.5–147.0	544.0–544.5	2020
74982A	3A above contact	*Paramylodon harlani*	Dermal ossicle	2	205	2	12.14–12.19	148.0–148.5	546.0–546.5	2020
93901A	3A above contact	*Paramylodon harlani*	Dermal ossicle	2	403	12	12.08–12.13	146.5–147.0	545.5–546.0	2020
93915A	3A above contact	*Paramylodon harlani*	Dermal ossicle	2	205	4	12.24–12.29	148.5–149.0	546.0–546.5	2020
93921A	3A above contact	*Paramylodon harlani*	Dermal ossicle	1	403	13	12.13–12.18	146.5–147.0	545.5–546.0	2020
93932A	3A above contact	*Paramylodon harlani*	Dermal ossicle	1	403	13	12.13–12.18	147.18	545.12	2020
93942A	3A above contact	*Paramylodon harlani*	Dermal ossicle	1	402	16	12.21–12.26	146.5–147.0	544.0–544.5	2020
93946A	3A above contact	*Paramylodon harlani*	Dermal ossicle	1	403	14	12.18–12.24	146.0–146.5	545.5–546.0	2020
93976A	3A above contact	*Paramylodon harlani*	Dermal ossicle	2	403	15	12.24–12.3	146.5–147.0	545.5–546.0	2020
74752A	Contact of 3A/2	*Paramylodon harlani*	Dermal ossicle	3	17	4	12.96–exp.E2	148.5–149.0	544.0–544.5	2020
74573A	Stratum 3A, south profile	*Paramylodon harlani*	Molariform tooth	1	—	sup.	sup.	—	—	2020

Table D.10. Lower Bonebed, *Tapirus haysii*

INAH bag number	*Stratigraphic context*	*Taxon*	*Element*	*Portion of element*	*Side*	*E*	*N*	*Year Collected*	*Comments*
46256	Contact of 3A/2	*Tapirus haysii*	M3 (lower)	Complete	Left	163.34	542.65	2007	Identified by J.A.C., Eileen Johnson

Table D.11. Lower Bonebed, Family Camelidae

INAH Bag Mumber	*Stratigraphic Context*	*Taxon*	*Element*	*Portion of Element*	*Side*	*Unit*	*Level*	*Depth Below Datum (m)*	*E*	*N*	*Year Collected*	*Comments*
59345	Eroded from Loc. 1 landform	Camelidae	Astragalus	Incomplete	Right	—	sup.	sup.	160.9–161.1	542	2008	Identified by Alejandro Jiménez

Table D.12. Lower Bonebed, Order Proboscidea

INAH Bag Number	Stratigraphic Context	Taxon	Element	Portion of Element	Side	Mastodont Tooth Wear Stage	State of Fusion	Weathering Stage	Orientation to Grid N	Unit	Level	Depth Below Datum (m)	E	N	Year Collected
33481	Contact of 3A/2	Proboscidea	1st or 2nd phalanx							8	6	12.465–12.515	153.24–153.38	545.54–545.63	2018
62540	Contact of 3A/2	Proboscidea	3rd phalanx							235-236		12.46–12.55	150–150.1	546.11–546.22	2008
33081	3A above contact	Proboscidea	Carpal or tarsal							8	2	12.25–12.31	152.98–153.07	545–545.08	2018
33540	3A above contact	Proboscidea	Carpal or tarsal							5	7	12.395–12.442	153.4–153.53	546.1–546.25	2018
33300	Contact of 3A/2	Proboscidea	Flat bone	Fragment						9	8	12.61–12.64	150.87–150.91	544.59–544.67	2018
33250	Contact of 3A/2	Proboscidea	Flat bone	Fragment						9	7	12.58–12.6	151.25–151.38	544.86–545	2018
-	Contact of 3A/2	Proboscidea	Humerus	Long diaphysis (tube)	Left			4	5	235	6	12.48–12.59	150.5	546.3	2008
62618	Contact of 3A/2	Proboscidea	Long bone	Shaft fragment				4		231	8	12.65–12.75	161.97–162	545.98–546.02	2008
33366	Contact of 3A/2	Proboscidea	Long bone	Fragment						208	6	12.45–12.49	149.8–149.9	545.85–545.97	2018
33484	Contact of 3A/2	Proboscidea	Long bone	Fragment						8	6	12.47–12.516	153.1–153.22	545.55–545.76	2018
74515A	Contact of 3A/2	Proboscidea	M3 (upper or lower)			4				8	7	12.53–12.6	152.43–152.62	545.42–545.58	2018
33053	3A above contact	Proboscidea	Molar	Fragment						8	2	12.27	153.98	545.48	2018
33068	3A above contact	Proboscidea	Molar	Fragment						9	2	12.305	151.09–544.08		2018
33183	3A above contact	Proboscidea	Molar	Fragment						10	3	12.355–12.36	153.51–153.53	544.12–544.14	2018
33301	Contact of 3A/2	Proboscidea	Molar	Fragment						9	8	12.6–12.615	150.89–150.91	544.52–544.54	2018
33340	3A above contact	Proboscidea	Molar	Fragment						8	5	12.455–12.465	153.67–153.71	545.49–545.52	2018
33511	Contact of 3A/2	Proboscidea	Molar	Fragment		4				10	7	12.52–12.53	152.68–152.7	544.3–544.35	2018
33536	Contact of 3A/2	Proboscidea	Molar	Fragment						8	7	12.54–12.55	152.24–152.25	545.50–545.52	2018

(continued)

Table D.12. (continued)

INAH Bag Number	*Stratigraphic Context*	*Taxon*	*Element*	*Portion of Element*	*Side*	*Mastodont Tooth Wear Stage*	*State of Fusion*	*Weathering Stage*	*Orientation to Grid N*	*Unit*	*Level*	*Depth Below Datum (m)*	*E*	*N*	*Year Collected*
46187	Contact of 3A/2	Proboscidea	Molar	Crown fragment						2	12		158.5–159	543–543.5	2007
46192	Contact of 3A/2	Proboscidea	Molar	Crown fragment		4				2	12	12.89	159.55	543.42	2007
59677	Contact of 3A/2	Proboscidea	Molar	Crown fragment						227	12	12.85–12.87	162.32	544.6	2008
62398	Contact of 3A/2	Proboscidea	Molar							205	6	12.5–12.53	150.22	546.22	2008
62417	Contact of 3A/2	Proboscidea	Molar	Rot fragment						227B	11	12.95–12.98	163.53–163.66	543.17	2008
62550	Contact of 3A/2	Proboscidea	Molar	Fragment						231	9	12.92–12.95	161–162	543–544	2008
62613	Contact of 3A/2	Proboscidea	Molar	Root fragment						231	8	12.65–12.75	161.73–161.85	545.7–545.79	2008
62619	Contact of 3A/2	Proboscidea	Molar	Crown fragment						231	8	10.51–10.52	161.87–161.91	545.74–545.84	2008
74556	Contact of 3A/2	Proboscidea	Molar	Fragment						8	7	12.526–12.545	152.84–152.88	545.49–545.55	2018
74514A	Contact of 3A/2	Proboscidea	Molar			3				8	7	12.535–12.62	152.42–152.52	545.55–545.68	2018
74571A	Stratum 3A, north profile	Proboscidea	Molar	Fragment							sup.	sup.			2020
74950A	3A above contact	Proboscidea	Molar	Fragment						402	14	12.11–.16	147–147.5	544–544.5	2020
33490	3A above contact	Proboscidea	Flat bone	Fragment						10	6	12.34–12.42	152.7–152.79	544.93–545.03	2018
33520	Contact of 3A/2	Proboscidea	Phalanx or metapodial	Epiphysis						10	7	12.5–12.545	153.19–153.24	544.13–544.18	2018
33373	3A above contact	Proboscidea	Rib	Medial shaft <1/2					170	7		12.33–12.40	150.38–150.52	545.36–545.60	2018
33487	Contact of 3A/2	Proboscidea	Rib	Shaft fragment					70	8	6	12.44–12.515	153.38–153.52	545.65–545.73	2018
74509A	Contact of 3A/2	Proboscidea	Rib	Proximal shaft					100	8	7	12.565–12.615	152.17–152.45	545.22–545.33	2018
74696A	3A above contact	Proboscidea	Rib						95	17	2	12.355–.39	148.03–148.19	544.84–544.88	2020

Table D.12. (continued)

INAH Bag Number	Stratigraphic Context	Taxon	Element	Portion of Element	Side	Mastodont Tooth Wear Stage	State of Fusion	Weathering Stage	Orientation to Grid N	Unit	Level	Depth Below Datum (m)	E	N	Year Collected
74706A	Stratum 3A, north profile	Proboscidea	Rib	Distal < 1/2							sup.	sup.			2020
62396	Contact of 3A/2	Proboscidea	Scapula	Nearly complete	Right		Fully fused	4	135	227	6	12.62–12.78	162	544.5	2008
33483	Contact of 3A/2	Proboscidea	Carpal or tarsal							8	6	12.468–12.516	153.03–153.09	545.71–545.77	2018
33512	Contact of 3A/2	Proboscidea	Carpal or tarsal							7	8	12.595–12.635	151.32–151.41	545.15–545.26	2018
33258	3A above contact	Proboscidea	Thoracic vertebra	Body						208	4	12.315–12.38	148–148.15	545.21–545.34	2018
62621	Contact of 3A/2	Proboscidea	Thoracic vertebra	Complete			Fully fused	3	90	236	4		149.53–150	546.11–546.31	2008
59691	Contact of 3A/2	Proboscidea	Tusk	Distal < 1/2				4	165	227	12	12.8–12.87	162.3	545	2008
33063	3A above contact	Proboscidea	Tusk	Fragment						9	2	12.365	150.57–150.64	544.54–544.56	2018
33069	3A above contact	Proboscidea	Tusk	Fragment						8	2	12.265–12.3	152.54	545.23	2018
33155	3A above contact	Proboscidea	Tusk	Fragment						10	3	12.33–12.34	152.2–152.24	544.3–544.38	2018
33173	3A above contact	Proboscidea	Tusk	Fragment						8	3	12.33–12.38	153.1–153.19	545.38–545.49	2018
33246	3A above contact	Proboscidea	Tusk	Fragment						8	4	12.350	152.74–152.78	545.54–545.6	2018
33255	3A above contact	Proboscidea	Tusk	Fragment						8	5	12.44–12.45	152.94–152.98	545.28–545.31	2018
33256	3A above contact	Proboscidea	Tusk	Fragment						7	4	12.42–12.43	151–151.06	545.22–545.77	2018
33265	Contact of 3A/2	Proboscidea	Tusk	Fragment					80	7	5	12.475–12.48	150.93–151.02	545.74–545.78	2018
33296	Contact of 3A/2	Proboscidea	Tusk	Fragment						8	5	12.45–12.46	152.76–152.78	545.61–545.64	2018
33364	3A above contact	Proboscidea	Tusk	Fragment						10	5	12.415–12.42	152.35–152.41	544.18–544.24	2018
33480	Contact of 3A/2	Proboscidea	Tusk	Fragment						8	6	12.512–12.515	152.77–152.84	545.47–545.52	2018

(continued)

Table D.12. (continued)

INAH Bag Number	*Stratigraphic Context*	*Taxon*	*Element*	*Portion of Element*	*Side*	*Mastodont Tooth Wear Stage*	*State of Fusion*	*Weathering Stage*	*Orientation to Grid N*	*Unit*	*Level*	*Depth Below Datum (m)*	*E*	*N*	*Year Collected*
33482	Contact of 3A/2	Proboscidea	Tusk	Fragment						10	6	12.473–12.518	153.27–153.32	544.79–544.82	2018
33486	Contact of 3A/2	Proboscidea	Tusk	Fragment						8	6	12.48–12.504	153.15–153.21	545.79–545.85	2018
46175	Contact of 3A/2	Proboscidea	Tusk	Distal < 1/2				4	5	2	12	12.82	158.94	543	2007
62385	Contact of 3A/2	Proboscidea	Tusk							234	7	12.46–12.58	152	546.5	2008
74512A	Contact of 3A/2	Proboscidea	Tusk	Fragment						8	7	12.525–12.594	152.25–152.45	545.47–545.6	2018
74513A	Contact of 3A/2	Proboscidea	Tusk	Fragment						8	7	12.57–12.584	152.26–152.36	545.57–545.66	2018
74518A	Contact of 3A/2	Proboscidea	Tusk	Fragment						8	7	12.52–12.557	152.41–152.55	545.33–545.43	2018
74519A	Contact of 3A/2	Proboscidea	Tusk	Fragment						8	7	12.53–12.566	152.52–152.58	545.59–545.63	2018
74551A	Contact of 3A/2	Proboscidea	Tusk	Fragment						8	7	12.546–12.556	152.55–152.6	545.62–545.66	2018
74561A	3A above contact	Proboscidea	Vertebra	Body fragment			unfused		80	17	2-3	12.28–12.325	148.72–149.84	544.83–544.87	2018

Table D.13. Lower Bonebed, Family Leporidae

INAH Bag Number	*Stratigraphic Context*	*Taxon*	*Element*	*Portion of Element*	*Side*	*Unit*	*Level*	*Depth Below Datum (m)*	*E*	*N*	*Year Collected*
33209	3A above contact	Leporidae	humerus	distal	left	8	4	12.395–12.405	153.01–153.02	545.11	2018

Table D.14. Lower Bonebed, Order Rodentia

INAH Bag Number	Stratigraphic Context	Taxon	Element	Portion of Element	Side	Unit	Level	Depth Below Datum (m)	E	N	Year Collected	Comments
59889	Bioturbated context	Rodentia	Cranium	Complete		231	6	12.09–12.19	161.45	543.3	2008	
46209	Bioturbated context	Rodentia: *Spermophilus* spp.	Mandible	Complete	Right	11	3	13.03–13.13	150–152	514–516	2007	Identified by Eileen Johnson

Table D.15. Lower Bonebed, Order Testudines

INAH Bag Number	Stratigraphic Context	Taxon	Element	Portion of Element	Count	Unit	Level	Depth Below Datum (m)	E	N	Year Collected
46051	3A above contact	Testudines	Carapace	Fragment	1	2	7	12.1–12.2	158.5–159	543–544	2007
74976A	3A above contact	Testudines	Carapace	Fragment	1	401	14	12.11–.16	146–146.5	543.5–544	2020
74901A	3A above contact	Testudines	Plastron	Fragment	2	402	13	12.06–.11	147.5–148	544.5–545	2020
74931A	3A above contact	Testudines	Plastron	Fragment	1	402	13	12.06–12.11	146–146.5	544–544.5	2020
74953A	3A above contact	Testudines	Plastron	Fragment	1	402	14	12.11–.16	147–147.5	544.5–545	2020
74700A	Contact of 3A/2	Testudines	Carapace	Fragment	1	17	3	12.41–.46	149.5–150	544–544.5	2020
74758A	Contact of 3A/2	Testudines	Carapace	Fragment	3	17	4	12.46–exp.E2	148–148.5	544–544.5	2020
33400	Contact of 3A/2	Testudines	Plastron	Fragment	1	5	5	12.335	153.04	546.45	2018
74699A	Contact of 3A/2	Testudines	Plastron	Fragment	1	17	3	12.41–.46	149–149.5	544.5–545	2020
74744A	Contact of 3A/2	Testudines	Plastron	Fragment	1	17	3	12.41–.46	148	544–544.5	2020
74754A	Contact of 3A/2	Testudines	Plastron	Fragment	2	17	4	12.46–exp.E2	148.5–149	544.5–545	2020

Diatom Data

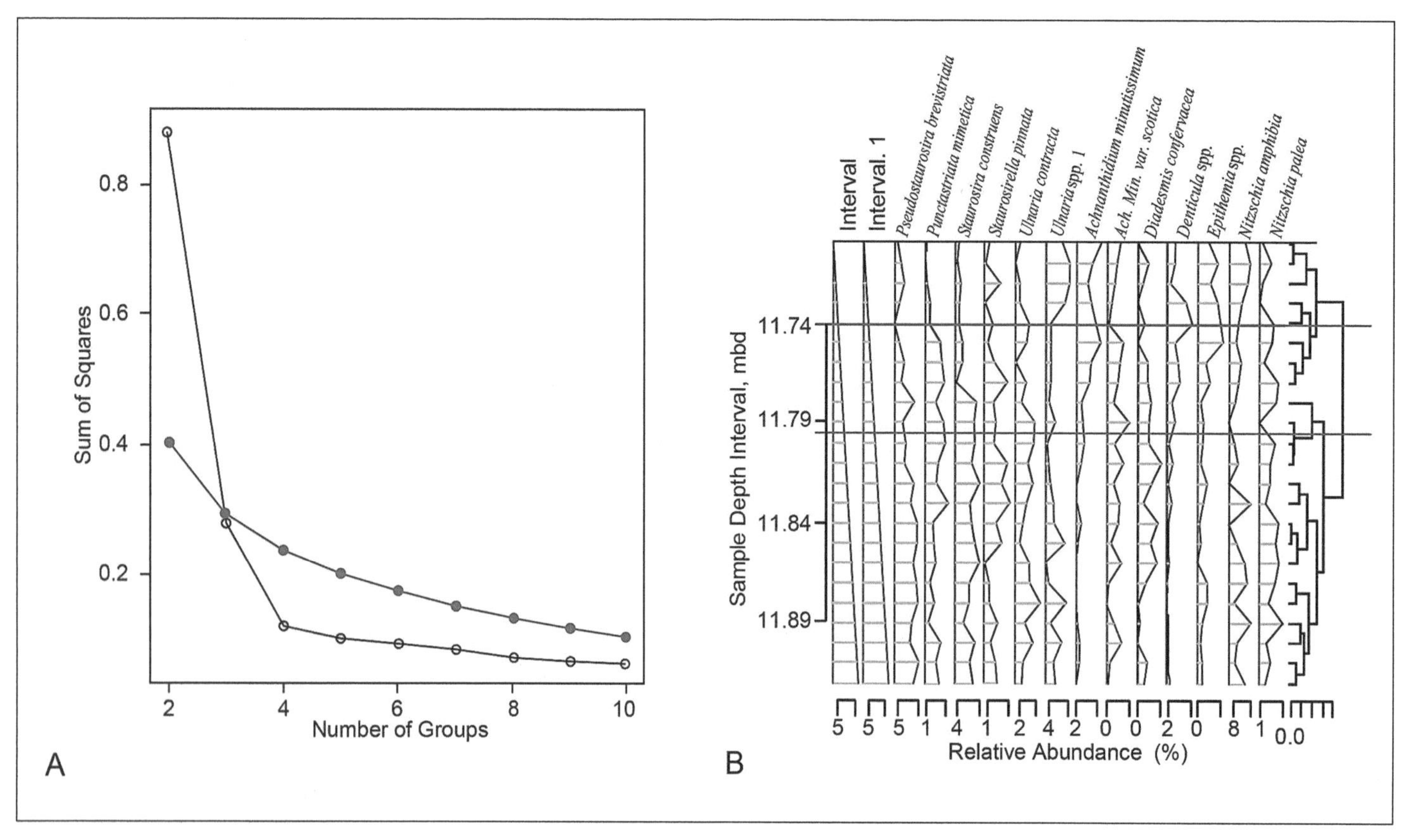

Figure E.1. Broken stick cluster model and superposition in Profile 10-4 stratigraphy. (A) Broken stick cluster analysis identified two significant shifts in the diatom assemblages (R software). The point at which the two black lines cross indicates significant shifts in the record (at least two, but no more than three). Neither axis reflects depth below the surface. (B) When the broken stick model is superimposed upon the stratigraphy in Profile 10-4, the two significant shifts are illustrated with the two horizontal lines. Dr. Victoria Chraibi, Tarleton University conducted the broken stick analysis for Profile 10-4. Profile 08-1 did not provide enough information. Drawing by Jim Abbott.

In the tables that follow, the dominant species are shown in **bold**.

In the tables, Profile 08-1 represents Strata 5, 4de, plus one sample from 3B, all collected in 5 cm segments (Stratum 4d was 5cm thick, unstratified, and not examined). Profile 10-4 represents lower 4de, all of 4d, and upper 3B, sampled in 5mm segments.

Table E.1. Total and Relative Abundance of Diatoms of Profiles 08-1 and 10-4, Assemblage I

Profile	Sample ID	Depth from site datum in meters below datum	Stratum	Diatoms	Microspheres	Chrysophytes	Diatoms Concentration / (# x 10^8 mL $^{-1}$ gdw)	*Achnanthidium levanderii*		***Achnanthidium minutissimum***		*Achnanthidium minutissimum var. scotica*		***Achnanthidium*** **spp. 1 (pot belly)**		*Amphora copulata-ovalis*		*Cocconeis placentula*		*Diploneis elliptica*		*Halamphora elongata*		*Halamphora veneta*		***Navicula spp* 1 (cf. *N. cryptocephela*)**		***Nitzschia amphibia***	
	(#)	(m bd)		(#)	(#)	(#)		(#)	(%)	(#)	(%)	(#)	(%)	(#)	(%)	(#)	(%)	(#)	(%)	(#)	(%)	(#)	(%)	(#)	(%)	(#)	(%)	(#)	(%)
08-1	1	10.98–11.03	5	208	7	36	3.630	0	0.0	21	10.1	1	0.5	1	0.5	0	0.0	0	0.0	1	0.5	0	0.0	0	0.0	4	1.9	4	1.9
	2	11.03–11.08		277	13	42	4.835	0	0.0	31	11.2	2	0.7	0	0.0	1	0.4	0	0.0	1	0.4	1	0.4	2	0.7	4	1.4	10	3.6
	3	11.08–11.13		204	24	42	3.560	0	0.0	12	5.9	3	1.5	0	0.0	0	0.0	0	0.0	1	0.5	0	0.0	0	0.0	3	1.5	4	2.0
	4	11.13–11.18		300	21	10	5.236	0	0.0	22	7.3	3	1.0	1	0.3	1	0.3	0	0.0	4	1.3	0	0.0	2	0.7	6	2.0	3	1.0
	5	11.18–11.23		339	9	27	5.917	0	0.0	44	13.0	0	0.0	0	0.0	7	2.1	0	0.0	2	0.6	0	0.0	1	0.3	5	1.5	13	3.8
	6	11.23–11.28		217	16	33	3.787	0	0.0	14	6.5	2	0.9	0	0.0	4	1.8	0	0.0	0	0.0	1	0.5	1	0.5	2	0.9	4	1.8
	7	11.28–11.33		310	2	2	5.411	0	0.0	57	18.4	0	0.0	0	0.0	2	0.6	0	0.0	2	0.6	0	0.0	2	0.6	2	0.6	11	3.5
	8	11.33–11.38	4de	346	5	14	1.132	0	0.0	47	13.6	1	0.3	0	0.0	5	1.4	0	0.0	0	0.0	1	0.3	2	0.6	1	0.3	13	3.8
	9	11.38–11.43		340	2	25	1.113	0	0.0	70	20.6	1	0.3	0	0.0	1	0.3	0	0.0	0	0.0	1	0.3	1	0.3	3	0.9	17	5.0
	10	11.43–11.38		331	12	59	1.195	0	0.0	49	14.8	0	0.0	0	0.0	2	0.6	1	0.3	1	0.3	4	1.2	1	0.3	4	1.2	11	3.3
	11	11.48–11.53		414	33	36	1.667	2	0.5	54	13.0	2	0.5	0	0.0	5	1.2	2	0.5	4	1.0	2	0.5	4	1.0	3	0.7	14	3.4
	12	11.53–11.58		350	6	25	1.527	0	0.0	42	12.0	2	0.6	0	0.0	3	0.9	0	0.0	2	0.6	2	0.6	3	0.9	0	0.0	37	10.6
	13	11.58–11.63		305	12	56	1.101	0	0.0	31	10.2	0	0.0	0	0.0	1	0.3	0	0.0	1	0.3	0	0.0	0	0.0	2	0.7	63	20.7
	14	11.63–11.68		318	11	25	2.775	0	0.0	57	17.9	0	0.0	0	0.0	1	0.3	0	0.0	1	0.3	0	0.0	0	0.0	4	1.3	87	27.4
	15	11.68–11.73		303	12	29	2.144	0	0.0	44	14.5	1	0.3	0	0.0	1	0.3	0	0.0	0	0.0	1	0.3	4	1.3	5	1.7	49	16.2
	16	11.88–11.93	3B	356	12	24	2.485	0	0.0	28	7.9	0	0.0	0	0.0	3	0.8	3	0.8	0	0.0	4	1.1	12	3.4	6	1.7	71	19.9

(continued)

Table E.1. (continued)

Profile	Sample ID	Depth from site datum in meters below datum	Stratum	Diatoms	Microspheres	Chrysophytes	Diatoms Concentration / (# x 10^8 mL $^{-1}$ gdw)	*Achnanthidium levanderii*		***Achnanthidium minutissimum***		*Achnanthidium minutissimum var. scotica*		***Achnanthidium*** **spp. 1 (pot belly)**		*Amphora copulata-ovalis*		*Cocconeis placentula*		*Diploneis elliptica*		*Halamphora elongata*		*Halamphora veneta*		***Navicula* spp 1 (cf. *N. cryptocephela*)**		***Nitzschia amphibia***	
	(#)	(m bd)		(#)	(#)	(#)		(#)	(%)	(#)	(%)	(#)	(%)	(#)	(%)	(#)	(%)	(#)	(%)	(#)	(%)	(#)	(%)	(#)	(%)	(#)	(%)	(#)	(%)
10-4	1	11.790–11.795	4de	306	15	13	1.842	1	0.3	47	15.4	10	3.3	1	0.3	0	0.0	1	0.3	1	0.3	3	1.0	0	0.0	4	1.3	36	11.8
	2	11.795–11.800		308	17	23	1.628	0	0.0	33	10.7	8	2.6	0	0.0	5	1.6	2	0.6	0	0.0	1	0.3	2	0.6	3	1.0	43	14.0
	3	11.800–11.805		304	15	0	1.817	0	0.0	25	8.2	6	2.0	0	0.0	4	1.3	1	0.3	3	1.0	0	0.0	5	1.6	1	0.3	39	12.8
	4	11.805–11.810		304	18	17	1.519	0	0.0	30	9.9	4	1.3	0	0.0	4	1.3	3	1.0	6	2.0	1	0.3	2	0.7	0	0.0	32	10.5
	5	11.810–11.815		304	17	21	1.613	0	0.0	39	12.8	2	0.7	0	0.0	2	0.7	1	0.3	1	0.3	2	0.7	5	1.6	1	0.3	29	9.5
	6	11.815–11.820		302	17	10	1.595	0	0.0	48	15.9	13	4.3	0	0.0	3	1.0	0	0.0	2	0.7	0	0.0	1	0.3	2	0.7	27	8.9
	7	11.820–11.825		308	26	9	1.058	0	0.0	30	9.7	10	3.2	0	0.0	3	1.0	2	0.6	0	0.0	3	1.0	1	0.3	2	0.6	32	10.4
	8	11.825–11.830		319	17	0	1.695	0	0.0	29	9.1	10	3.1	0	0.0	3	0.9	4	1.3	1	0.3	0	0.0	3	0.9	0	0.0	31	9.7
	9	11.830–11.835	Mixed 4d & de	312	4	0	6.991	1	0.3	14	4.5	6	1.9	0	0.0	4	1.3	0	0.0	1	0.3	0	0.0	2	0.6	1	0.3	27	8.7
	10	11.835–11.840	4d	302	4	8	6.775	0	0.0	17	5.6	19	6.3	0	0.0	5	1.7	1	0.3	5	1.7	0	0.0	2	0.7	0	0.0	19	6.3
	11	11.840–11.845		306	14	2	1.958	0	0.0	21	6.9	7	2.3	0	0.0	2	0.7	6	2.0	2	0.7	0	0.0	2	0.7	0	0.0	26	8.5
	12	11.845–11.850		320	0	2	0.0	0	0.0	15	4.7	14	4.4	0	0.0	6	1.9	1	0.3	3	0.9	0	0.0	9	2.8	1	0.3	30	9.4
	13	11.850–11.855		308	4	3	6.926	0	0.0	6	1.9	5	1.6	0	0.0	1	0.3	3	1.0	1	0.3	0	0.0	4	1.3	1	0.3	20	6.5
	14	11.855–11.860		317	1	2	2.834	0	0.0	6	1.9	11	3.5	0	0.0	4	1.3	4	1.3	0	0.0	1	0.3	0	0.0	2	0.6	43	13.6
	15	11.860–11.865		309	6	0	4.627	1	0.3	15	4.9	9	2.9	0	0.0	1	0.3	0	0.0	1	0.3	0	0.0	1	0.3	1	0.3	20	6.5
	16	11.865–11.870		307	3	4	9.132	0	0.0	8	2.6	4	1.3	0	0.0	2	0.7	0	0.0	1	0.3	0	0.0	2	0.7	0	0.0	29	9.4
	17	11.870–11.875		304	4	3	6.801	0	0.0	6	2.0	12	3.9	0	0.0	4	1.3	1	0.3	1	0.3	1	0.3	3	1.0	0	0.0	36	11.8
	18	11.875–11.880		301	6	8	4.461	0	0.0	6	2.0	2	0.7	0	0.0	5	1.7	2	0.7	1	0.3	0	0.0	4	1.3	2	0.7	38	12.6
	19	11.880–11.885		300	5	8	5.286	0	0.0	6	2.0	0	0.0	0	0.0	3	1.0	1	0.3	2	0.7	0	0.0	3	1.0	0	0.0	25	8.3
	20	11.885–11.890		310	4	7	6.846	0	0.0	9	2.9	7	2.3	0	0.0	0	0.0	8	2.6	5	1.6	0	0.0	5	1.6	0	0.0	45	14.5
	21	11.890–11.895	Mixed 3 & 4	307	5	3	4.816	0	0.0	12	3.9	12	3.9	0	0.0	4	1.3	2	0.7	0	0.0	0	0.0	2	0.7	2	0.7	29	9.4
	22	11.895–11.900		308	13	0	1.580	0	0.0	13	4.2	3	1.0	0	0.0	4	1.3	2	0.6	0	0.0	1	0.3	0	0.0	0	0.0	26	8.4
	23	11.900–11.905	3	302	12	5	1.122	0	0.0	6	2.0	1	0.3	0	0.0	2	0.7	0	0.0	0	0.0	0	0.0	3	1.0	1	0.3	35	11.6

Table E.1. (continued)

Profile	Sample ID	Depth from site datum in meters below datum	Stratum	***Nitzschia palea***		***Pinnularia borealis***		***Pinnularia* sp. cf. *P. saprophila***		***Pinnularia* spp. 2 (large)**		*Planothidium lanceolatum*		***Staurosirella pinnata***		***Ulnaria contracta***		**Assemblage I**
	(#)	(m bd)		(#)	(%)	(#)	(%)	(#)	(%)	(#)	(%)	(#)	(%)	(#)	(%)	(#)	(%)	(%)
08-1	1			6	2.9	1	0.5	5	2.4	0	27.0	0	0.0	2	1.0	10	4.8	27.0
	2	10.98–11.03	5	18	6.5	1	0.4	4	1.4	1	32.9	4	1.4	5	1.8	6	2.2	32.9
	3	11.03–11.08		4	2.0	7	3.4	0	0.0	0	22.2	2	1.0	2	1.0	7	3.4	22.2
	4	11.08–11.13		11	3.0	1	0.3	1	0.3	0	24.0	4	1.3	5	1.4	11	3.7	24.0
	5	11.13–11.18		13	3.8	1	0.3	5	1.5	0	28.9	1	0.3	1	0.3	5	1.5	28.9
	6	11.18–11.23		7	3.2	0	0.0	4	1.8	2	20.6	0	0.0	0	0.0	4	1.8	20.6
	7	11.23–11.28		10	3.2	2	0.6	9	2.9	0	33.2	1	0.3	1	0.3	5	1.6	33.2
	8	11.28–11.33		10	2.9	2	0.6	7	2.0	1	31.7	3	0.9	4	1.2	12	3.5	31.7
	9	11.33–11.38	4de	7	2.1	0	0.0	15	4.4	3	38.4	1	0.3	2	0.6	8	2.4	38.4
	10	11.38–11.43		18	5.4	0	0.0	5	1.5	0	31.0	1	0.3	2	0.6	4	1.2	31.0
	11	11.43–11.38		14	3.4	1	0.2	7	1.7	4	34.9	2	0.5	5	1.2	19	4.6	34.9
	12	11.48–11.53		15	4.3	1	0.3	3	0.9	1	34.8	1	0.3	3	0.9	6	1.7	34.8
	13	11.53–11.58		9	3.0	0	0.0	1	0.3	2	40.4	1	0.3	0	0.0	12	3.9	40.4
	14	11.58–11.63		7	2.2	0	0.0	7	2.2	2	55.0	1	0.3	0	0.0	8	2.5	55.0
	15	11.63–11.68		7	2.3	2	0.7	3	1.0	0	44.3	4	1.3	1	0.3	12	4.0	44.3
	16	11.68–11.73	3B	11	3.1	0	0.0	10	2.8	5	49.7	5	1.4	2	0.6	17	4.8	49.7

(continued)

Table E.1. (continued)

Profile	Sample ID	Depth from site datum in meters below datum	Stratum	***Nitzschia palea***		***Pinnularia borealis***		***Pinnularia* sp. cf. *P. saprophila***		***Pinnularia* spp. 2 (large)**		*Planothidium lanceolatum*		***Staurosirella pinnata***		***Ulnaria contracta***		**Assemblage I**
	(#)	(m bd)		#	%	#	%	#	%	#	%	#	%	#	%	#	%	%
10-4	1	11.790–11.795	4de	36	11.8	4	1.3	1	0.3	4	1.3	8	2.6	6	2.0	4	1.3	25
	2	11.795–11.800		43	14.0	9	2.9	0	0.0	2	0.6	14	4.5	6	1.9	1	0.3	33
	3	11.800–11.805		39	12.8	4	1.3	0	0.0	3	1.0	2	0.7	4	1.3	14	4.6	32
	4	11.805–11.810		32	10.5	2	0.7	0	0.0	3	1.0	2	0.7	6	2.0	1	0.3	27
	5	11.810–11.815		29	9.5	10	3.3	0	0.0	4	1.3	8	2.6	6	2.0	6	2.0	14
	6	11.815–11.820		27	8.9	9	3.0	0	0.0	4	1.3	3	1.0	3	1.0	3	1.0	14
	7	11.820–11.825		32	10.4	5	1.6	0	0.0	3	1.0	4	1.3	0	0.0	8	2.6	15
	8	11.825–11.830		31	9.7	13	4.1	0	0.0	4	1.3	9	2.8	2	0.6	19	6.0	15
	9	11.830–11.835	Mixed 4d & de	27	8.7	12	3.8	0	0.0	2	0.6	4	1.3	4	1.3	8	2.6	13
	10	11.835–11.840	4d	19	6.3	2	0.7	1	0.3	1	0.3	0	0.0	3	1.0	9	3.0	19
	11	11.840–11.845		26	8.5	11	3.6	1	0.3	4	1.3	2	0.7	12	3.9	7	2.3	11
	12	11.845–11.850		30	9.4	9	2.8	0	0.0	0	0.0	9	2.8	3	0.9	18	5.6	14
	13	11.850–11.855		20	6.5	8	2.6	0	0.0	3	1.0	1	0.3	3	1.0	14	4.5	14
	14	11.855–11.860		43	13.6	6	1.9	0	0.0	3	0.9	1	0.3	7	2.2	20	6.3	18
	15	11.860–11.865		20	6.5	13	4.2	0	0.0	2	0.6	3	1.0	1	0.3	9	2.9	18
	16	11.865–11.870		29	9.4	12	3.9	0	0.0	2	0.7	1	0.3	2	0.7	14	4.6	28
	17	11.870–11.875		36	11.8	14	4.6	0	0.0	0	0.0	3	1.0	1	0.3	2	0.7	10
	18	11.875–11.880		38	12.6	11	3.7	0	0.0	0	0.0	0	0.0	8	2.7	4	1.3	13
	19	11.880–11.885		25	8.3	7	2.3	0	0.0	0	0.0	1	0.3	1	0.3	4	1.3	28
	20	11.885–11.890		45	14.5	16	5.2	0	0.0	0	0.0	1	0.3	3	1.0	10	3.2	15
	21	11.890–11.895	Mixed 3 & 4	29	9.4	7	2.3	0	0.0	0	0.0	3	1.0	7	2.3	5	1.6	26
	22	11.895–11.900		26	8.4	8	2.6	1	0.3	1	0.3	0	0.0	2	0.6	9	2.9	17
	23	11.900–11.905	3	35	11.6	6	2.0	0	0.0	1	0.3	3	1.0	2	0.7	10	3.3	19

Table E.2. Total and Relative Abundance of Diatoms of Profiles 08-1 and 10-4, Assemblage II

Profile	Sample ID	Depth from site datum in meters below datum	Stratum	Diatoms	Microspheres	Chrysophytes	Diatoms Concentration / (# x 10^8 mL^{-1} gdw)	*Achnanthidium exiguum*		*Amphora pediculus*		***Aulacoseira islandica***		***Aulacoseira italica***		***Cyclotella atomus***		*Cyclotella comensis*		*Cyclostephanos* spp.		*Cymbopleura lapponica*		*Diadesmis confervacea*		*Diadesmis* spp. 1 (oval)		***Eunotia arcus/bilunaris***	
	(#)	(m bd)		(#)	(#)	(#)		(#)	(%)	(#)	(%)	(#)	(%)	(#)	(%)	(#)	(%)	(#)	(%)	(#)	(%)	(#)	(%)	(#)	(%)	(#)	(%)	(#)	(%)
	1	10.98–11.03		208	7	36	3.630	1	0.5	0	0.0	2	1.0	2	1.0	0	0.0	0	0.0	0	0.0	0	0.0	0	0.0	0	0.0	10	4.8
	2	11.03–11.08		277	13	42	4.835	0	0.0	2	0.7	2	0.7	1	0.4	0	0.0	0	0.0	1	0.4	0	0.0	2	0.7	0	0.0	13	4.7
	3	11.08–11.13		204	24	42	3.560	0	0.0	4	2.0	0	0.0	7	3.4	3	1.5	0	0.0	0	0.0	0	0.0	1	0.5	0	0.0	6	2.9
	4	11.13–11.18	5	300	21	10	5.236	0	0.0	2	0.7	16	5.3	3	1.0	6	2.0	0	0.0	2	0.7	0	0.0	4	1.3	2	0.7	3	1.0
	5	11.18–11.23		339	9	27	5.917	0	0.0	2	0.6	2	0.6	4	1.2	0	0.0	0	0.0	0	0.0	1	0.3	1	0.3	3	0.9	13	3.8
	6	11.23–11.28		217	16	33	3.787	0	0.0	7	3.2	0	0.0	0	0.0	0	0.0	0	0.0	0	0.0	0	0.0	1	0.5	0	0.0	13	6.0
	7	11.28–11.33		310	2	2	5.411	1	0.3	3	1.0	1	0.3	2	0.6	0	0.0	0	0.0	0	0.0	0	0.0	3	1.0	2	0.6	12	3.9
	8	11.33–11.38		346	5	14	1.132	1	0.3	2	0.6	4	1.2	1	0.3	0	0.0	0	0.0	0	0.0	0	0.0	0	0.0	1	0.3	11	3.2
08-1	9	11.38–11.43		340	2	25	1.113	1	0.3	1	0.3	0	0.0	1	0.3	0	0.0	0	0.0	0	0.0	2	0.6	0	0.0	1	0.3	10	2.9
	10	11.43–11.38		331	12	59	1.195	0	0.0	0	0.0	0	0.0	0	0.0	0	0.0	0	0.0	0	0.0	0	0.0	1	0.3	2	0.6	10	3.0
	11	11.48–11.53		414	33	36	1.667	0	0.0	1	0.2	2	0.5	5	1.2	3	0.7	0	0.0	1	0.2	0	0.0	0	0.0	2	0.5	10	2.4
	12	11.53–11.58	4de	350	6	25	1.527	1	0.3	0	0.0	2	0.6	5	1.4	0	0.0	0	0.0	0	0.0	0	0.0	0	0.0	0	0.0	29	8.3
	13	11.58–11.63		305	12	56	1.101	3	1.0	0	0.0	1	0.3	15	4.9	1	0.3	0	0.0	1	0.3	0	0.0	0	0.0	0	0.0	11	3.6
	14	11.63–11.68		318	11	25	2.775	0	0.0	0	0.0	1	0.3	16	5.0	0	0.0	0	0.0	0	0.0	0	0.0	1	0.3	0	0.0	15	4.7
	15	11.68–11.73		303	12	29	2.144	3	1.0	2	0.7	2	0.7	6	2.0	0	0.0	0	0.0	0	0.0	0	0.0	6	2.0	0	0.0	13	4.3
	16	11.88–11.93	3B	356	12	24	2.485	1	0.3	3	0.8	0	0.0	1	0.3	0	0.0	0	0.0	2	0.6	1	0.3	2	0.6	3	0.8	7	2.0

(continued)

Table E.2. (continued)

Profile	Sample ID	Depth from site datum in meters below datum	Stratum	Diatoms	Microspheres	Chrysophytes	Diatoms Concentration / (# x 10^8 mL^{-1} gdw)	*Achnanthidium exiguum*		*Amphora pediculus*		***Aulacoseira islandica***		***Aulacoseira italica***		***Cyclotella atomus***		*Cyclotella comensis*		*Cyclostephanos* spp.		*Cymbopleura lapponica*		*Diadesmis confervacea*		*Diadesmis* spp. 1 (oval)		***Eunotia arcus/bilunaris***	
	(#)	(m bd)		(#)	(#)	(#)		(#)	(%)	(#)	(%)	(#)	(%)	(#)	(%)	(#)	(%)	(#)	(%)	(#)	(%)	(#)	(%)	(#)	(%)	(#)	(%)	(#)	(%)
10-4	1	11.790–11.795	4de	306	15	13	1.842	0	0.0	0	0.0	14	3.6	11	4.6	5	1.6	0	0.0	0	0.0	0	0.0	0	0.0	0	0.0	3	1.0
	2	11.795–11.800		308	17	23	1.628	1	0.3	1	0.3	1	1.6	5	0.3	2	0.6	0	0.0	0	0.0	0	0.0	7	2.3	2	0.6	1	0.3
	3	11.800–11.805		304	15	0	1.817	6	2.0	2	0.7	1	2.3	7	0.3	0	0.0	0	0.0	0	0.0	2	0.7	3	1.0	0	0.0	5	1.6
	4	11.805–11.810		304	18	17	1.519	2	0.7	4	1.3	3	3.0	9	1.0	0	0.0	0	0.0	0	0.0	0	0.0	0	0.0	0	0.0	3	1.0
	5	11.810–11.815		304	17	21	1.613	0	0.0	6	2.0	1	1.0	3	0.3	0	0.0	0	0.0	0	0.0	0	0.0	5	1.6	0	0.0	2	0.7
	6	11.815–11.820		302	17	10	1.595	1	0.3	1	0.3	1	1.7	5	0.3	0	0.0	0	0.0	0	0.0	1	0.3	0	0.0	0	0.0	8	2.6
	7	11.820–11.825		308	26	9	1.058	0	0.0	4	1.3	1	1.9	6	0.3	1	0.3	0	0.0	0	0.0	0	0.0	7	2.3	1	0.3	2	0.6
	8	11.825–11.830		319	17	0	1.695	1	0.3	1	0.3	0	0.0	10	0.0	3	0.9	1	0.3	0	0.0	0	0.0	7	2.2	1	0.3	2	0.6
	9	11.830–11.835	4d Mixed 4d & de	312	4	0	6.991	0	0.0	9	2.9	0	0.0	3	1.0	0	0.0	0	0.0	0	0.0	0	0.0	9	2.9	0	0.0	1	0.3
	10	11.835–11.840	4d	302	4	8	6.775	3	1.0	6	2.0	1	0.7	2	0.6	0	0.0	0	0.0	0	0.0	0	0.0	8	2.6	0	0.0	0	0.0
	11	11.840–11.845		306	14	2	1.958	1	0.3	15	4.9	0	0.0	6	2.0	0	0.0	0	0.0	0	0.0	0	0.0	8	2.6	0	0.0	3	1.0
	12	11.845–11.850		320	0	2	0.0	4	1.3	4	1.3	2	0.6	0	0.0	0	0.0	0	0.0	0	0.0	1	0.3	17	5.3	0	0.0	4	1.3
	13	11.850–11.855		308	4	3	6.926	4	1.3	13	4.2	0	0.0	2	0.6	0	0.0	0	0.0	0	0.0	0	0.0	9	2.9	0	0.0	1	0.3
	14	11.855–11.860		317	1	2	2.834	2	0.6	4	1.3	1	0.3	0	0.0	0	0.0	0	0.0	0	0.0	0	0.0	4	1.3	1	0.3	0	0.0
	15	11.860–11.865		309	6	0	4.627	10	3.2	11	3.6	1	0.3	0	0.0	0	0.0	0	0.0	0	0.0	0	0.0	14	4.5	0	0.0	0	0.0
	16	11.865–11.870		307	3	4	9.132	2	0.7	2	0.7	2	1.6	0	0.0	0	0.0	0	0.0	0	0.0	0	0.0	9	2.9	0	0.0	0	0.0
	17	11.870–11.875		304	4	3	6.801	3	1.0	7	2.3	0	0.0	2	0.7	0	0.0	0	0.0	0	0.0	0	0.0	13	4.3	0	0.0	0	0.0
	18	11.875–11.880		301	6	8	4.461	2	0.7	6	2.0	3	1.0	3	1.0	0	0.0	0	0.0	0	0.0	0	0.0	6	2.0	0	0.0	2	0.7
	19	11.880–11.885		300	5	8	5.286	2	0.7	4	1.3	1	0.3	1	06	0	0.0	2	0.7	0	0.0	0	0.0	0	0.0	0	0.0	0	0.0
	20	11.885–11.890		310	4	7	6.846	5	1.6	7	2.3	4	1.6	5	1.8	0	0.0	0	0.0	0	0.0	0	0.0	3	1.0	0	0.0	3	1.0
	21	11.890–11.895	Mixed 3 & 4	307	5	3	4.816	1	0.3	8	2.6	2	0.7	2	0.7	1	0.3	1	0.3	0	0.0	0	0.0	0	0.0	0	0.0	1	0.3
	22	11.895–11.900		308	13	0	1.580	2	0.6	10	3.2	7	2.3	7	2.3	0	0.0	0	0.0	1	0.3	0	0.0	7	2.3	0	0.0	4	1.3
	23	11.900–11.905	3	302	12	5	1.122	1	0.3	6	2.0	4	3.3	10	4.2	0	0.0	0	0.0	0	0.0	1	0.3	5	1.7	0	0.0	1	0.3

Table E.2. (continued)

Profile	Sample ID	Depth from site datum in meters below datum	Stratum	*Eunotia* spp.		*Gomphonema acuminatum*		***Gomphonema consector***		*Gomphonema* spp. 1 (cf. *G. mexicanum?*)		*Psammothidium didyum*		***Pseudostaurosira brevistriata***		***Punctastriata mimetica***		*Stauroneis smithii*		***Staurosira construens***		***Ulnaria* spp. 1 (short, pinched waist)**		**Assemblage II**
	(#)	(m bd)		(#)	(%)	(#)	(%)	(#)	(%)	(#)	(%)	(#)	(%)	(#)	(%)	(#)	(%)	(#)	(%)	(#)	(%)	(#)	(%)	(%)
08-1	1	10.98–11.03	5	0.0	0.0	0	0.0	3	1.4	0	0.0	0	0.0	1	0.5	3	1.4	1	0.5	0	0.0	0	0.0	11.0
	2	11.03–11.08		0.0	0.0	0	0.0	1	0.4	0	0.0	0	0.0	1	0.4	1	0.4	0	0.0	2	0.7	0	0.0	9.5
	3	11.08–11.13		0.0	0.0	0	0.0	0	0.0	1	0.3	0	0.0	0	0.0	6	2.9	0	0.0	2	1.0	1	0.5	14.7
	4	11.13–11.18		1.0	0.3	0	0.0	0	0.0	0	0.0	0	0.0	10	3.3	9	3.0	3	1.0	1	0.3	1	0.3	20.9
	5	11.18–11.23		0.0	0.0	0	0.0	4	1.2	0	0.0	0	0.0	2	0.6	5	1.5	0	0.0	1	0.3	0	0.0	11.3
	6	11.23–11.28		0.0	0.0	0	0.0	0	0.0	0	0.0	0	0.0	1	0.5	2	0.9	0	0.0	0	0.0	0	0.0	11.1
	7	11.28–11.33		0.0	0.0	0	0.0	3	1.0	0	0.0	0	0.0	0	0.0	0	0.0	1	0.3	0	0.0	0	0.0	9.1
	8	11.33–11.38	4de	0.0	0.0	0	0.0	7	2.0	0	0.0	0	0.0	15	0.3	5	1.4	0	0.0	1	0.3	0	0.0	9.8
	9	11.38–11.43		0.0	0.0	0	0.0	0	0.0	0	0.0	0	0.0	1	0.3	0	0.0	0	0.0	1	0.3	0	0.0	5.3
	10	11.43–11.38		0.0	0.0	0	0.0	0	0.0	0	0.0	0	0.0	0	0.0	4	1.2	0	0.0	0	0.0	0	0.0	5.1
	11	11.48–11.53		0.0	0.0	0	0.0	0	0.0	0	0.0	0	0.0	1	0.2	4	1.0	0	0.0	7	1.7	0	0.0	8.6
	12	11.53–11.58		0.0	0.0	0	0.0	4	1.1	0	0.0	0	0.0	1	0.3	1	0.3	0	0.0	0	0.0	0	0.0	12.3
	13	11.58–11.63		0.0	0.0	0	0.0	0	0.0	0	0.0	0	0.0	1	0.3	1	0.3	0	0.0	0	0.0	0	0.0	11.0
	14	11.63–11.68		0.0	0.0	0	0.0	5	1.6	0	0.0	0	0.0	2	0.6	3	0.9	0	0.0	0	0.0	0	0.0	13.9
	15	11.68–11.73		0.0	0.0	0	0.0	2	0.7	0	0.0	1	0.3	5	1.7	7	2.3	1	0.3	1	0.3	0	0.0	16.2
	16	11.88–11.93	3B	0.0	0.0	0	0.0	1	0.3	0	0.0	0	0.0	12	3.4	7	2.0	0	0.0	3	0.8	4	1.1	13.3

(continued)

Table E.2. (continued)

Profile	Sample ID	Depth from site datum in meters below datum	Stratum	*Eunotia* spp.		*Gomphonema acuminatum*		***Gomphonema consector***		*Gomphonema* spp. 1 (cf. *G. mexicanum?*)		*Psammothidium didyum*		***Pseudostaurosira brevistriata***		***Punctastriata mimetica***		*Stauroneis smithii*		***Staurosira construens***		***Ulnaria* spp. 1 (short, pinched waist)**		**Assemblage II**
	(#)	(m bd)		(#)	(%)	(#)	(%)	(#)	(%)	(#)	(%)	(#)	(%)	(#)	(%)	(#)	(%)	(#)	(%)	(#)	(%)	(#)	(%)	(%)
10-4	1	11.790–11.795	4de	0	0.0	0	0.0	0	0.0	0	0.0	1	0.3	16	5.2	3	1.0	0	0.0	14	4.6	6	2.0	23.9
	2	11.795–11.800		0	0.0	0	0.0	2	0.6	0	0.0	0	0.0	28	9.1	2	0.6	0	0.0	12	3.9	2	0.6	21.4
	3	11.800–11.805		0	0.0	0	0.0		0.0	0	0.0	0	0.0	46	15.1	4	1.3	0	0.0	16	5.3	6	2.0	32.2
	4	11.805–11.810		0	0.0	0	0.0	1	0.3	0	0.0	0	0.0	26	8.6	7	2.3	0	0.0	15	4.9	6	2.0	25.0
	5	11.810–11.815		0	0.0	0	0.0	1	0.3	1	0.3	0	0.0	2	0.7	6	2.0	0	0.0	14	4.6	18	5.9	19.4
	6	11.815–11.820		0	0.0	0	0.0	2	0.7	2	0.7	0	0.0	23	7.6	17	5.6	0	0.0	18	6.0	14	4.6	30.8
	7	11.820–11.825		2	0.6	0	0.0	1	0.3	0	0.0	0	0.0	43	14.0	18	5.8	0	0.0	21	6.8	3	1.0	35.7
	8	11.825–11.830		0	0.0	0	0.0		0.0	0	0.0	0	0.0	30	9.4	20	6.3	1	0.3	8	2.5	15	4.7	31.3
	9	11.830–11.835	Mixed 4d & de	0	0.0	0	0.0	3	1.0	0	0.0	0	0.0	86	27.6	12	3.8	0	0.0	44	14.1	8	2.6	56.1
	10	11.835–11.840	4d	0	0.0	0	0.0	6	2.0	0	0.0	0	0.0	39	12.9	19	6.3	0	0.0	42	13.9	22	7.3	49.0
	11	11.840–11.845		0	0.0	0	0.0	1	0.3	5	1.6	0	0.0	52	17.0	19	6.2	0	0.0	41	13.4	22	7.2	56.5
	12	11.845–11.850		0	0.0	0	0.0	5	1.6	0	0.0	0	0.0	43	13.4	14	4.4	1	0.3	38	11.9	16	5.0	46.6
	13	11.850–11.855		0	0.0	0	0.0	1	0.3	0	0.0	0	0.0	83	26.9	12	3.9	0	0.0	50	16.2	21	6.8	63.6
	14	11.855–11.860		0	0.0	0	0.0	1	0.3	2	0.6	0	0.0	77	24.3	25	7.9	0	0.0	34	10.7	14	4.4	52.1
	15	11.860–11.865		0	0.0	0	0.0	5	1.6	0	0.0	0	0.0	103	33.3	10	3.2	0	0.0	37	12.0	10	3.2	65.0
	16	11.865–11.870		0	0.0	0	0.0	2	0.7	0	0.0	0	0.0	97	31.6	11	3.6	0	0.0	42	13.7	7	2.3	57.7
	17	11.870–11.875		0	0.0	0	0.0	2	0.7	2	0.7	0	0.0	87	28.6	12	3.9	0	0.0	48	15.8	17	5.6	63.5
	18	11.875–11.880		0	0.0	3	1.0		0.0	0	0.0	0	0.0	98	32.6	6	2.0	1	0.3	30	10.0	16	5.3	57.5
	19	11.880–11.885		0	0.0	3	1.0	5	1.7	0	0.0	0	0.0	106	35.3	11	3.7	1	0.3	30	10.0	28	9.3	64.3
	20	11.885–11.890		0	0.0	2	0.6	1	0.3	0	0.0	0	0.0	80	25.8	5	1.6	0	0.0	24	7.7	12	3.9	47.4
	21	11.890–11.895	Mixed 3 & 4	0	0.0	0	0.0		0.0	3	1.0	0	0.0	70	22.8	17	5.5	1	0.3	42	13.7	23	7.5	55.4
	22	11.895–11.900		0	0.0	0	0.0	1	0.3	0	0.0	0	0.0	109	35.4	12	3.9	0	0.0	33	10.7	10	3.2	63.6
	23	11.900-11.905	3	0	0.0	0	0.0	9	3.0	0	0.0	0	0.0	82	27.2	15	5.0	0	0.0	37	12.3	9	3.0	58.3

Table E.3. Total and Relative Abundance of Diatoms of Profiles 08-1 and 10-4, Assemblage III

Profile	Sample ID	Depth from site datum in meters below datum	Stratum	Diatoms	Microspheres	Chrysophytes	Diatoms Concentration / (# x 10^8 mL $^{-1}$ gdw)	*Anomoeneis* spp.		*Craticula* spp.		*Cymbella mexicana*		*Cymbella microcephala*		*Cymbella* spp. 1		*Cymbella* spp. 2		*Gomphonema kobayasii*		*Gomphonema turgidum*		*Gyrosigma* spp.		*Hantzschia* spp.		*Karayevia* spp.	
	(#)	(m bd)		(#)	(#)	(#)		(#)	(%)	(#)	(%)	(#)	(%)	(#)	(%)	(#)	(%)	(#)	(%)	(#)	(%)	(#)	(%)	(#)	(%)	(#)	(%)	(#)	(%)
	1	10.98–11.03		208	7	36	3.630	0	0.0	0	0.0	0	0.0	0	0.0	0	0.0	0	0.0	0	0.0	0	0.0	0	0.0	8	3.8	1	0.5
	2	11.03–11.08		277	13	42	4.835	0	0.0	2	0.4	0	0.0	1	0.4	0	0.0	0	0.0	0	0.0	0	0.0	0	0.0	19	6.9	2	0.7
	3	11.08–11.13		204	24	42	3.560	0	0.0	0	0.0	0	0.0	0	0.0	0	0.0	0	0.0	0	0.0	0	0.0	0	0.0	13	6.4	0	0.0
	4	11.13–11.18	5	300	21	10	5.236	0	0.0	3	1.0	0	0.0	2	0.7	0	0.0	0	0.0	0	0.0	1	0.3	0	0.0	13	4.3	0	0.0
	5	11.18–11.23		339	9	27	5.917	0	0.0	1	0.3	0	0.0	0	0.0	0	0.0	0	0.0	0	0.0	0	0.0	0	0.0	18	5.3	1	0.3
	6	11.23–11.28		217	16	33	3.787	0	0.0	0	0.0	0	0.0	0	0.0	0	0.0	1	0.3	0	0.0	0	0.0	0	0.0	8	3.7	0	0.0
	7	11.28–11.33		310	2	2	5.411	0	0.0	0	0.0	0	0.0	0	0.0	0	0.0	0	0.0	0	0.0	0	0.0	0	0.0	5	1.6	0	0.0
08-1	8	11.33–11.38		346	5	14	1.132	1	0.3	1	0.3	0	0.0	0	0.0	0	0.0	0	0.0	0	0.0	0	0.0	0	0.0	19	5.5	2	0.6
	9	11.38–11.43		340	2	25	1.113	0	0.0	0	0.0	0	0.0	0	0.0	0	0.0	0	0.0	0	0.0	0	0.0	0	0.0	11	3.2	2	0.6
	10	11.43–11.38		331	12	59	1.195	0	0.0	3	0.9	1	0.3	1	0.3	0	0.0	1	0.3	0	0.0	0	0.0	0	0.0	15	4.5	0	0.0
	11	11.48–11.53	4de	414	33	36	1.667	2	0.7	1	0.2	0	0.0	0	0.0	0	0.0	0	0.0	1	0.2	0	0.0	0	0.0	5	1.2	3	0.7
	12	11.53–11.58		350	6	25	1.527	0	0.0	2	0.6	0	0.0	0	0.0	0	0.0	0	0.0	0	0.0	0	0.0	0	0.0	9	2.6	0	0.0
	13	11.58–11.63		305	12	56	1.101	1	0.3	1	0.3	0	0.0	0	0.0	0	0.0	1	0.3	0	0.0	0	0.0	0	0.0	15	4.9	0	0.0
	14	11.63–11.68		318	11	25	2.775	0	0.0	1	0.3	0	0.0	0	0.0	0	0.0	0	0.0	0	0.0	0	0.0	0	0.0	2	0.6	0	0.0
	15	11.68–11.73		303	12	29	2.144	0	0.0	2	0.7	0	0.0	0	0.0	0	0.0	0	0.0	0	0.0	0	0.0	2	0.7	17	5.6	0	0.0
	16	11.88–11.93	3B	356	12	24	2.485	0	0.0	4	1.1	0	0.0	0	0.0	0	0.0	0	0.0	0	0.0	0	0.0	0	0.0	10	2.8	1	0.3

(continued)

Table E.3. (continued)

Profile	Sample ID	Depth from site datum in meters below datum	Stratum	Diatoms	Microspheres	Chrysophytes	Diatoms Concentration / (# x 10^8 mL $^{-1}$ gdw)	*Anomoeneis* spp.		*Craticula* spp.		*Cymbella mexicana*		*Cymbella microcephala*		*Cymbella* spp. 1		*Cymbella* spp. 2		*Gomphonema kobayasii*		*Gomphonema turgidum*		*Gyrosigma* spp.		*Hantzschia* spp.		*Karayevia* spp.	
	(#)	(m bd)		(#)	(#)	(#)		(#)	(%)	(#)	(%)	(#)	(%)	(#)	(%)	(#)	(%)	(#)	(%)	(#)	(%)	(#)	(%)	(#)	(%)	(#)	(%)	(#)	(%)
10-4	1	11.790–11.795	4de	306	15	13	1.842	1	0.3	1	0.3	0	0.0	4	1.3	0	0.0	0	0.0	0	0.0	0	0.0	1	0.3	5	1.6	5	1.6
	2	11.795–11.800		308	17	23	1.628	2	0.6	2	0.6	1	0.3	0	0.0	0	0.0	1	0.3	2	0.6	0	0.0	1	0.3	0	0.0	0	0.0
	3	11.800–11.805		304	15	0	1.817	2	0.7	3	1.0	2	0.7	0	0.0	0	0.0	0	0.0	1	0.3	0	0.0	0	0.0	4	1.3	0	0.0
	4	11.805–11.810		304	18	17	1.519	2	0.7	5	1.6	4	1.3	2	0.7	0	0.0	0	0.0	0	0.0	2	0.7	1	0.3	0	0.0	2	0.7
	5	11.810–11.815		304	17	21	1.613	4	1.3	6	2.0	2	0.7	3	1.0	0	0.0	0	0.0	0	0.0	0	0.0	0	0.0	1	0.3	0	0.0
	6	11.815–11.820		302	17	10	1.595	1	0.3	4	1.3	1	0.3	0	0.0	0	0.0	0	0.0	0	0.0	0	0.0	0	0.0	1	0.3	2	0.7
	7	11.820–11.825		308	26	9	1.058	2	0.6	2	0.6	3	1.0	1	0.3	0	0.0	0	0.0	0	0.0	5	1.6	1	0.3	2	0.6	2	0.6
	8	11.825–11.830		319	17	0	1.695	0	0.0	0	0.0	0	0.0	0	0.0	0	0.0	0	0.0	2	0.6	0	0.0	0	0.0	1	0.3	0	0.0
	9	11.830–11.835	Mixed 4d & de	312	4	0	6.991	0	0.0	0	0.0	2	0.6	2	0.6	0	0.0	0	0.0	0	0.0	0	0.0	0	0.0	0	0.0	1	0.3
	10	11.835–11.840	4d	302	4	8	6.775	1	0.3	1	0.3	1	0.3	0	0.0	0	0.0	0	0.0	0	0.0	1	0.3	0	0.0	4	1.3	4	1.3
	11	11.840–11.845		306	14	2	1.958	0	0.0	1	0.3	1	0.3	0	0.0	0	0.0	0	0.0	0	0.0	0	0.0	0	0.0	0	0.0	2	0.7
	12	11.845–11.850		320	0	2	0.0	0	0.0	2	0.6	3	0.9	0	0.0	0	0.0	0	0.0	0	0.0	1	0.3	0	0.0	2	0.6	0	0.0
	13	11.850–11.855		308	4	3	6.926	1	0.3	0	0.0	0	0.0	0	0.0	0	0.0	0	0.0	0	0.0	0	0.0	1	0.3	0	0.0	0	0.0
	14	11.855–11.860		317	1	2	2.834	0	0.0	0	0.0	1	0.3	0	0.0	1	0.3	0	0.0	1	0.3	0	0.0	0	0.0	2	0.6	0	0.0
	15	11.860–11.865		309	6	0	4.627	0	0.0	2	0.6	0	0.0	0	0.0	0	0.0	0	0.0	0	0.0	0	0.0	0	0.0	0	0.0	0	0.0
	16	11.865–11.870		307	3	4	9.132	1	0.3	0	0.0	0	0.0	0	0.0	0	0.0	0	0.0	2	0.7	0	0.0	0	0.0	0	0.0	2	0.7
	17	11.870–11.875		304	4	3	6.801	0	0.0	0	0.0	0	0.0	0	0.0	0	0.0	0	0.0	0	0.0	0	0.0	0	0.0	0	0.0	0	0.0
	18	11.875–11.880		301	6	8	4.461	1	0.3	1	0.3	0	0.0	0	0.0	0	0.0	0	0.0	1	0.3	0	0.0	0	0.0	1	0.3	0	0.0
	19	11.880–11.885		300	5	8	5.286	2	0.7	1	0.3	3	1.0	0	0.0	1	0.3	0	0.0	0	0.0	0	0.0	0	0.0	2	0.7	0	0.0
	20	11.885–11.890		310	4	7	6.846	1	0.3	2	0.6	3	1.0	0	0.0	0	0.0	2.0	0.6	1	0.3	3	1.0	0	0.0	0	0.0	0	0.0
	21	11.890–11.895	Mixed 3 & 4	307	5	3	4.816	1	0.3	0	0.0	1	0.3	0	0.0	5	1.6	0	0.0	0	0.0	2	0.7	0	0.0	0	0.0	0	0.0
	22	11.895–11.900		308	13	0	1.580	1	0.3	2	0.6	1	0.3	0	0.0	0	0.0	0	0.0	0	0.0	0	0.0	0	0.0	0	0.0	0	0.0
	23	11.900–11.905	3	302	12	5	1.122	1	0.3	1	0.3	3	1.0	1	0.3	0	0.0	0	0.0	0	0.0	2	0.7	0	0.0	2	0.7	1	0.3

Table E.3. (continued)

Profile	Sample ID	Depth from site datum in meters below datum	Stratum	*Luticola* spp.		***Neidium* sp. cf. *N. dubium***		***Pinnularia* spp. 1 (glassy striae)**		**Assemblage III**
	(#)	(m bd)		(#)	(%)	(#)	(%)	(#)	(%)	(%)
08-1	1	10.98–11.03	5	0	0.0	1	0.5	3	1.4	6.2
	2	11.03–11.08		0	0.0	3	1.1	4	1.4	10.9
	3	11.08–11.13		3	1.5	0	0.0	0	0.0	7.9
	4	11.13–11.18		1	0.3	1	0.3	0	0.0	6.9
	5	11.18–11.23		0	0.0	3	0.9	1	0.3	7.1
	6	11.23–11.28		0	0.0	0	0.0	1	0.5	4.5
	7	11.28–11.33		0	0.0	1	0.3	4	1.3	3.2
	8	11.33–11.38	4de	1	0.3	0	0.0	3	0.9	7.9
	9	11.38–11.43		0	0.0	3	0.9	4	1.2	5.9
	10	11.43–11.38		2	0.6	0	0.0	5	1.5	8.4
	11	11.48–11.53		0	0.0	2	0.5	6	1.4	4.9
	12	11.53–11.58		0	0.0	1	0.3	7	2.0	5.5
	13	11.58–11.63		0	0.0	1	0.3	5	1.6	7.7
	14	11.63–11.68		0	0.0	2	0.6	5	1.6	3.1
	15	11.68–11.73		2	0.7	2	0.7	7	2.3	10.7
	16	11.88–11.93	3B	0	0.0	4	1.1	8	2.2	7.5

Profile	Sample ID	Depth from site datum in meters below datum	Stratum	*Luticola* spp.		***Neidium* sp. cf. *N. dubium***		***Pinnularia* spp. 1 (glassy striae)**		**Assemblage III**
	(#)	(m bd)		(#)	(%)	(#)	(%)	(#)	(%)	(%)
10-4	1	11.790–11.795	4de	2	0.7	0	0.0	0	0.0	6.2
	2	11.795–11.800		0	0.0	2	0.6	6	1.9	5.5
	3	11.800–11.805		0	0.0	2	0.7	0	0.0	4.6
	4	11.805–11.810		0	0.0	3	1.0	0	0.0	7.0
	5	11.810–11.815		0	0.0	2	0.7	1	0.3	6.3
	6	11.815–11.820		0	0.0	1	0.3	2	0.7	4.0
	7	11.820–11.825		0	0.0	1	0.3	5	1.6	7.8
	8	11.825–11.830		0	0.0	4	1.3	3	0.9	3.1
	9	11.830–11.835	Mixed 4d & de	0	0.0	2	0.6	2	0.6	2.8
	10	11.835–11.840	4d	0	0.0	3	1.0	0	0.0	5.0
	11	11.840–11.845		1	0.3	0	0.0	0	0.0	1.6
	12	11.845–11.850		0	0.0	2	0.6	0	0.0	3.1
	13	11.850–11.855		0	0.0	2	0.6	2	0.6	1.9
	14	11.855–11.860		0	0.0	1	0.3	1	0.3	2.2
	15	11.860–11.865		0	0.0	0	0.0	0	0.0	0.6
	16	11.865–11.870		0	0.0	0	0.0	0	0.0	1.6
	17	11.870–11.875		0	0.0	0	0.0	0	0.0	0.0
	18	11.875–11.880		0	0.0	0	0.0	0	0.0	1.3
	19	11.880–11.885		0	0.0	0	0.0	1	0.3	3.3
	20	11.885–11.890		0	0.0	0	0.0	2	0.6	4.5
	21	11.890–11.895	Mixed 3 & 4	0	0.0	0	0.0	1	0.3	3.3
	22	11.895–11.900		0	0.0	1	0.3	2	0.6	2.3
	23	11.900–11.905	3	0	0.0	0	0.0	1	0.3	3.9

Table E.4. Total and Relative Abundance of Diatoms of Profiles 08-1 and 10-4, Assemblage IV

Profile	Sample ID	Depth from site datum in meters below datum	Stratum	Diatoms	Microspheres	Chrysophytes	Diatoms Concentration / (# x 10^8 mL $^{-1}$ gdw)	***Denticula* sp. cf. *D. tenuis***		***Epithemia* sp. cf. *E. adnata***		*Epithemia* spp. 2 (small)		***Fragilaria tenera***		***Gomphonema johnsonii***		***Nitzschia* spp. 1 (large, prominent struts)**		***Nitzschia spp.* 2 (small, capitate, no striae)**		*Orthoseira* spp.		*Planothidium abbreviatum*		***Rhopalodia gibba***		***Rhopalodia musculus***	
	(#)	(m bd)		(#)	(#)	(#)		(#)	(%)	(#)	(%)	(#)	(%)	(#)	(%)	(#)	(%)	(#)	(%)	(#)	(%)	(#)	(%)	(#)	(%)	(#)	(%)	(#)	(%)
08-1	1	10.98–11.03	5	208	7	36	3.630	42	20.2	11	5.3	0	0.0	8	3.8	3	1.4	5	2.4	4	1.9	0	0.0	0	0.0	28	13.5	12	5.8
	2	11.03–11.08		277	13	42	4.835	42	15.2	18	6.5	0	0.0	13	4.7	4	1.4	2	0.7	5	1.8	0	0.0	0	0.0	23	8.3	13	4.7
	3	11.08–11.13		204	24	42	3.560	44	21.6	5	2.5	0	0.0	10	4.9	0	0	3	1.5	4	2.0	0	0.0	0	0.0	26	12.7	16	7.8
	4	11.13–11.18		300	21	10	5.236	37	12.3	15	5	1	0.3	18	6.0	0	0	11	3.7	3	1.0	3	1.0	1	0.3	15	5.0	28	9.3
	5	11.18–11.23		339	9	27	5.917	62	18.3	39	11.5	0	0.0	4	1.2	2	0.6	5	1.5	4	1.2	0	0.0	0	0.0	31	9.1	26	7.7
	6	11.23–11.28		217	16	33	3.787	53	24.4	18	8.3	0	0.0	5	2.3	4	1.8	4	1.8	4	1.8	0	0.0	0	0.0	19	8.8	21	9.7
	7	11.28–11.33		310	2	2	5.411	68	21.9	9	2.9	0	0.0	12	3.9	4	1.3	5	1.6	2	0.6	0	0.0	0	0.0	22	7.1	39	12.6
	8	11.33–11.38	4de	346	5	14	1.132	56	16.2	26	7.5	0	0.0	15	4.3	2	0.6	3	0.9	7	2.0	0	0.0	0	0.0	36	10.4	23	6.6
	9	11.38–11.43		340	2	25	1.113	64	18.8	19	5.6	0	0.0	8	2.4	4	1.2	2	0.6	4	1.2	0	0.0	0	0.0	34	10.0	34	10.0
	10	11.43–11.38		331	12	59	1.195	56	16.9	20	6	0	0.0	3	0.9	2	0.6	5	1.5	0	0.0	0	0.0	0	0.0	47	15.2	40	12.1
	11	11.48–11.53		414	33	36	1.667	68	16.4	33	8	0	0.0	16	3.9	0	0	0	0.0	4	1.0	0	0.0	0	0.0	61	15.7	19	4.6
	12	11.53–11.58		350	6	25	1.527	45	12.9	13	3.7	0	0.0	12	3.4	5	1.4	3	0.9	4	1.1	0	0.0	0	0.0	48	13.7	30	8.6
	13	11.58–11.63		305	12	56	1.101	19	6.2	10	3.3	0	0.0	8	2.6	1	0.3	9	3.0	1	0.3	0	0.0	0	0.0	64	21.0	11	3.6
	14	11.63–11.68		318	11	25	2.775	17	5.3	14	4.4	0	0.0	1	0.3	4	1.3	3	0.9	2	0.6	0	0.0	0	0.0	22	6.9	22	6.9
	15	11.68–11.73		303	12	29	2.144	18	5.9	17	5.6	0	0.0	2	0.7	2	0.7	4	1.7	5	1.7	0	0.0	0	0.0	28	9.2	6	2.0
	16	11.88–11.93	3B	356	12	24	2.485	27	7.6	26	7.3	0	0.0	5	1.4	2	0.6	2	0.6	2	0.6	0	0.0	0	0.0	15	4.2	11	3.1

Table E.4. (continued)

Profile	Sample ID	Depth from site datum in meters below datum	Stratum	Diatoms	Microspheres	Chrysophytes	Diatoms Concentration / (# x 10^8 mL^{-1} gdw)	***Denticula* sp. cf. *D. tenuis***		***Epithemia* sp. cf. *E. adnata***		*Epithemia* spp. 2 (small)		***Fragilaria tenera***		***Gomphonema johnsonii***		***Nitzschia* spp. 1 (large, prominent struts)**		***Nitzschia spp.* 2 (small, capitate, no striae)**		*Orthoseira* spp.		*Planothidium abbreviatum*		***Rhopalodia gibba***		***Rhopalodia musculus***	
	(#)	(m bd)		(#)	(#)	(#)		(#)	(%)	(#)	(%)	(#)	(%)	(#)	(%)	(#)	(%)	(#)	(%)	(#)	(%)	(#)	(%)	(#)	(%)	(#)	(%)	(#)	(%)
	1	11.790–11.795		306	15	13	1.842	18	5.9	9	2.9	0	0.0	1	0.3	0	0.0	0	0.0	0	0.0	3	1.0	1	0.3	11	3.6	8	2.6
	2	11.795–11.800		308	17	23	1.628	18	5.8	16	5.2	3	1.0	0	0.0	0	0.0	5	1.6	0	0.0	0	0.0	0	0.0	13	4.2	6	1.9
	3	11.800–11.805		304	15	0	1.817	9	3.0	10	3.3	0	0.0	0	0.0	0	0.0	3	1.0	1	0.3	0	0.0	0	0.0	7	2.3	14	4.6
	4	11.805–11.810	4de	304	18	17	1.519	35	11.5	15	4.9	0	0.0	0	0.0	0	0.0	1	0.3	3	1.0	0	0.0	0	0.0	15	4.9	9	3.0
	5	11.810–11.815		304	17	21	1.613	46	15.1	18	5.9	0	0.0	0	0.0	0	0.0	1	0.3	0	0.0	0	0.0	0	0.0	14	4.6	12	3.9
	6	11.815–11.820		302	17	10	1.595	17	5.6	20	6.6	0	0.0	0	0.0	0	0.0	2	0.7	0	0.0	0	0.0	0	0.0	10	3.3	12	4.0
	7	11.820–11.825		308	26	9	1.058	24	7.8	6	1.9	1	0.3	0	0.0	0	0.0	2	0.6	0	0.0	3	1.0	0	0.0	11	3.6	5	1.6
	8	11.825–11.830		319	17	0	1.695	26	8.2	10	3.1	0	0.0	2	0.6	0	0.0	2	0.6	0	0.0	0	0.0	0	0.0	10	3.1	12	3.8
	9	11.830–11.835	Mixed 4d & de	312	4	0	6.991	12	3.8	3	1.0	0	0.0	0	0.0	0	0.0	2	0.6	0	0.0	0	0.0	0	0.0	6	1.9	5	1.6
	10	11.835–11.840		302	4	8	6.775	17	5.6	5	1.7	0	0.0	0	0.0	1	0.3	0	0.0	0	0.0	0	0.0	0	0.0	6	2.0	5	1.7
	11	11.840–11.845		306	14	2	1.958	7	2.3	4	1.3	0	0.0	0	0.0	0	0.0	1	0.3	0	0.0	0	0.0	0	0.0	0	0.0	2	0.7
10-4	12	11.845–11.850		320	0	2	0.0	16	5.0	2	0.6	0	0.0	0	0.0	0	0.0	0	0.0	0	0.0	0	0.0	0	0.0	6	1.9	3	0.9
	13	11.850–11.855		308	4	3	6.926	8	2.6	7	2.3	0	0.0	0	0.0	2	0.6	1	0.3	0	0.0	0	0.0	0	0.0	2	0.6	1	0.3
	14	11.855–11.860		317	1	2	2.834	13	4.1	5	1.6	0	0.0	0	0.0	0	0.0	0	0.0	0	0.0	0	0.0	0	0.0	1	0.3	0	0.0
	15	11.860–11.865	4d	309	6	0	4.627	7	2.3	1	0.3	0	0.0	0	0.0	0	0.0	0	0.0	0	0.0	0	0.0	0	0.0	1	0.3	1	0.3
	16	11.865–11.870		307	3	4	9.132	8	2.6	2	0.7	0	0.0	0	0.0	0	0.0	0	0.0	0	0.0	0	0.0	0	0.0	5	1.6	2	0.7
	17	11.870–11.875		304	4	3	6.801	10	3.3	0	0.0	0	0.0	0	0.0	0	0.0	0	0.0	2	0.7	0	0.0	0	0.0	3	1.0	1	0.3
	18	11.875–11.880		301	6	8	4.461	7	2.3	8	2.7	0	0.0	0	0.0	1	0.3	0	0.0	0	0.0	0	0.0	0	0.0	4	1.3	4	1.3
	19	11.880–11.885		300	5	8	5.286	5	1.7	7	2.3	0	0.0	0	0.0	0	0.0	0	0.0	0	0.0	0	0.0	0	0.0	3	1.0	1	0.3
	20	11.885–11.890		310	4	7	6.846	6	1.9	4	1.3	0	0.0	0	0.0	1	0.3	1	0.3	0	0.0	0	0.0	0	0.0	3	1.0	4	1.3
	21	11.890–11.895	Mixed 3 & 4	307	5	3	4.816	6	2.0	2	0.7	0	0.0	0	0.0	0	0.0	2	0.7	0	0.0	0	0.0	0	0.0	4	1.3	1	0.3
	22	11.895–11.900		308	13	0	1.580	7	2.3	3	1.0	0	0.0	0	0.0	2	0.6	0	0.0	0	0.0	0	0.0	0	0.0	3	1.0	0	0.0
	23	11.900–11.905	3	302	12	5	1.122	9	3.0	3	1.0	0	0.0	0	0.0	0	0.0	2	0.7	0	0.0	2	0.7	0	0.0	4	1.3	2	0.7

(continued)

Table E.4. (continued)

Profile	Sample ID	Depth from site datum in meters below datum	Stratum	***Sellaphora saugarresii* (aka *Navicula minima*)**		*Sellaphora* spp. 1 (cf. *pupula*)		*Sellaphora* spp. 2 (cf. *laevis/laevissima*)		***Surirella* sp. cf. *S. ovalis***		**Assemblage IV**
	(#)	(m bd)		(#)	(%)	(#)	(%)	(#)	(%)	(#)	(%)	(%)
08-1	1	10.98–11.03	5	0	0.0	0	0.0	0	0.0	3	1.4	55.7
	2	11.03–11.08		6	2.2	0	0.0	0	0.0	4	1.4	46.9
	3	11.08–11.13		0	0.0	2	1.0	0	0.0	2	1.0	55.0
	4	11.13–11.18		6	2.0	3	1.0	0	0.0	2	0.7	47.6
	5	11.18–11.23		3	0.9	0	0.0	0	0.0	2	0.6	52.6
	6	11.23–11.28		9	4.1	0	0.0	0	0.0	2	0.9	63.9
	7	11.28–11.33		2	0.6	0	0.0	0	0.0	4	1.3	53.8
	8	11.33–11.38	4de	3	0.9	0	0.0	0	0.0	4	1.2	50.6
	9	11.38–11.43		0	0.0	0	0.0	0	0.0	3	0.9	50.7
	10	11.43–11.38		7	2.1	0	0.0	0	0.0	6	1.8	57.1
	11	11.48–11.53		1	0.2	0	0.0	0	0.0	11	2.7	52.5
	12	11.53–11.58		2	0.6	0	0.0	0	0.0	5	1.4	47.7
	13	11.58–11.63		0	0.0	0	0.0	0	0.0	1	0.3	40.6
	14	11.63–11.68		3	0.9	0	0.0	0	0.0	2	0.6	28.1
	15	11.68–11.73		2	0.7	0	0.0	0	0.0	3	1.0	29.2
	16	11.88–11.93	3B	2	0.9	0	0.0	0	0.0	13	3.7	30.0

Profile	Sample ID	Depth from site datum in meters below datum	Stratum	***Sellaphora saugarresii* (aka *Navicula minima*)**		*Sellaphora* spp. 1 (cf. *pupula*)		*Sellaphora* spp. 2 (cf. *laevis/laevissima*)		***Surirella* sp. cf. *S. ovalis***		**Assemblage IV**
	(#)	(m bd)		(#)	(%)	(#)	(%)	(#)	(%)	(#)	(%)	(%)
10-4	1	11.790-11.795	4de	0	0.0	3	0.9	0	0.0	1	0.3	20.7
	2	11.795-11.800		0	0.0	0	0.0	0	0.0	2	0.6	9.6
	3	11.800-11.805		0	0.0	0	0.0	0	0.0	1	0.3	11.6
	4	11.805-11.810		0	0.0	0	0.0	0	0.0	0	0.0	4.6
	5	11.810-11.815		0	0.0	1	0.3	0	0.0	1	0.3	9.1
	6	11.815-11.820		0	0.0	0	0.0	0	0.0	1	0.3	7.1
	7	11.820-11.825		0	0.0	0	0.0	0	0.0	0	0.0	6.0
	8	11.825-11.830		0	0.0	1	0.3	0	0.0	0	0.0	3.6
	9	11.830-11.835	Mixed 4d & de	0	0.0	2	0.7	0	0.0	1	0.3	6.5
	10	11.835-11.840	4d	0	0.0	1	0.3	0	0.0	0	0.0	5.6
	11	11.840-11.845		1	0.3	0	0.0	2	0.7	1	0.3	9.3
	12	11.845-11.850		0	0.0	0	0.0	0	0.0	0	0.0	5.3
	13	11.850-11.855		1	0.3	2	0.6	0	0.0	3	1.0	8.1
	14	11.855-11.860		0	0.0	0	0.0	0	0.0	1	0.3	5.2
	15	11.860-11.865		0	0.0	2	0.6	0	0.0	1	0.3	5.8
	16	11.865-11.870		0	0.0	0	0.0	0	0.0	3	1.0	8.3
	17	11.870-11.875		0	0.0	3	0.9	0	0.0	1	0.3	20.7
	18	11.875-11.880		0	0.0	0	0.0	0	0.0	2	0.6	9.6
	19	11.880-11.885		0	0.0	0	0.0	0	0.0	1	0.3	11.6
	20	11.885-11.890		0	0.0	0	0.0	0	0.0	0	0.0	4.6
	21	11.890-11.895	Mixed 3 & 4	0	0.0	1	0.3	0	0.0	1	0.3	9.1
	22	11.895-11.900		0	0.0	0	0.0	0	0.0	1	0.3	7.1
	23	11.900-11.905	3	0	0.0	0	0.0	0	0.0	0	0.0	6.0

References Cited

Abrantes, Fatima, Isabelle M. Gil, Cristina Lopes, and M. Castro

2005 Quantitative Diatom Analyses—a Faster Cleaning Procedure. *Deep-Sea Research* I 52:189–198.

Adamiec, G., and M. J. Aitken

1998 Dose Rate Conversion Factors: Update. *Ancient TL* 16:37–50.

AGI

1982 AGI Data Sheets. American Geological Institute. Falls Church, Virginia.

Aitken, M. J.,

1985, *Thermoluminescence Dating*, Academic Press, London.

Alberdi, María T., and Eduardo Corona-M

2005 Revisión de los gonfoterios en el Cenozoico tardío de México. *Revista Mexicana de Ciencias Geológicas* 22:246–260.

Alberdi, María T., and José L. Prado

2022 Diversity of the Fossil Gomphotheres from South America, *Historical Biology* 34:1685–1691. doi: 10.1080/08912963.2022.2067754.

Alberdi, María T., Javier Juárez-Woo, Oscar J. Polaco, and Joaquín Arroyo-Cabrales

2009 Description of the Most Complete Skeleton of *Stegomastodon* (Mammalia, Gomphotheriidae) Recorded for the Mexican Late Pleistocene. *Neues Jahrbuch für Geologie und Paläontologie-Abhandlungen*, 251:239–255.

Albert, Rosa M., Ofer Lavi, Lara Estroff, Steve Weiner, Alexander Tsatskin, Avraham Ronen, and Simcha Lev-Yadun

1999 Mode of Occupation of Tabun Cave, Mt Carmel, Israel During the Mousterian Period: A Study of the Sediments and Phytoliths. *Journal of Archaeological Science* 26:1249–1260.

Allaun D'Lopez, Sarah

2019 *Spatial Analysis of the Upland Occupation of El Fin del Mundo, Sonora, Mexico.* Master's Thesis, Department of Anthropology, University of Wyoming, Laramie.

Alonso-Zarza, Ana María, and V. P. Wright

2010 Palustrine Carbonates. *Developments in Sedimentology* 61:103–131.

Anderson, David G.

1996 Models of Paleoindian and Early Archaic Settlement in the Lower Southeast. In *The Paleoindian and Early Archaic Southeast*, edited by David G. Anderson and Kenneth E. Sassaman, pp. 29–57. University of Alabama Press, Tuscaloosa.

Andrefsky, William

2005 *Lithics: Macroscopic Approaches to Analysis*, second edition. Cambridge University Press, Cambridge.

2009 The Analysis of Stone Tool Procurement, Production and Maintenance. *Journal of Archaeological Research* 17:65–103.

Antevs, Ernst
1953 Artifacts with Mammoth Remains, Naco Arizona. Age of the Clovis Fluted Points with the Naco Mammoth. *American Antiquity* 19(1):15–18.
1955 Geologic-Climatic Dating in the West. *American Antiquity* 20(4):317–335.

Arroyo-Cabrales, Joaquín, Ana Luisa Carreño, and Socorro Lozano-García
2008 La diversidad en el pasado. La diversidad en el pasado. In *Capital natural de México,* Vol. I: *Conocimiento actual de la biodiversidad,* edited by Jorge Soberón, Gonzalo Halfter, and Jorge Llorente Bousquets, pp. 227–262. Comisión Nacional para el Conocimiento y Uso de la Biodiversidad, México City.

Arroyo-Cabrales, Joaquín, Oscar J. Polaco, Eileen Johnson, and A. F. Guzmán
2003 The Distribution of the Genus *Mammuthus* in Mexico. *Deinsea 9*(1):27–40.

Arroyo-Cabrales, Joaquín, Oscar J. Polaco, César Laurito, Eileen Johnson, María T. Alberdi, and Ana Lucía V. Zamora
2007 The Proboscideans (Mammalia) from Mesoamerica. *Quaternary International* 169:17–23.

Ashley, Gail M.
2001 Archaeological Sediments in Springs and Wetlands. In *Sediments in Archaeological Context,* edited by Julie K. Stein, and William R. Farrand, pp. 183–210. University of Utah Press, Salt Lake City.

Aslan, Andres, and Anna K. Behrensmeyer
1996 Taphonomy and Time Resolution of Bone Assemblages in a Contemporary Fluvial System; the East Fork River, Wyoming. *Palaios* 11(5):411–421.

Auclair, M., M. Lamothe, and S. Huot
2003 Measurement of Anomalous Fading for Feldspar IRSL Using SAR. *Radiation Measurements,* 37:487–492.

Ballenger, Jesse A.M.
2015 Densest Concentration on Earth? Quantifying Human-Mammoth Associations in the San Pedro Basin, Southeastern Arizona, USA. In *Clovis: On the Edge of a New Understanding,* edited by Ashley M. Smallwood and Thomas A. Jennings, pp. 183–204. Texas A&M University Press, College Station.

Ballenger, Jesse A.M., Vance T. Holliday, Andrew L. Kowler, William T. Reitze, Mary M. Prasciunas, D. Shane Miller, and Jason D. Windingstad
2011 Evidence for Younger Dryas Global Climate Oscillation and Human Response in the American Southwest. *Quaternary International* 242:502–519.

Bartolini, C., Paul E. Damon, M. Shafiqullah, and Mariano Morales M.
1995 Geochronologic Contributions to the Tertiary Sedimentary-Volcanic Sequences ("Baucarit Formation") in Sonora, México. *Geofisica Internacional, 34*(1):67–77. doi:10.22201/igeof.00167169p.1995.34.1.1286.

Barton, C. Michael, and Julien Riel-Salvatore
2014 The Formation of Lithic Assemblages. *Journal of Archaeological Science* 46:334–352.

Battarbee, Richard W.
1986 Diatom Analysis. In *Handbook of Holocene Paleoecology and Palaeohydrology,* edited by Björn E. Berglund, pp. 527–570. John Wiley and Sons, New York.

Battarbee, Richard W., Gavin L. Simpson, Helen Bennion, and Christopher Curtis
2011 A Reference Typology of Low Alkalinity Lakes in the UK Based on Pre-acidification Diatom Assemblages from Lake Sediment Cores. *Journal of Paleolimnology* 45:489–505.

Behrensmeyer, Anna K.
1978 Taphonomic and Ecologic Information from Bone Weathering. *Paleobiology* 4:140–162.
1982 Time Resolution in Fluvial Vertebrate Assemblages. *Paleobiology* 8:211–227.
1991 Terrestrial Vertebrate Accumulations. In *Taphonomy: Releasing the Data Locked in the Fossil Record,* edited by Peter A. Allison and Derek E. G. Briggs, pp. 291–335. Plenum Press, New York.
2007 Bonebeds through Time. In *Bonebeds: Genesis, Analysis, and Paleobiological Significance,* edited by Raymond R. Rogers, David A. Eberth, and Anthony R. Fiorillo, pp. 65–102. University of Chicago Press, Chicago.

Bement, Leland C.
2009 Clovis Sites, Gut Piles, and Environmental Reconstructions in Northwest Oklahoma. *Plains Anthropologist* 54:325–331.

Bere, Taurai
2014 Ecological Preferences of Benthic Diatoms in a Tropical River System in Sao Carlos-SP, Brazil. *Tropical Ecology* 55(1):47–61.

Betancourt, Julio L., Thomas R. Van Devender, and Paul S. Martin (editors)
1990 *Packrat Middens—The Last 40,000 Years of Biotic Change.* University of Arizona Press, Tucson.

Binford, Lewis R.
1979 Organization and Formation Processes: Looking at Curated Technologies. *Journal of Anthropological Research* 35(3):255–273.
1980 Willow Smoke and Dog's Tails: Hunter-Gatherer Settlement System and Archaeological Site Formation. *American Antiquity* 45(1):4–20.

Birkeland, Peter W.
1999 Soils and Geomorphology, third edition. Oxford University Press, New York.

Bleed, Peter
1986 The Optimal Design of Hunting Weapons: Maintainability or Reliability. *American Antiquity* 51(4):737–747.

Boldurian, Anthony T., and John L. Cotter
1999 *Clovis Revisited. New Perspectives on Paleoindian Adaptations from Blackwater Draw, New Mexico.* University of Pennsylvania Museum, Philadelphia.

Bousman, C. Britt, and Eric Oksanen
2012 The Protoarchaic in Central Texas and Surrounding Areas. In *From the Pleistocene to the Holocene. Human Organization and Cultural Transformations in Prehistoric North America*, edited by C. Britt Bousman and Bradley J. Vierra, pp. 197–232. Texas A & M University Press, College Station.

Bousman, C. Britt, Barry W. Baker, and Anne C. Kerr
2004 Paleoindian Archaeology in Texas. In *The Prehistory of Texas*, edited by Timothy Perttula, pp. 15–97. Texas A&M University Press, College Station.

Bradley, Bruce A., Michael B. Collins, and Andrew Hemmings
2010 *Clovis Technology*. Michigan Archaeological Series No. 17. International Monographs in Prehistory, Ann Arbor.

Brambilla, Luciano, Marcelo Javiér Toledo, José Augusto Haro, and José Luis Aguilar
2019 New Osteoderm Morphotype (Xenarthra, Mylodontidae) from the Middle Pleistocene of Argentina. *Journal of South American Earth Sciences*, 95:102298.

Bremond, Laurent, Anne Alexandre, Odile Peyron, and Joël Guiot
2004 Grass Water Stress Estimated from Phytoliths in West Africa. *Journal of Biogeography* 31:1–17.

Bright, Jordan, Darrell S. Kaufman, Steven L. Forman, William C. McIntosh, Jim I. Mead, and Arturo Baez
2010 Comparative Dating of a Bison-Bearing Late-Pleistocene Deposit, Térapa, Sonora, Mexico. *Quaternary Geochronology* 5:631–643.

Broster, John B., and Mark R. Norton
1993 The Carson-Conn-Short Site (40BN190): An Extensive Clovis Habitation in Benton County Tennessee. *Current Research in the Pleistocene* 10:3–4.

Broster, John B., Mark R. Norton, Dennis J. Stanford, C. Vance Haynes, Jr., and Margaret A. Jodry
1996 Stratified Fluted Point Deposits in the Western Valley of Tennessee. In *Proceedings of the 14th Annual Mid-South Archaeological Conference*, edited by R. Walling, C. Wharey, and C. Stanley, pp. 1–11. Special Publications No. 1. Panamerican Consultants, Inc., Memphis.

Brunswig, Robert H.
2007 New Interpretations of the Dent Mammoth site. In *Frontiers in Colorado Paleoindian Archaeology*, edited by Robert H. Brunswig and Bonnie L. Pitblado, pp. 87–122. University Press of Colorado, Boulder.

Buchanan, Briggs, J. David Kilby, Jason M. LaBelle, Todd A. Surovell, Jacob Holland-Lulewicz, and Marcus J. Hamilton
2022 Bayesian Modeling of the Clovis and Folsom Radiocarbon Records Indicates a 200-Year Multigenerational Transition. *American*

Buchanan, Briggs, J. David Kilby, Jason M. LaBelle, Todd A. Surovell, Jacob Holland-Lulewicz, and Marcus J. Hamilton (continued)
Antiquity 87(3):567–580. doi:10.1017/aaq.2021.153

Buckley, James, and Eric H. Willi
1972 Isotopes' Radiocarbon Measurements IX. *Radiocarbon* 14:114–139.

Bull, William B.
1991 *Geomorphic Responses to Climatic Change.* Oxford University Press, New York.

Burford, Euan P., Stephen Hillier, and Geoffrey M. Gadd
2006 Biomineralization of Fungal Hyphae with Calcite ($CaCO_3$) and Calcium Ocalate Mono- and Dehydrate in Carboniferous Limestone Microcosms. *Geomicrobiology Journal,* 23:599–611.

Búrquez, Alberto, Angela Martínez, A., and Paul S. Martin
1992 From the High Sierra Madre to the Coast: Changes in Vegetation along Highway 16, Maycoba–Hermosillo. In *Geology and Mineral Resources of the Northern Sierra Madre Occidental, Mexico*, edited by K. F. Clark, J. Roldín Quintana, and R. H. Schmidt, pp. 239–232. El Paso Geological Society, El Paso.

Cabanes, Dan, and Ruth Shahack-Gross
2015 Understanding Fossil Phytolith Preservation: The Role of Partial Dissolution in Paleoecology and Archaeology. *PloS One* 10:e0125532.

Cantonati, Marco, and Horst Lange-Bertalot
2010 Diatom Biodiversity of Springs in the Berhtesgaden National Park (North-Eastern Alps, Germany), with the Ecological and Morphological Characterization of Two Species New to Science. *Diatom Research* 25(2):251–280.

Cantonati, Marco, Nicola Angeli, Ermanno Bertuzzi, and Daniel Spitale
2012 Diatoms in Springs of the Alps: Spring Types, Environmental Determinants, and Sub-Stratum. *Freshwater Science* 31(2):499–524.

Cantonati, Marco, Nicola Angeli, Laura Virtanen, Agata Z. Wojtal, Jacopo Gabrieli, Elisa Falasco, Isabelle Lavoie, Soizic Morin, Aldo Marchetto, Claude Fortin, and Svetlana Smirnova
2013 *Achnanthidium minutissimum* (Bacillariophyta) Valve Deformities as Indicators of Metal Enrichment in Diverse Widely-Distributed Freshwater Habitats. *Science of the Total Environment* 475:201–215. http://dx.doi.org/10.1016/j.scitotenv.2013.10.018.

Cantonati, Marco, Silvia Scola, Nicola Angeli, Graziano Guella, and Rita Frassanito
2009 Environmental Controls of Epilithic Diatom Depth-Distribution in an Oligotrophic Lake Characterized by Marked Water-Level Fluctuations. *European Journal of Phycology* 44(1):15–29, doi: 10.1080/09670260802079335.

Cantonati, Marco, Stefano Segadelli, Daniel Spitale, Jacopo Gabrieli, Reinhard Gerecke, Nicola Angeli, Maria Teresa De Nardo, Kei Ogata, and John D. Wehr
2020 Geological and Hydrochemical Prerequisites of Unexpectedly High Biodiversity in Spring Ecosystems at the Landscape Level. *Science of the Total Environment* 740:140–157.

Cohen, Andrew S.
2003 *Paleolimnology: The History and Evolution of Lake Systems.* Oxford University Press, New York.

Collins, Michael B.
1999 *Clovis Blade Technology.* University of Texas Press, Austin.

Collins, Michael B., Gail L. Bailey, C. Britt Bousman, Susan W. Dial, Paul Goldberg, Jan Guy, Vance T. Holliday, C. E. Mear, and Paul R. Takac
1998 *Wilson-Leonard: An 11,000-year Archeological Record of Hunter-Gatherers in Central Texas,* Vol. I: *Introduction, Background, and Syntheses.* Index of Texas Archaeology: Open Access Gray Literature from the Lone Star State 1998(1), https://scholarworks.sfasu.edu/ita/vol1998/iss1/24/.

Converse, H. H.
1973 A Pleistocene Vertebrate Fauna from Palm Beach County, Florida. *Plaster Jacket* 21:1–12.

Corona-M, Eduardo, and María T. Alberdi
2006 Two New Records of Gomphotheriidae (Mammalia, Proboscidea) in Southern Mexico and Some Biogeographic Implications. *Journal of Paleontology* 80:357–366.

Cotter, John L.
1937 The Occurrence of Flints and Extinct Animals in Pluvial Deposits near Clovis, New Mexico,

Part IV: Report on Excavation at the Gravel Pit, 1936. *Proceedings of the Philadelphia Academy of Natural Sciences* 90:2–16.

1938 The Occurrence of Flints and Extinct Animals in Pluvial Deposits near Clovis, New Mexico, Part VI: Report on the Field Season of 1937. *Proceedings of the Philadelphia Academy of Natural Sciences* 90:113–117.

Cruz-y-Cruz, Tamara, Guadalupe Sánchez, Sergey Sedov, Alejando Terrazas-Mata, Elizabeth Solleiro-Rebolledo, Rosa Elena Tovar-Liceaga, and John Carpenter

2015 Spatial Variability of Late Pleistocene-Early Holocene Soil Formation and Its Relation to Early Human Paleoecology in Northwest Mexico. *Quaternary International* 365:135–149.

Cruz-y-Cruz, Tamara, Sergey Sedov, Guadalupe Sánchez, T. Pi-Puig, K. Pustovoytov, H. Barceinas-Cruz, B. Ortega-Guerrero, and Elizabeth Solleiro-Rebolledo

2014 Late Pleistocene-Holocene Paleosols in the North of Sonora, Mexico: Chronostratigraphy, Pedogenesis, and Implications for Environmental History. *European Journal of Soil Science* 65:455-469.

Cruz-y-Cruz, Tamara, Victor Adrián Pérez-Crespo, P. Morales-Puente, Sergey Sedov, R.E. Tovar-Liceaga, , Joaquín Arroyo-Cabrales, A. Terrazas-Mata, and Guadalupe Sánchez-Miranda

2016 Paleosol (Organic Matter and Pedogenic Carbonates) and Paleontological δ13C Records Applied to the Paleoecology of Late Pleistocene -Holocene in Mexico: *Quaternary International* 418:147–164.

da Silva, Cássia F.M., Lezilda Carvalho-Torgan, and Fabiana Schneck

2019 Temperature and Surface Runoff Affect Community of Periphytic Diatoms and Have Distinct Effects on Functional Groups: Evidence of a Mesocosms Experiment. *Hydrobiologia* 839(1):37–50.

Dello-Russo, Robert D.

2010 *Archaeological Testing at the Water Canyon Site (LA134764), Socorro County, New Mexico: Interim Report for the 2008 and 2009 Field Seasons*. ERG Report No. 2009-09. Submitted by Escondida Research Group, to the New Mexico Historic Preservation Division, Department of Cultural Affairs, Santa Fe.

Dudley, J. P.

1996 Mammoths, Gomphotheres, and the Great American Faunal Interchange. In *The Proboscidea: Evolution and Paleoecology of Elephants and Their Relatives*, edited by Jeheskel Shoshani and Pascal Tassy, pp. 289–295. Oxford University Press, New York.

Duffy, Kevin

1984 *Children of the Forest: Africa's Mbuti Pygmies*. Waveland Press, Longrove, Illinois.

Duller, Geoff. A.T.

2008 *Luminescence Dating: Guidelines on Using Luminescence Dating in Archaeology*. English Heritage, Swindon.

Durand, Nicolas, H. Curtis Monger, Matthew G. Canti, and Eric P. Verrecchia

2018 Calcium Carbonate Features. In *Interpretation of Micromorphological Features of Soils and Regoliths,* second edition, edited by Georges Stoops, Vero Marcelino, and Florios Mees, pp. 205–258. Elsevier, New York.

Eren, Metin I.

2008 Paleoindian Stability during the Younger Dryas in the North American Lower Great Lakes. In *Transitions in Prehistory: Papers in Honor of Ofer Bar-Yosef*, edited by John J. Shea and Daniel E. Lieberman, pp. 389–422. American School of Prehistoric Research Press, Oxbow Books, Oakville, Connecticut.

Eren, Metin I., and Brain G. Redmond

2011 Clovis Blades at Paleo Crossing (33ME274), Ohio. *Midcontinental Journal of Archaeology*, 36(2):173–194.

Eren, Metin I., David J. Meltzer, Brett Story, Briggs Buchanan, Don Yeager, and Michelle R. Bebber

2021 On the Efficacy of Clovis Fluted Points for Hunting Proboscideans. *Journal of Archaeological Science Reports* 39:103166. doi.org/10.1016/j.jasrep.2021.103166.

2022 Not Just for Proboscidean Hunting: On the Efficacy and Functions of Clovis Fluted Points. *Journal of Archaeological Science Reports* 45:103601. doi.org/10.1016/j.jasrep.2022.103601.

Faegri, Knut, and Johs. Iversen

1989 *Textbook of Pollen Analysis*. John Wiley & Sons, Amsterdam.

Feathers, James K., and Hugo G. Nami
2018 Luminescence Dating of late Pleistocene and Holocene Sediments in Uruguay. *Latin American* Antiquity 29(3):495–513.

Feathers, James K., María Nieves Zedeño, Lawrence C. Todd, and Stephen Aaberg
2015 Dating Stone Alignments by Luminescence. *Advances in Archaeological Practice* 3:378–396.

Fedoroff, Nicolas, Marie-Agnès Courty, and Zhentang Guo
2010 Palaeosoils and Relict Soils. In *Interpretation of Micromorphological Features of Soils and Regoliths,* edited by Georges Stoops, Vera Marcelino, and Florias Mees, pp. 623–662. Elsevier, Amsterdam.

Ferring, C. Reid
2001 *The Archaeology and Paleoecology of the Aubrey Clovis Site (41DN479) Denton County, Texas.* Center for Environmental Archaeology, University of North Texas, Denton.
2012 *The "Long" Clovis Chronology: Evidence from the Aubrey and Friedkin Sites, Texas.* Society for American Archaeology 77th Annual Meeting, Memphis.

Figgins, Jesse D.
1933 A Further Contribution to the Antiquity of Man in America. *Proceedings of the Colorado Museum of Natural History* 12(2):4–10.

Fisher, Daniel C., and David L. Fox
2007 Season of Death of the Dent Mammoths. In *Frontiers in Colorado Paleoindian Archaeology,* edited by Robert H. Brunswig and Bonnie L. Pitblado, pp. 123–154. University Press of Colorado, Boulder.

Fisher, John W. W., Jr.
1992 Observations on the Late Pleistocene Bone Assemblage from the Lamb Spring Site, Colorado. *Ice Age Hunters of the Rockies,* edited by Dennis J. Stanford and Jane S. Day, pp. 51–81. Denver Museum of Natural History and University Press of Colorado, Boulder.

Freytet, Pierre, and Eric P. Verrecchia
2002 Lacustrine and Palustrine Carbonate Petrography: An Overview. *Journal of Paleolimnology* 27(2):221–237.

Fritz, Sherilyn C., Brian F. Cumming, François. Gasse, and Kathleen R. Laird
1999 Diatoms as Indicators of Hydrologic and Climatic Change in Saline Lakes. In *The Diatoms: Applications for the Environmental and Earth Sciences,* edited by Eugene F. Stoermer and John P. Smol, pp. 41–72. Cambridge University Press, Cambridge.

Gaines, Edmund P.
2006 *Paleoindian Geoarchaeology of the Upper San Pedro Valley, Sonora, Mexico.* Master's thesis, School of Anthropology, University of Arizona, Tucson.

Gaines, Edmund P., Guadalupe Sánchez, and Vance T. Holliday
2009 Paleoindian Archaeology in Northern and Central Sonora, Mexico (A Review and Update). *Kiva* 74(3):305–335.

Galbraith, R. F., and Richard G. Roberts
2012 *Statistical Aspects of Equivalent Dose and Error Calculation and Genera.* Cambridge University Press, Cambridge.

Gamez-Rascon, Thanairi
2016 *Paleovegetación y paleoambientes del Pleistoceno Tardío y Holoceno del sitio arqueológico "La Playa," Municipio de Trincheras, Sonora.* Tesis de Licenciatura en Ecología, Universidad Estatal de Sonora. Hermosillo, Sonora.

Gardner, William M.
1977 Flint Run Paleoindian Complex and its Implications for Eastern North American Prehistory. *Annals of the New York Academy of Science* 288(1):257–263.

Gile, Leland H., J. W. Hawley, and R. B. Grossman
1981 *Soils and Geomorphology in the Basin and Range Area of Southern New Mexico—Guidebook to the Desert Project.* Memoir No. 39. New Mexico Bureau of Mines and Mineral Resources, Socorro.

González-León, Carlos M., and Gilberto S. Moreno-Hurtado
2021 Geologic Map and Geochronology Data Base of Sonora, Mexico. *Terra Digitalis* 5(1):1–7.

Goodyear, Albert C.
2000 The Topper Site 2000: Results of the 2000 Allendale Paleoindian Expedition. *Legacy* 5 (2):18–25.

Goodyear, Albert C., and Kenn Steffy
2003 Evidence of a Clovis Occupation at the Topper Site, 38AL23, Allendale County, South Carolina. *Current Research in the Pleistocene* 20:23–25.

Grayson, Donald K., and David J. Meltzer
2015 Revisiting Paleoindian Exploitation of Extinct North American Mammals. *Journal of Archaeological Science* 56:177–193.

Green, F. Earl
1962 The Lubbock Reservoir Site. *The Museum Journal* 6:83–123. West Texas Museum Association, Lubbock.

Green, Jeremy L.
2002 *Mammut americanum* (Kerr, 1792). *Fossil Species of Florida* 1:1–11.

Green, Jeremy L., and Richard C. Hulbert, Jr.
2005 The Deciduous Premolars of *Mammut americanum* (Mammalia, Proboscidea). *Journal of Vertebrate Paleontology* 25(3):702–715.

Grimm, Eric C.
2004 *TILIA Graph Computer Program.* Illinois State Museum, Research and Collections Center, Springfield.

Guo, Meie, Dongmei Jie, Hongmei Liu, Shuai Luo, and Na Li
2012 Phytolith Analysis of Selected Wetland Plants from Changbai Mountain Region and Implications of Palaeoenvironment. *Quaternary International* 250:119–128.

Gutiérrez-Castorena, María del Carmen, and William R. Effland
2010 Pedogenic and Biogenic Siliceous Features. In *Interpretation of Micromorphological Features of Soils and Regoliths*, edited by Georges Stoops, Vero Marcelino, and Florios Mees, pp. 471–496. Elsevier, Amsterdam.

Hall, Stephen A., William L. Penner, Manuel R. Palacios-Fest, Artie L. Metcalf, and Susan J. Smith
2012 Cool, Wet Conditions Late in the Younger Dryas in Semi-arid New Mexico. *Quaternary Research* 77:87–95.

Hamilton, Marcus J., Briggs Buchanan, Bruce B. Huckell, Vance T. Holliday, M. Steven Shackley, and Matthew E. Hill
2013 Clovis Paleoecology and Lithic Technology in the Central Rio Grande Rift Region, New Mexico. *American Antiquity* 78(2):248–265.

Harris-Parks, Erin
2016 The Micromorphology of Younger Dryas-Aged Black Mats from Nevada, Arizona, Texas, and New Mexico. *Quaternary Research* 85(1):94–106.

Haury, Emil W.
1953 Artifacts with Mammoth Remains, Naco, Arizona. *American Antiquity* 19(1):1–24.

Haury, Emil W., E. B. Sayles, and William W. Wasley
1959 The Lehner Mammoth Site, Southeastern Arizona. *American Antiquity* 25(1):2–30.

Haynes, C. Vance, Jr.
1975 Pleistocene and Recent Stratigraphy. In *Late Pleistocene Environments of the Southern High Plains,* edited by Fred Wendorf and James J. Hester, pp. 57–96. Publication No. 9. Fort Burgwin Research Center, Taos.
1982 Archaeological Investigations at the Lehner site, Arizona, 1974-1975. *National Geographic Society Research Reports* 14:325–334.
1991 Geoarchaeological and Paleohydrological Evidence for a Clovis-age Drought in North America and Its Bearing on Extinction. *Quaternary Research* 35:438–450.
1992 Contributions of Radiocarbon Dating to the Geochronology of the Peopling of the New. In *Radiocarbon after Four Decades: An Interdisciplinary Perspective*, edited by Ervin Taylor, Austin Long, and Renee S. Kra, pp. 355–374. New York, Springer-Verlag.
1995 World Geochronology of Paleoenvironmental Change, Clovis Type Site, Blackwater Draw, New Mexico. *Geoarchaeology,* 10(5):317–388.
2007 Clovis Investigations in the San Pedro Valley. In *Murray Springs. A Clovis Site with Multiple Activity Areas in the San Pedro Valley, Arizona,* edited by C. Vance Haynes and Bruce B. Huckell, pp. 1–15. University of Arizona Press, Tucson.
2008 Younger Dryas "Black Mats" and the Rancholabrean Termination in North America. *Proceedings of the National Academy of Sciences, U.S.A.* 105:6520–6525.

Haynes, C. Vance, Jr., and George A. Agogino
1966 Prehistoric Springs and Geochronology of the Clovis Site, New Mexico. *American Antiquity,* 31(6):812–821.

Haynes, C. Vance, and Bruce B. Huckell (editors)
2007 *Murray Springs. A Clovis Site with Multiple Activity Areas in the San Pedro Valley, Arizona.* Anthropological Paper No. 71. University of Arizona Press, Tucson.

Haynes, Gary
1987 Proboscidean Die-offs and Die-outs: Age Profiles in Fossil Collections. *Journal of Archaeological Science* 14(6):659–668.
1991 *Mammoths, Mastodonts, and Elephants: Biology, Behavior, and the Fossil Record.* Cambridge University Press, Cambridge.
2002 *The Early Settlement of North America: The Clovis Era.* Cambridge University Press, Cambridge.

Haynes, Gary, and Janis Klimowicz
2015 Recent Elephant-Carcass Utilization as a Basis for Interpreting Mammoth Exploitation. *Quaternary International* 359:19–37.

Hedges R.E.M., Julia A. Lee-Thorp and Noreen C. Tuross
1995 Is Tooth Enamel Carbonate a Suitable Material for Radiocarbon Dating? *Radiocarbon* 37(2):285–290.

Hemmings, E. Thomas
2007 Buried Animal Kills and Processing Localities In *Murray Springs. A Clovis Site with Multiple Activity Areas in the San Pedro Valley, Arizona,* edited by C. Vance Haynes and Bruce B. Huckell, pp. 83–137. Anthropological Paper No. 71. University of Arizona Press, Tucson.

Hemmings, Thomas, and C. Vance Haynes
1969 The Escapule Mammoth and Associated Projectile Points, San Pedro Valley, Arizona. *Journal of the Arizona Academy of Science* 5(3):184–188.

Hester, James J., Ernest L. Lundelius, Jr., and Roald Fryxell
1972 *Blackwater Locality No. 1: A Stratified Early Man Site in Eastern New Mexico.* Fort Burgwin Research Center, Taos.

Hester, Thomas R.
2015 *The Golondrina Expansion.* Electronic document, Accessed January 17, 2023 (http://www.texasbeyond history.net/st-plains/prehistory/images/golondrina.html).
2017 The St. Mary's Hall Type: The History, Chronology, and Distribution of St. Mary's Hall Projectile Points. In *Plainview: The Enigmatic Paleoindian Artifact Style of the Great Plains,* edited by Vance T. Holliday, Eileen Johnson, and Ruthann Knudson, pp. 145–173. University of Utah Press, Salt Lake City.

Heusser, Calvin J.
1971 *Pollen and Spores of Chile: Modern Types of the Pteridophyta, Gymnospermae. and Angiospermae.* University of Arizona Press, Tucson.

Hibbard, Claude W.
1955 Pleistocene Vertebrates from the Upper Becerra (Becerra Superior) Formation, Valley of Tequixquiac, México, with Notes on Other Pleistocene Forms. *Contributions from the Museum of Paleontology, University of Michigan* 12(5):47–96.

Holliday, Vance T.
1985 Archaeological Geology of the Lubbock Lake Site, Southern High Plains of Texas. *Geological Society of America Bulletin,* 96(12):1483–1492.
1995 *Stratigraphy and Paleoenvironments of Late Quaternary Valley Fills on the Southern High Plains.* Memoir No. 186. Geological Society of America, Boulder.
1997 *Paleoindian Geoarchaeology of the Southern High Plains.* University of Texas Press, Austin.
2000a The Evolution of Paleoindian Geochronology and Typology on the Great Plains. *Geoarchaeology* 15:227–290.
2000b Folsom Drought and Episodic Drying on the Southern High Plains from 10,900–10,200 ^{14}C yr B.P. *Quaternary Research,* 53:1–12.
2004 *Soils in Archaeological Research.* Oxford University Press, New York.
2015 Clovis Landscapes in the Greater Southwest of North America. In *Clovis: On the Edge of a New Understanding,* edited by Ashley M. Smallwood and Thomas A. Jennings, pp. 205–241. Texas A&M University Press, College Station.

Holliday, Vance T., and David J. Meltzer
2010 The 12.9ka Impact Hypothesis and North American Paleoindians. *Current Anthropology* 51:575–585.

Holliday, Vance T., Robert D. Dello-Russo, and Susan M. Mentzer
2020 Geoarchaeology of the Water Canyon Paleoindian site, West-Central New Mexico.

Geoarchaeology 35:112-140. doi: 10.1002/gea.21765.

Holliday, Vance T., Thomas C. Gustavson, and Susan D. Hovorka
1996 Stratigraphy and Geochronology of Playa Fills on the Southern High Plains. *Geological Society of America Bulletin* 108:953–965.

Holliday, T. Vance, C. Vance Haynes Jr., and Bruce B. Huckell
in press Clovis Archaeology along the Upper San Pedro River Valley, Southeastern Arizona. In *The Paleoindian Southwest,* edited by David Kilby and Bruce Huckell. University of Utah Press, Salt Lake City.

Holliday, Vance T., James H. Mayer, and Glen Fredlund
2008 Geochronology and Stratigraphy of Playa Fills on the Southern High Plains. *Quaternary Research*, 70:11–25.

Holliday, Vance T., Peter C. Condon, Matthew Cuba, Brendan Fenerty, and David Bustos
in press Paleoindian Archaeology of the Tularosa Basin, South-Central New Mexico. In *Paleoindian Archaeology of the Southwest*, edited by David Kilby and Bruce Huckell. University of Utah Press, Salt Lake City.

Holliday, Vance T., Allison Harvey, Matthew T. Cuba, and Aimee M. Weber
2019 Paleoindians, Paleo-Lakes and Paleo-Playas: Landscape Geoarchaeology of the Tularosa Basin, New Mexico. *Geomorphology* 331:92–106.

Holliday, Vance T., C. Vance Haynes, Jr., Jack L. Hofman, and David J. Meltzer
1994 Geoarchaeology and Geochronology of the Miami (Clovis) Site, Southern High Plains of Texas. *Quaternary Research* 41(2):234–244.

Holliday, Vance T., Bruce B. Huckell, J. M. Mayer, and S. L. Forman
2006 Geoarchaeology of the Boca Negra Wash Area, Albuquerque Basin, New Mexico. *Geoarchaeology* 21:765–802.

Holliday, Vance T., Bruce B. Huckell, Marcus Hamilton, Robert H. Weber, William R. Reitze, and James H. Mayer
2009 Geoarchaeology of the Mockingbird Gap (Clovis) Site, Jornada del Muerto, New Mexico. *Geoarchaeology* 24:438–370. doi .org/10.1002/gea.20265.

Holliday, Vance T., Andrew Kowler, Todd Lange, Susan M. Mentzer, Gregory Hodgins, Natalia Martínez-Tagüeña, Edmund P. Gaines, Joaquín Arroyo-Cabrales, Guadalupe Sánchez, and Ismael Sánchez-Morales
2014 Supporting Information: A Human (Clovis)/Gomphothere (*Cuvieronius*) Association ~13,390 cal years B.P in Sonora, Mexico. *Proceedings of the National Academy of Sciences* 111. doi/10.1073/pnas.1404546111.

Holmgren, Camille A., Julio L. Betancourt, and Kate A. Rylander
2006 A 36,000-yr Vegetation History from the Peloncillo Mountains, Southeastern Arizona, USA. *Palaeogeography, Palaeoclimatology, Palaeoecology* 240:405–422.

Holmgren Camille A., Jodi Norris, and Julio L. Betancourt
2007 Inferences about Winter Temperatures and Summer Rains from the Late Quaternary Record of C4 Perennial Grasses and C3 Desert Shrubs in the Northern Chihuahuan Desert. *Journal of Quaternary Science* 22:141–161.

Holmgren, Camille A., M. Cristina Peñalba, Kate Aasen Rylander, and Julio L. Betancourt
2003 A 16,000 14C yr B.P. Packrat Midden Series from the USA– Mexico Borderlands. *Quaternary Research* 60, 319–329.

Hoppe, Kathryn A.
2004 Late Pleistocene Mammoth Herd Structure, Migration Patterns, and Clovis Hunting Strategies Inferred from Isotopic Analyses of Multiple Death Assemblages. *Paleobiology* 30(1):129–145.

Huckell, Bruce B.
2007 Clovis Lithic Technology: A View from the Upper San Pedro Valley. In *Murray Springs: A Clovis Site with Multiple Activities Areas in the San Pedro Valley, Arizona,* edited by C. Vance Haynes and Bruce B. Huckell, pp. 170–213. Anthropological Papers No. 71. University of Arizona Press, Tucson.

Huntley, D. J., and M. Lamothe
2001 Ubiquity of Anomalous Fading in K-feldspars, and Measurement and Correction for It in Optical Dating. *Canadian Journal of Earth Sciences* 38:1093–1106.

Ibarra, Georgina, Elizabeth Solleiro-Rebolledo, Guadalupe Sánchez, Sergey Sedov, Ismael Sánchez-Morales, John Carpenter, and Bruno Chávez-Vergara
2020 Response of Surface Processes to the Holocene Landscape Changes in Sonora: Evidences from the Paleosol-Sedimentary Sequences at the Archaeological Sites El Fin del Mundo and El Gramal. *Journal of South American Earth Sciences* 104:102947. 10.1016/j.jsames.2020.102947.

Jackson Donald, Cèsar Méndez, and Patricio de Souza
2004 Poblamiento Paleoindio en el norte-centro de Chile: Evidencias, problemas y perspectivas de studio. *Complutum* 15:165–176.

Jennings, Thomas, A., Ashley M. Smallwood, and John Greer
2016 Redating the Late Paleoindian/Early Archaic Golondrina Component at Baker Cave, Texas and Implications for the Dalton/Golondrina Expansion. *PaleoAmerica,* 2(4):332–342, doi: 10.1080/20555563.2016.1212159.

Jennings, Thomas A., Ashley M. Smallwood, and Michael R. Waters
2015 Exploring Late Paleoindian and Early Archaic Unfluted Lanceolate Point Classification in the Southern Plains. *North American Archaeologist* 36:243–265. doi: 10.1177/0197693115572763.

Johnson, Eileen (editor)
1987 *Lubbock Lake: Late Quaternary Studies on the Southern High Plains.* Texas A&M University, College Station.

Johnson, LeRoy, Jr.
1964 *The Devil's Mouth Site: A Stratified Campsite at Amistad Reservoir, Val Verde County, Texas.* Archaeology Series 6. Austin: University of Texas, Austin.
1989 *Great Plains Interlopers in the Eastern Woodlands During Late Paleo-Indian Times.* Office of the State Archeologist Report 36. Texas Historical Commission, Austin.

Juggins, Steve
2014 *C2 Version 1.7.2. Software for Ecological and Palaeoecological Data Analysis and Visualisation.* Newcastle University, Newcastle upon Tyne, UK. Craticula.ncl.ac.uk.
2015 *Rioja: Analysis of Quaternary Science Data.* R package version 0.9-5. http://www.staff.ncl.ac.uk/stephen.juggins/.

Katz, Ofir, Dan Cabanes, Stephen Weiner, Aren M. Maeir, Elisabetta Boaretto, and Ruth Shahack-Gross
2010 Rapid Phytolith Extraction for Analysis of Phytolith Concentrations and Assemblages during an Excavation: An Application at Tell es-Safi/Gath, Israel. *Journal of Archaeological Science* 37:1557–1563.

Kellogg, Douglas C., Daniel G. Roberts, and Arthur Spiess
2003 The Neal Garrison Paleoindian Site, York County, Maine. *Archaeology of Eastern North America* 31:73–131.

Kelly, Robert L.
1988 The Three Sides of a Biface. *American Antiquity* 53(4):717–734.

Kelly, Robert L., and Larry C. Todd
1988 Coming into the Country: Early Paleoindian Mobility and Hunting. *American Antiquity* 53(2):231–244.

Kelly, Thomas C.
1982 Criteria for Classification of Plainview and Golondrina Projectile Points. *La Tierra* 9:2–25.

Kilby, David, Todd Surovell, Bruce B. Huckell, C. Ringstaff, M. Hamilton, and C. Vance Haynes
2022 Evidence Supports the Efficacy of Clovis Points for Hunting Proboscideans. *Journal of Archaeological Science Reports* 45:103600. doi.org/10.1016/j.jasrep.2022.103600.

Kim, Yong Jin, and Ok Min Lee
2017 A Study of Low Temperature and Mountain Epilithic Diatom Community in Mountain Stream at the Han River System, Korea. *Journal of Ecology and Environment* 41(28):1–10.

Koch, Paul L., Kathryn A. Hoppe, and S. David Webb
1998 The Isotopic Ecology of Late Pleistocene Mammals in North America, Part 1. *Florida. Chemical Geology* 152:119–138.

Koch, Paul L., Noreen Tuross, and Marilyn L. Fogel
1997 The Effects of Sample Treatment and Diagenesis on the Isotopic Integrity of Carbonate in Biogenic Hydroxylapatite. *Journal of Archaeological Science* 24(5):417–426.

Kooyman, Brian, Len V. Hills, Paul McNeil, and Shayne Tolman
2006 Late Pleistocene Horse Hunting at the Wally's Beach Site (DhPg-8), Canada. *American Antiquity* 71(1):101–121.

Kooyman, Brian, Len V. Hills, Shayne Tolman, and Paul McNeil
2012 Late Pleistocene Western Camel (*Camelops hesternus*) Hunting in Southwestern Canada. *American Antiquity* 77(1):15–124.

Kooyman, Brian, Margaret E. Newman, Christine Cluney, Murray Lobb, Shayne Tolman, Paul McNeil, and Len V. Hills
2001 Identification of Horse Exploitation by Clovis Hunters Based on Protein Analysis. *American Antiquity* 66(4):686–691.

Kuhn, Steve L.
1994 A Formal Approach to the Design and Assembly of Mobile Toolkits. *American Antiquity* 59(3):426–442.

Lambert, W. D.
1996 The Biogeography of the Gomphoteriid Proboscidean of North America. In *The Proboscidea: Evolution and Paleoecology of Elephants and Their Relatives*, edited by Jeheskal Shoshani, and Pascal Tassy, pp. 143–148. Oxford University Press, New York.

Leira, Manel, Rosa Meijide-Failde, and Enrique Torres
2017 Diatom Communities in the Thermo-Mineral Springs of Galicia (NW Spain). *Diatom Research* 32(1):29–42.

Liutkus, Cynthia M., and Gail M. Ashley
2003 Facies Model of a Semiarid Freshwater Wetland, Olduvai Gorge, Tanzania. *Journal of Sedimentary Research* 73(5):691–705.

Lu, Houyuan, and Kam-Biu Liu
2003 Phytoliths of Common Grasses in the Coastal Environments of Southeastern USA. *Estuarine, Coastal and Shelf Science* 58:587–600.

Lucas, Spencer G.
2008 *Cuvieronius* (Mammalia, Proboscidea) from the Neogene of Florida. *Neogene Mammals: Bulletin* 44:31.

Lucas, Spencer G., Gary S. Morgan, John W. Estep, Greg H. Mack, and John W. Hawley
1999 Co-occurrence of the Proboscideans *Cuvieronius, Stegomastodon*, and *Mammuthus* in the Lower Pleistocene of Southern New Mexico. *Journal of Vertebrate Paleontology* 19(3):595–597, doi: 10.1080/02724634.1999.10011169.

Luchsinger, Heidi Marie
2002 *Micromorphological Analysis of the Sediment and Soils from the Gault Site, a Clovis Site in Central Texas.* M.A. thesis, Department of Anthropology, Texas A & M University, College Station.

Lundelius, Ernest L., Jr., Kenneth J. Thies, Russell W. Graham, Christopher J. Bell, Gregory James Smith, and Larisa R.G. DeSantis
2019 Proboscidea from the Big Cypress Creek Fauna, Deweyville Formation, Harris County, Texas. *Quaternary International* 530:59–68.

MacDonald, George F.
1985 *Debert: A Paleo-Indian Site in Central Nova Scotia.* Persimmon Press, Buffalo, New York.

Machette, Michael N.
1988 *Quaternary Movement along the La Jencia Fault, Central New Mexico.* U.S. Geological Survey Professional Paper No. 1440. U.S. Government Printing Office, Washington, D.C.

Macphail, Richard I., and Joseph M. McAvoy
2008 A Micromorphological Analysis of Stratigraphic Integrity and Site Formation at Cactus Hill, an Early Paleoindian and Hypothesized Pre-Clovis Occupation in South-Central Virginia, USA. *Geoarchaeology* 23(5):675–694.

Madella, Marco, Anne Alexandre, and Terry Ball
2005 International Code for Phytolith Nomenclature 1.0. *Annals of Botany* 96, 253–260.

Mallol, Carolina
2006 What's in a Beach? Soil Micromorphology of Sediments from the Lower Paleolithic Site of 'Ubeidiya, Israel. *Journal of Human Evolution* 51(2):185–206.

Mallol, Carolina, Susan M. Mentzer, and Patrick J. Wrinn
2009 A Micromorphological and Mineralogical Study of Site Formation Processes at the lLte Pleistocene Site of Obi-Rakhmat, Uzbekistan. *Geoarchaeology: An International Journal* 24(5):548–575.

Mannion, A. M.
1987 Fossil Diatoms and Their Significance in Archaeological Research. *Oxford Journal of Archaeology* 6(2):131–147.

Martin, Paul S.
1963 *The Last 10,000 years: A Fossil Pollen Record of the American Southwest.* University of Arizona Press, Tucson.

Martínez Ramírez, Júpiter
2010 *Informe del Proyecto de Salvamento Arqueológico Gamma, Hermosillo. Reviviendo el Desierto del Hombre Nómada.* Versión digital, Centro INAH Sonora, Hermosillo.

Mason, I. M., Myszka A.J. Guzkowska, and Chris G. Rapley
1994 The Response of Lake Levels and Areas to Climatic Change. *Climatic Change* 27:16–197.

McDonald, H. Gregory
2018 An Overview of the Presence of Osteoderms in Sloths: Implications for Osteoderms as a Plesiomorphic Character of the Xenarthra. *Journal of Mammalian Evolution* 25(4):485–493.

Mead, Jim I., Joaquín Arroyo-Cabrales, and Sandra L. Swift
2019 Late Pleistocene Mammuthus and Cuvieronius (Proboscidea) from Térapa, Sonora, Mexico. *Quaternary Science Reviews* 223:105949.

Mead, Jim I., C. Vance Haynes, and Bruce B. Huckell
1979 A Late Pleistocene Mastodon (*Mammut americanum*) from the Lehner Site, Southeastern Arizona. *The Southwestern Naturalist* 24(2):231–238.

Mead, Jim I., Arturo Baez, Sandra L. Swift, Mary C. Carpenter, Marci Hollenshead, Nicolas J. Czaplewski, David W. Steadman, Jordan Bright, and Joaquín Arroyo-Cabrales
2006 Tropical Marsh and Savanna of the Late Pleistocene in Northeastern Sonora, Mexico. *Southwestern Naturalist* 51:226–239.

Mead, Jim I., Richard S. White, Arturo Baez, Marci G. Hollenshead, Sandra L. Swift, and Mary C. Carpenter
2010 Late Pleistocene (Rancholabrean) *Cynomys* (Rodentia, Sciuridae: prairie dog) from Northwestern Sonora, Mexico. *Quaternary International* 217(1-2):138–142.

Mehringer, Peter J., and C. Vance Haynes, Jr.
1965 The Pollen Evidence for the Environment of Early Man and Extinct Mammals at the Lehner Mammoth Site, Southeastern Arizona. *American Antiquity* 31(1):17–23.

Meltzer, David J.
2021 *First Peoples in a New World: Populating Ice Age America,* second edition. Cambridge University Press, New York. https://doi.org/10.1017/9781108632867.

Meltzer, David J., and Vance T. Holliday
2010 Would North American Paleoindian Have Noticed Younger Dryas Age Climate Change? *Journal of World Prehistory* 23:1–41.

Mentzer, Susan M.
2017 Micro XRF. In *Archaeological Soil and Sediment Micromorphology,* edited by C. Nicosia and Georges Stoops, pp. 431–440. John Wiley and Sons, Hoboken, New Jersey.

Merriam, John C.
1906 Recent Discoveries of Quaternary Mammals in Southern California. *Science* 24(608):248–250.

Metcalfe, Jessica Z.
2011 *Late Pleistocene Climate and Proboscidean Paleoecology in North America: Insights from Stable Isotope Compositions of Skeletal Remains.* Ph.D. dissertation, University of Western Ontario, London, Ontario, Canada.

Metcalfe, S., Allison Say, Stuart Black, Robert McCulloch, and Sarah O'Hara
2002 Wet Conditions during the Last Glaciation in the Chihuahuan Desert, Alta Babicora Basin, Mexico. *Quaternary Research* 57:91–101.

Miller, C. E., and P. Goldberg
2009 Micromorphology and Paleoenvironments. In *Hell Gap: A Stratified Paleoindian Campsite at the Edge of the Rockies,* edited by Mary Lou Larson, Marcel Kornfeld, George C. Frison, pp. 71–89. University of Utah Press, Salt Lake City.

Miller, D. Shane, Vance T. Holliday, and Jordon Bright
2013 Clovis Across the Continent. In *Paleoamerican Odyssey*, edited by Kelly E. Graf, C. V. Ketron, and Michael R. Waters, pp. 207–220. Texas A&M University Press, College Station.

Miller, D. Shane, Ashley M. Smallwood, and Jesse W. Tune (editors)
2022 *The American Southeast at the End of the Ice Age.* University of Alabama Press, Tuscaloosa.

Montané Martí, Julio C.
1996 Desde los Orígenes hasta 3000 años antes del presente. In *Historia general de Sonora.* Tomo I *Periodo prehistórico y prehispánico*, second edition, pp. 151–195. Instituto Sonorense de Cultura, Gobierno del Estado de Sonora, Hermosillo.

Montellano-Ballesteros, Marisol
2002 New *Cuvieronius* Finds from the Pleistocene of Central Mexico. *Journal of Paleontology* 76:578–583.

Morales, E.
2010 *Staurosira construens.* In *Diatoms of the United States.* Retrieved August 18, 2016, from http://diatoms.org/species/staurosira_construens.

Moreno, Francisco Pascasio, and Arthur Smith Woodward
1899 On a Portion of Mammalian Skin, Named *Neomylodon listai,* from a Cavern near Consuelo Cove, Last Hope Inlet, Patagonia. *Proceedings of the Zoological Society of London* 67(1):144–156.

Morrow, Juliet E.
1997 Endscraper Morphology and Use-Life: An Approach for Studying Paleoindian Lithic Technology and Mobility. *Lithic Technology* 22(1):70–85.

Murray, A. S., and A. G. Wintle
2000 Luminescence Dating of Quartz Using an Improved Single-Aliquot Regenerative-Dose Protocol. *Radiation Measurements* 32:57–73.

Nelson, Margaret C.
1991 The Study of Technological Organization. In *Archaeological Method and Theory,* edited by Michael B. Schiffer, pp. 57–100. University of Arizona Press, Tucson.

Neuendorf, Klaus K.E., James P. Mehl, Jr., and Julia A. Jackson
2005 *Glossary of Geology,* fifth edition. American Geological Institute, Alexandria, Virginia.

Neumann Katharina, Caroline Strömberg, Terry Ball, R. M. Albert, Luc Vrydaghs, and Linda S. Cummings
2019 International Code for Phytolith Nomenclature (ICPN) 2.0. *Annals of Botany* 124:189–199.

Novello, Doris Barboni, Laure Berti-Equille, Jean-Charles Mazur, Pierre Poilecot, and Patrick Vignaud
2012 Phytolith Signal of Aquatic Plants and Soils in Chad, Central Africa. *Review of Palaeobotany and Palynology* 178:43–58.

Nunez, Elvis E., Bruce J. Macfadden, Jim I. Mead, and Arturo Baez
2010 Ancient Forests and Grasslands in the Desert: Diet and Habitat of Late Pleistocene Mammals from Northcentral Sonora, Mexico. *Palaeogeography, Palaeoclimatology, Palaeoecology* 297(2):391–400.

Ochoa-Castillo, Patricia, Mario Pérez-Campa, Ana L. Martín del Pozzo, and Joaquín Arroyo-Cabrales
2003 New Excavations in Valsequillo, Puebla, México. *Current Research in the Pleistocene* 20:61–63.

Ochoa D'Aynés, Sarahí
2004 *La industria lítica de bifaciales y puntas de proyectil en el ditio de La Playa, Sonora.* B.A. thesis, Departamento de Antropología, Universidad de Las Américas, Puebla.

Ortega-Rosas, Carmen Isela, M. C. Peñalba, and Joel Guiot
2008 Holocene Altitudinal Shifts in Vegetation Belts and Environmental Changes in the Sierra Madre Occidental, Northwestern Mexico, Based on Modern and Fossil Pollen Data. *Review of Palaeobotany and Palynology* 151:1–20.

Ortega-Rosas, Carmen Isela, Jesús Roberto R. Vidal-Solano, D. Williamson, M. C. Peñalba, and Joel Guiot
2017 Geochemical and Magnetic Evidence of a Change from Winter to Summer Rainfall Regimes at 9.2 ka BP in Northwestern Mexico. *Palaeogeography, Palaeoclimatology, Palaeoecology* 465:64-78. doi.org/10.1016/j.palaeo.2016.10.017.

Palacios-Fest, Manuel R., and Vance T. Holliday
2017 Paleoecology of a Cienega at the Mockingbird Gap Site, Chupadera Draw, New Mexico.

Palacios-Fest, Manuel R., and Vance T. Holliday (continued)
Quaternary Research 89:1-15; doi.org/10.1017/qua2017.82

Palacios-Fest, Manuel R., Daron Duke, D. Craig Young, Jason D. Kirk, and Charles G. Oviatt
2021 Paleo-lake and Wetland Paleoecology and Human Use of the Distal Old River Bed Delta at the Pleistocene-Holocene Transition in the Bonneville Basin, Utah. *Quaternary Research* 106:5–93. doi.org/10.1017/qua.2021.49.

Pasenko, Michael R., and Spencer G. Lucas
2011 A Review of Gomphotheriid (Proboscidea, Mammalia) Remains from the Pliocene Benson and Curtis Ranch Local Faunas, Southern Arizona, with a Discussion of Gomphotheriids in Arizona. *New Mexico Museum of Natural History and Science, Bulletin* 53:592–600.

Patrick, Ruth
1938 The Occurrence of Flints and Extinct Animals in Pluvial Deposits near Clovis, New Mexico, Part V. Diatom Evidence from the Mammoth Pit. *Proceedings of the Philadelphia Academy of Natural Science* 90:15–24.

Pearson, Georges A.
2017 Bridging the Gap: An Updated Overview of Clovis across Middle America and its Techno-Cultural Relation with Fluted Point Assemblages from South America, *PaleoAmerica* 3(3):203–230, doi: 10.1080/20555563.2017.1328953.

Pérez-Crespo, Victor Adrián, Joaquín Arroyo-Cabrales, Eduardo Corona-M, Pedro Morales-Puente, Edith Cienfuegos-Alvarado, and Francisco J. Otero
2015 Diet of Rincothere (Proboscidea, Gomphotheriidae, Rhynchotherium sp.) of Taxco, Guerrero, Mexico. *Southwestern Naturalist* 60:97–98. https://doi.org/10.1894/SWNAT-D-14-00008R1.1.

Pérez-Crespo, Victor Adrián, Joaquín Arroyo-Cabrales, Guadalupe Sánchez, Vance T. Holliday, Pedro Morales-Puente, Edith Cienfuegos-Alvarado, and Francisco J. Otero
2016 Información isotópica de los gonfoterios, los mastodontes y el tapir de El Fin del Mundo, un sitio Clovis mexicano. In *El poblamiento temprano en América,* edited by José Concepción Jiménez-López, Carlos Serrano-Sánchez, Felisa Aguilar-Arellano, and Arturo González-González,, pp. 211–220. Museo del Desierto, México.

Pérez-Crespo, Victor Adrián, Joaquín Arroyo-Cabrales, L. M. Alva-Valdivia, Pedro Morales-Puente, Edith Cienfuegos-Alvarado, and Francisco J. Otero
2012 Estado actual de la aplicación de los marcadores biogeoquímicos en paleoecología de mamíferos del pleistoceno tardío de México. *Archaeobios* (6):53–65.

Pérez-Crespo, Victor Adrián, Gerardo Carbot-Chanona, Pedro Morales-Puente, Edith Cienfuegos-Alvarado, and Francisco J. Otero
2015 Paleoambiente de la Depresion Central de Chiapas, con base a los isotopos estables de carbono y oxígeno. Rev. Mex. *Ciencias Geol.* 32:272–282.

Pérez-Crespo, Victor Adrián, Pedro Morales-Puente, Joaquín Arroyo-Cabrales, and Patricia Ochoa-Castillo
2019 Paleoambiente en cuatro sitios mexicanos del Pleistoceno tardío con actividad humana inferidos a partir de la fauna. *Boletín de la Sociedad Geológica Mexicana,* 71(2):343–358.

Pérez-Crespo, Victor Adrián, Josè Luis Prado, María T. Alberdi, Joaquín Arroyo-Cabrales, and Eileen Johnson
2020 Feeding Ecology of the Gomphotheres (Proboscidea, Gomphotheriidae) of America. *Quaternary Science Reviews* 229:106–126.

Pigati, Jeffrey S., Jason A. Rech, and Jeffrey C. Nekol
2010 Radiocarbon Dating of Small Terrestrial Gastropod Shells in North America. *Quaternary Geochronology* 5 (2010) 519–532.

Pigati, Jeffrey S., Jay Quade, Timothy M. Shanahan, and C. Vance Haynes Jr.
2004 Radiocarbon Dating of Minute Gastropods and New Constraints on the Timing of Spring-Discharge Deposits in Southern Arizona, USA. *Palaeogeography, Palaeoclimatology, Palaeoecology* 204:33–45.

Pimentel, N. L., V. P. Wright, and Teresa Mira Azevedo
1996 Distinguishing Early Groundwater Alteration Effects from Pedogenesis in Ancient Alluvial Basins: Examples from the Palaeogene of Southern Portugal. *Sedimentary Geology* 105(1-2):1–10.

Piperno, Dolores R.
1991 The Status of Phytolith Analysis in the American Tropics. *Journal of World Prehistory* 5:155–191.
2006 *Phytoliths: A Comprehensive Guide for Archaeologists and Paleoecologists.* Altamira Press, Lanham, Maryland.

Polaco, Oscar J., Joaquín Arroyo-Cabrales, Eduardo Corona-M, and J. G. López-Oliva
2001 The American Mastodon *Mammut americanum* in Mexico. In *The World of Elephants-International Congress, Rome*, edited by G. Cavarretta, P. Gioia, M. Mussi, and M. R. Palombo, pp. 237–242. Consiglio Nazionale delle Ricerche, Roma.

Potapova, Marina
2009 *Achnanthidium minutissimum.* In *Diatoms of the United States.* Retrieved August 18, 2016, from http://diatoms.org/species/*Achnanthidium_minutissimum.*

Potapova, Marina, and Paul Hamilton
2007 Morphological and Ecological Variation within the *Achnanthidium minutissimum* (Bacillariophyta) species complex. *Journal of Phycology* 43:561–575.

Prado, Josè Luis, María T. Alberdi, Beatriz Azanza, Begonia. Sánchez, and Daniel Frassinetti
2005 The Pleistocene Gomphotheres (Proboscidea) from South America. *Quaternary International* 126-128:21–30.

Prado, Josè Luis, Joaquín Arroyo-Cabrales, Eileen Johnson, María T. Alberdi, and Oscar J. Polaco
2015 New World Proboscidean Extinctions: Comparisons between North and South America. *Archaeological and Anthropological Sciences* 7:277–288 https://doi.org/10.1007/s12520-012-0094-3.

Prasciunas, Mary M., and Todd A. Surovell
2015 Reevaluating the Duration of Clovis: The Problem of Non-representative Radiocarbon. In *Clovis: On the Edge of a New Understanding,* edited by Ashley M. Smallwood, and Thomas A. Jennings, pp. 21–35. Texas A&M University Press, College Station.

Prescott, J. R., and J. T. Hutton
1994 Cosmic Ray Contributions to Dose Rates for Luminescence and ESR Dating. *Radiation Measurements* 23:497–500.

Reimer Paula, William E.N. Austin, Edouard Bard, Aliss Bayliss, and others
2020 The IntCal20 Northern Hemisphere Radiocarbon Age Calibration Curve (0-55 cal kB) *Radiocarbon* 62. doi: 10.1017/RDC.2020.41.

Robles, O.M.
1974 Distribucion de Artefactos Clovis en Sonora, *Boletín del Instituto Nacional de Antropología e Historia* (segunda época) 9:25–32.

Robles Ortiz, Manuel, and Francisco Manzo Taylor
1972 Clovis Fluted Points from Sonora, Mexico. *Kiva* 37(4):199–206.

Rodrigues, Kathleen, W. Jack Rink, M. B. Collins, Thomas J. Williams, Amanda Keen-Zebert, and Gloria I. López
2016 OSL Ages of the Clovis, Late Paleoindian, and Archaic Components at Area 15 of the Gault Site, Central Texas, U.S.A. *Journal of Archaeological Science Reports* 7:94-103 http://dx.doi.org/10.1016/j.jasrep.2016.03.014.

Rogers, Raymond R., and Susan M. Kidwell
2007 A Conceptual Framework for the Genesis and Analysis of Vertebrate Skeletal Concentrations. In *Bonebeds: Genesis, Analysis, and Paleobiological Significance,* edited by Raymond R. Rogers, David A. Eberth, and Anthony R. Fiorillo, pp. 1–63. University of Chicago Press, Chicago.

Round, F.E., R. M. Crawford, and D. G. Mann
1990 *The Diatoms: Biology and Morphology of the Genera.* Cambridge University Press, Cambridge.

Roy, Priyardarsi D., Muthuswamy P. Jonathan, Ligia L. Pérez-Cruz, María M. Sánchez-Córdova, Jesús D. Quiroz-Jiménez, and Francisco M. Romero
2012 A Millennial-scale Late Pleistocene–Holocene Palaeoclimatic Record from the Western Chihuahua Desert, Mexico. *Boreas* 41:707–717. doi.org/10.1111/j.1502-3885.2012.00266.x. ISSN 0300-9483.

Roy, Priyardarsi D., Axel Rivero-Navarette, Nayeli Lopez-Balbiaux, Ligia L. Pérez-Cruz, Sarah E. Metcalfe, G. Muthu Sankar, and José Luis Sánchez-Zavala
2013 A Record of Holocene Summer-Season Palaeohydrological Changes from the Southern Margin of Chihuahua. *The Holocene* 23(8):1105–1114.

Sánchez, Guadalupe
2010 *Los Primeros Mexicanos: Late Pleistocene/Early Holocene Archaeology of Sonora, Mexico.* Ph.D. dissertation, Department of Anthropology, University of Arizona, Tucson,
2016 *Los Primeros Mexicanos: Late Pleistocene and Early Holocene People of Sonora.* Anthropological Papers No. 76. University of Arizona Press, Tucson.

Sánchez, Guadalupe, and John P. Carpenter
2012 Paleoindian and Archaic Traditions in Sonora, Mexico. *From the Pleistocene to the Holocene. Human Organization and Cultural Transformations in Prehistoric North America,* edited by C. Britt Bousman and Bradley J. Vierra, pp. 125–147. Texas A & M University Press, College Station.

Sánchez, Guadalupe, Vance T. Holliday, John Carpenter, and Edmund Gaines
2015 Sonoran Clovis Groups: Lithic Technological Organization and Land Use. In *Clovis: On the Edge of a New Understanding,* edited by Ashley M. Smallwood and Thomas A. Jennings, pp. 243-261. Texas A&M University Press, College Station.

Sánchez, Guadalupe, Vance T. Holliday, Edmund P. Gaines, Joaquín Arroyo-Cabrales, Natalia Martínez-Tagüeña, Andrew Kowler, Todd Lange, Gregory W. L. Hodgins, Susan M. Mentzer, and Ismael Sánchez-Morales
2014 Human (Clovis) Gomphothere (*Cuvieronius sp.*) Association ~13,390 Calibrated yr BP in Sonora, Mexico *Proceedings of the National Academy of Sciences* 111: 10972–10977. doi/10.1073/pnas.1404546111.

Sánchez, Guadalupe, Vance Holliday, Edmund P. Gaines, and Ismael Sánchez-Morales
2008 *Informe de la Temporada de Campo Invierno 2007-2008 del Sitio Fin del Mundo.* Proyecto Geoarqueología y Tecnología Lítica de los Sitios Paleoindios de Sonora. Archivo Técnico del INAH, mecanoescrito, México City.

Sánchez-Morales, Ismael
2012 *Las industrias* líticas de *puntas de proyectil y bifaciales en los sitios arcaicos de Sonora.* BA thesis, Escuela Nacional de Antropología e Historia, Mexico City.
2018 The Clovis Lithic Assemblage from El Fin del Mundo, Sonora, Mexico: Evidence of Upland Campsite Localities. *PaleoAmerica, 4*(1):76–81.

Sánchez-Morales, Ismael, Guadalupe Sánchez, and Vance T. Holliday
2022 Clovis Stone Tools from El Fin del Mundo, Sonora, Mexico: Site Use and Associations between Localities. *American Antiquity* 87(3):523–543.

Sanders, Albert E.
2002 *Additions to the Pleistocene Mammal Faunas of South Carolina, North Carolina, and Georgia.* American Philosophical Society Transactions 92(5).

Saunders, Jeffrey J., and Edward B. Daeschler
1994 Descriptive Analyses and Taphonomical Observations of Culturally-Modified Mammoths Excavated at "The Gravel Pit," near Clovis, New Mexico in 1936. *Proceedings of the Academy of Natural Sciences of Philadelphia* 145:1–28.

Saunders, Jeffrey J.
2007 Processing Marks on Remains of *Mammuthus columbi* from the Dent Site, Colorado, in Light of Those from Clovis, New Mexico. In *Frontiers in Colorado Paleoindian Archaeology,* edited by Robert H. Brunswig and Bonnie L. Pitblado, pp. 155–184. University Press of Colorado, Boulder.

Sellards, Elias H.
1938 Artifacts Associated with Fossil Elephant. *Geological Society of America Bulletin* 49:999–1010.
1952 *Early Man in America.* University of Texas Press, Austin.

Sellards, Elias H., and Glen L. Evans
1960 The Paleo-Indian Cultural Succession in the Central High Plains of Texas and New Mexico. In *Men and Cultures,* edited by Anthony F.C. Wallace, pp. 639–649. University of Pennsylvania Press, Philadelphia.

Shott, Michael J.
1995 How Much is a Scraper? Curation, Use Rates and the Formation of Scraper Assemblages. *Lithic Technology* 20:53–72.

Singer, Michael J., and Peter Janitzky, P. (editors)
1986 *Field and Laboratory Procedures Used in a Soil Chronosequence Study.* Bulletin No. 1648. USDOI, U.S. Geological Survey, Washington D.C.

Smallwood, Ashley M.
2010 Clovis Biface Technology at the Topper Site, South Carolina: Evidence for Variation and Technological Flexibility. *Journal of Archaeological Science* 37:2413–2425.
2012 Clovis Technology and Settlement in the American Southeast: Using Biface Analysis to Evaluate Dispersal Models. *American Antiquity* 77(4):689–713.

Smallwood, Ashley M., and Thomas A. Jennings (editors)
2015 *Clovis: On the Edge of a New Understanding.* Texas A&M University Press, College Station.

Smedley, R. K., G.A.T. Duller, J. G. Pearce, and H. M. Roberts
2012 Determining the K-content of Single-Grains of Feldspar for Luminescence Dating. *Radiation Measurements* 47:790–796.

Smith, Gregory James, and Larisa R. G. DeSantis
2020 Extinction of North American *Cuvieronius* (Mammalia: Proboscidea: Gomphotheriidae) Driven by Dietary Resource Competition with Sympatric Mammoths and Mastodons. *Paleobiology* 46(1):41–57; doi: 10.1017/pab.2020.7.

Smith, Heather L., Ashley M. Smallwood, and Thomas J. DeWitt
2015 Defining the Normative Range of Clovis Fluted Point Shape Using Geographic Models of Geometric Morphometric Variation. In *Clovis: On the Edge of a New Understanding*, edited by Ashley M. Smallwood and Thomas A. Jennings, pp. 161–180. Texas A&M University Press, College Station.

Smol, John P.
2002 *Pollution of Lakes and Rivers: A Paleoenvironmental Perspective.* Arnold Publishers, London, co-published by Oxford University Press, New York.

Spaulding, S.
2010 *Epithemia.* In *Diatoms of the United States.* Retrieved August 18, 2016, from http://western diatoms.colorado.edu/taxa/genus/epithemia.

Spaulding, S., and M. Edlund
2008 *Denticula.* In *Diatoms of the United States.* Retrieved August 18, 2016, from http://westerndiatoms.colorado.edu/taxa/genus/Denticula.

Stafford, Thomas W., Jr.
1981 Alluvial Geology and Archaeological Potential of the Texas Southern High Plains. *American Antiquity* 46(3):548–565.

Starrat, Scott W.
2011 *Holocene Diatom Flora and Climate History of Medicine Lake, Northern California, USA.* Beihefte 141. Nova Hedwigia, Stuttgart.

Stoermer, Eugene F., and John P. Smol
1999 *The Diatoms: Applications for Environmental and Earth Sciences.* Cambridge University Press, Cambridge.

Stoops, Georges
2021 *Guidelines for Analysis and Description of Soil and Regolith Thin Sections*, second edition. Madison, Wisconsin and Wiley, New York.

Strauss, Lawrence G., and Ted Goebel
2011 Human and Younger Dryas: Dead End, Short Detour, or Open Road to the Holocene? *Quaternary International* 242(2): 259–584.

Street-Perrott, F. A., D. S. Marchand, N. Roberts, and S. P. Harrison
1989 *Global Lake-Level Variations from 18,000 to 0 Years Ago: A Palaeoclimatic Analysis.* Technical Report No 46. U.S. Department of Energy, Washington, D.C.

Stuiver, M., P. J. Reimer, and R. W. Reimer, R.W.
2021 CALIB 8.2 at http://calib.org, accessed 2021-3-1.

Surovell, Todd A.
2000 Radiocarbon Dating of Bone Apatite by Step Heating. *Geoarchaeology: An International Journal,* 15(6):591–608.

Surovell, Todd A., Joshua R. Boyd, C. Vance Haynes, Jr., and Gregory W. L. Hodgins
2016 On the Dating of the Folsom Complex and Its Correlation with the Younger Dryas, the End of Clovis, and Megafaunal Extinction. *PaleoAmerica,* 2(2):81–89, doi:10.1080/20555563.2016.1174559.

Thompson, R. S., C. Whitlock, P. J. Bartlein, S. P. Harrison, and W. G. Spaulding
1993 Climatic Changes in the Western United States Since 18,000 yr B.P. In *Global Climates Since the Last Glacial Maximum,* edited by H. E. Wright, Jr., J. E. Kutzbach, T. Webb III, W. F. Ruddiman, F. A. Street-Perrott, and P. J. Bartlein, pp. 468–513. University of Minnesota Press, Minneapolis.

Thulman, David K.
2022 Southeastern Late Paleoindian through Early Archaic Chronologies. In *The American Southeast at the End of the Ice Age,* edited by D. Shane Miller, Ashley M. Smallwood, and Jesse W. Tune, pp. 306–334. University of Alabama Press, Tuscaloosa.

Torrence, Robin
1983 Time Budgeting and Hunter-Gatherer Technology. In *Hunter-Gatherer Economy in Prehistory: A European Perspective,* edited by G. Bailey, pp. 11–22. Cambridge University Press, Cambridge.

Trumbore, Susan E.
2000 Radiocarbon Geochronology. In *Quaternary Geochronology: Methods and Applications,* edited by Jay Stratton Noller, Janet M. Sowers, and William R. Lettis, pp. 41–60. AGU Reference Shelf 4. American Geophysical Union, Washington, D.C.

Tune, Jess W., Thomas. A. Jennings, and Aaron Deter-Wolf
2022 Prismatic Blade Production at the Sinclair Site, Tennessee: Implications for Understanding Clovis Technological Organization. *American Antiquity* 87(3):601–610.

Turner Raymond, and David E. Brown
1994 Sonoran Desertscrub. In *Biotic Communities: Southwestern United States and Northwestern Mexico,* edited by David E. Brown, pp. 181–221. University of Utah Press, Salt Lake City.

Twiss, P. C., Erwin Suess, and R. M. Smith
1969 Morphological Classification of Grass Phytoliths. *Soil Genesis, Morphology, and Classification* 33:109–115.

Van Devender, Thomas R.
1990a Late Quaternary Vegetation and Climate of the Chihuahuan Desert, United States and Mexico. In *Packrat Middens: The Last 40,000 Years of Biotic Change,* edited by Julio L. Betancourt, Thomas R. Van Devender, and Paul S. Martin, pp. 104–133. University of Arizona Press, Tucson.
1990b Late Quaternary Vegetation and Climate of the Sonoran Desert, United States and Mexico. In *Packrat Middens: The Last 40,000 Years of Biotic Change,* edited by Julio L. Betancourt, Thomas Van Devender, and Paul S. Martin, pp. 134–165. University of Arizona Press, Tucson.

Vrydaghs, Luc, Terry B. Ball, and Yannick Devos
2016 Beyond Redundancy and Multiplicity: Integrating Phytolith Analysis and Micromorphology to the Study of Brussels Dark Earth. *Journal of Archaeological Science* 68:79–88.

Walthall, John A., and George R. Holley
1997 Mobility and Hunter Gatherer Toolkit Design: Analysis of a Dalton Lithic Cache. *Southeastern Archaeology* 16(2):152–162.

Waters, Michael R., and Thomas W. Stafford, Jr.
2007 Redefining the Age of Clovis: Implications for the Peopling of the Americas. *Science* 315:1122–1126.

Waters, Michael R., Charlotte D. Pevny, and David L. Carlson
2011 *Clovis Lithic Technology. Investigation of a Stratified Workshop at the Gault Site, Texas.* Texas A&M University Press, College Station.

Waters, Michael R., Thomas W. Stafford, and D. L. Carlson
2020 The Age of Clovis—13,050 to 12,750 cal yr BP. *Science Advances* 6, p.eaaz0455

Waters, Michael R., Joshua L. Keene, Steven L. Forman, Elton R. Prewitt, David L. Carlson, and James E. Wiederhold
2018 Pre-Clovis Projectile Points at the Debra L. Friedkin Site, Texas—Implications for the Late Pleistocene Peopling of the Americas. *Science Advances* 4:doi: 10.1126/sciadv.aat4505.

Waters, Michael R., Thomas W. Stafford, Brian Kooyman, and L. V. Hills
2015 Late Pleistocene Horse and Camel Hunting at the Southern Margin of the Ice-Free Corridor: Reassessing the Age of Wally's Beach, Canada. *Proceedings of the National Academy of Sciences* 112:4263–4267. doi:10.1073/pnas.1420650112.

Wells, Phillip V.
1966 Late Pleistocene and Degree of Pluvial Climate Change in the Chihuahuan Desert. *Science* 16:970–975.

Wessel, R.L., P. L. Eidenbach, L. M. Meyer, C. S. Comer, and B. Knight
1997 *From Playas to Highlands: PaleoIndian Adaptations to the Region of the Tularosa.* Human Systems Research, Las Cruces.

Wiant, Michael D., and Harold Hassen
1985 The Role of Lithic Availability and Accessibility in the Organization of Lithic Technology. In *Lithic Resource Procurement: Proceedings from the Second Conference on Prehistoric Chert Exploitation*, edited by Susan C. Vehik, pp. 101–114. Center for Archaeological Investigations Occasional Paper No. 4. Southern Illinois University, Carbondale.

Willard, Debra, Christopher E. Bernhardt, Lisa Weimer, Sherri R. Cooper, Desiré. Gamez, and Jennifer Jensen
2004 Atlas of Pollen and Spores of the Florida Everglades. *Palynology* 28:175–227.

Winsborough, Barbara M.
1995 Diatoms. In *Stratigraphy and Paleoenvironments of Late Quaternary Valley Fills on the Southern High Plains*, by Vance T. Holliday, pp. 67–82. Memoir No. 186. Geological Society of America, Boulder.

Winsborough, Barbara
2016 Diatom Paleoenvironmental Analysis of Sediments from the Water Canyon Paleoindian Site (LA134764). Winsborough Consulting, Technical Report prepared for the Office of Archaeological Studies, Museum of New Mexico. Santa Fe.

Wintle, Ann G., and Andrew S. Murray
2006 A Review of Quartz Optically Stimulated Luminescence Characteristics and Their Relevance in Single-Aliquot Regeneration Dating Protocols. *Radiation Measurements* 41:369–391.

Wormington, H. Marie
1957 *Ancient Man in North America.* Popular Series No. 4. Denver Museum of Natural History, Denver.

Yost, Chad L.
2016 *Phytolith Analysis of Late Pleistocene and Early Holocene Sediments from the Water Canyon Paleoindian Site, LA134764, New Mexico.* Technical report submitted to Office of Contract Archaeology, University of New Mexico, Albuquerque.

Yost, Chad L., and Mikhail S. Blinnikov
2011 Locally Diagnostic Phytoliths of Wild Rice (*Zizania palustris* L.) from Minnesota, USA: Comparison to Other Wetland Grasses and Usefulness for Archaeobotany and Paleoecological Reconstructions. *Journal of Archaeological Science* 38:1977–1991.

Zahajská, Petra, Sophie Opfergelt, Sherilyn C. Fritz, Johanna Stadmark, and Daniel J. Conley
2020 What Is Diatomite? *Quaternary Research* 96:48–52.

Zarzycka, Sandra E., Todd A. Surovell, Madeline E. Mackie, Spencer R. Pelton, Robert L. Kelly, Paul Goldberg, Janet Dewey, and Meghan Kent
2019 Long-Distance Transport of Red Ocher by Clovis Foragers. *Journal of Archaeological Science: Reports* 25:519–529.

Index

ABSTRACT

El Fin del Mundo is a multicomponent archaeological site that produced Clovis artifacts directly associated with the remains of two gomphotheres (*Cuvieronius*) buried in an ancient, low-energy stream in the final stages of alluviation. The archaeological context is indicative of a hunting event. An extensive lithic assemblage of Clovis and Archaic artifacts are scattered across the adjacent upland surface of a Pleistocene alluvial fan. Clovis hunters encountered the gomphotheres on a flat alluvial plain within a grassland. Below the Clovis bonebed is a rare paleontological accumulation of late Pleistocene gomphothere (*Cuvieronius* sp.), mastodon (*Mammut americanum*), and mammoth (*Mammuthus* sp.), representing a mixed steppe-grassland coincident with a perennial stream. Paleoenvironmental analysis revealed that the steppe-grassland during the Clovis occupation (≤13,415 cal yr B.P.) evolved into a warmer, drier grassland. The location containing the gomphothere remains became a diatomaceous lake, followed around 11,130 cal yr. B.P. by a marsh. El Fin del Mundo provides the first evidence for the association of Clovis hunters with gomphotheres; the first archaeological context for gomphotheres in North America; the northernmost post-LGM gomphotheres in North America; contains a rare association between a Clovis-megafauna feature and a campsite; and preserves a Late Pleistocene-Early Holocene paleoenvironmental record of the Sonoran Desert of Northwest Mexico. It is the first documented and excavated in situ Clovis site outside of the United States and is the first Clovis site to be excavated in Mexico and Central America.

RESUMEN

El Fin del Mundo es un sitio arqueológico multicomponente que ha producido artefactos líticos Clovis directamente asociados a los restos óseos de dos gonfoterios (*Cuvieronius*) enterrados en el canal de un arroyo antiguo de baja energía en las etapas finales de aluviación. Este contexto arqueológico es indicativo de un evento de cacería de megafauna. Un gran número de materiales líticos Clovis y Arcaicos se encuentra disperso sobre la superficie del abanico aluvial adyacente en áreas un poco más elevadas. Los cazadores Clovis enfrentaron a los gonfoterios en una planicie aluvial dentro de un pastizal. Por debajo de la cama de huesos de edad Clovis se encuentra una acumulación paleontológica del Pleistoceno tardío poco común que incluye restos de gonfoterio (*Cuvieronius* sp.), mastodonte (*Mammut americanum*), y mamut (*Mammuthus* sp.), representando un bioma mixto de pastizales y matorrales coincidente con un arroyo perenne. Análisis paleoambientales indican que el pastizal-matorral que caracterizó al sitio durante la ocupación Clovis (≤13,415 años cal. A.P.) dio lugar a un pastizal más seco y cálido. La localidad que contiene los restos de los gonfoterios derivó en un lago diatomáceo, seguido por una ciénaga alrededor de 11,130 años cal. AP. El Fin del Mundo ha producido la primera evidencia de la asociación de cazadores Clovis y gonfoterios; el primer contexto arqueológico de gonfoterios en Norte América; los gonfoterios post-último máximo glacial más norteños; contiene una asociación inusual de un contexto Clovis de cacería de megafauna y un campamento; y preserva un registro paleoambiental del Pleistoceno tardío-Holoceno temprano del Desierto sonorense del Noroeste de México. El Fin del Mundo es el primer sitio Clovis in situ documentado y excavado fuera de los Estados Unidos y es el primer sitio Clovis en ser excavado en México y Centroamérica.

ANTHROPOLOGICAL PAPERS OF THE UNIVERSITY OF ARIZONA

1. **Excavations at Nantack Village, Point of Pines, Arizona**
   David A. Breternitz. 1959.
2. **Yaqui Myths and Legends**
   Ruth W. Giddings. 1959. Now in book form.
3. **Marobavi: A Study of an Assimilated Group in Northern Sonora**
   Roger C. Owen. 1959.
4. **A Survey of Indian Assimilation in Eastern Sonora**
   Thomas B. Hinton. 1959.
5. **The Phonology of Arizona Yaqui with Texts**
   Lynn S. Crumrine. 1961.
6. **The Maricopas: An Identification from Documentary Sources**
   Paul H. Ezell. 1963.
7. **The San Carlos Indian Cattle Industry**
   Harry T. Getty. 1963.
8. **The House Cross of the Mayo Indians of Sonora, Mexico**
   N. Ross Crumrine. 1964.
9. **Salvage Archaeology in Painted Rocks Reservoir, Western Arizona**
   William W. Wasley and Alfred E. Johnson. 1965.
10. **An Appraisal of Tree-Ring Dated Pottery in the Southwest**
    David A. Breternitz. 1966.
11. **The Albuquerque Navajos**
    William H. Hodge. 1969.
12. **Papago Indians at Work**
    Jack O. Waddell. 1969.
13. **Culture Change and Shifting Populations in Central Northern Mexico**
    William B. Griffen. 1969.
14. **Ceremonial Exchange as a Mechanism in Tribal Integration Among the Mayos of Northwest Mexico**
    Lynn S. Crumrine. 1969.
15. **Western Apache Witchcraft**
    Keith H. Basso. 1969.
16. **Lithic Analysis and Cultural Inference: A Paleo-Indian Case**
    Edwin N. Wilmsen. 1970.
17. **Archaeology as Anthropology: A Case Study**
    William A. Longacre. 1970.
18. **Broken K Pueblo: Prehistoric Social Organization in the American Southwest**
    James N. Hill. 1970.
19. **White Mountain Redware: A Pottery Tradition of East-Central Arizona and Western New Mexico**
    Roy L. Carlson. 1970.
20. **Mexican Macaws: Comparative Osteology**
    Lyndon L. Hargrave. 1970.
21. **Apachean Culture History and Ethnology**
    Keith H. Basso and Morris E. Opler, eds. 1971.
22. **Social Functions of Language in a Mexican-American Community**
    George C. Barker. 1972.
23. **The Indians of Point of Pines, Arizona: A Comparative Study of Their Physical Characteristics**
    Kenneth A. Bennett. 1973.
24. **Population, Contact, and Climate in the New Mexico Pueblos**
    Ezra B. W. Zubrow. 1974.
25. **Irrigation's Impact on Society**
    Theodore E. Downing and McGuire Gibson, eds. 1974.
26. **Excavations at Punta de Agua in the Santa Cruz River Basin, Southeastern Arizona**
    J. Cameron Greenleaf. 1975.
27. **Seri Prehistory: The Archaeology of the Central Coast of Sonora, Mexico**
    Thomas Bowen. 1976.
28. **Carib-Speaking Indians: Culture, Society, and Language**
    Ellen B. Basso, ed. 1977.
29. **Cocopa Ethnography**
    William H. Kelly. 1977.
30. **The Hodges Ruin: A Hohokam Community in the Tucson Basin**
    Isabel Kelly, James E. Officer, and Emil W. Haury, collaborators; Gayle H. Hartmann, ed. 1978.
31. **Fort Bowie Material Culture**
    Robert M. Herskovitz. 1978.
32. **Wooden Ritual Artifacts from Chaco Canyon, New Mexico: The Chetro Ketl Collection**
    R. Gwinn Vivian, Dulce N. Dodgen, and Gayle H. Hartmann. 1978.
33. **Indian Assimilation in the Franciscan Area of Nueva Vizcaya**
    William B. Griffen. 1979.
34. **The Durango South Project: Archaeological Salvage of Two Late Basketmaker III Sites in the Durango District**
    John D. Gooding. 1980.
35. **Basketmaker Caves in the Prayer Rock District, Northeastern Arizona**
    Elizabeth Ann Morris. 1980.
36. **Archaeological Explorations in Caves of the Point of Pines Region, Arizona**
    James C. Gifford. 1980.
37. **Ceramic Sequences in Colima: Capacha, an Early Phase**
    Isabel Kelly. 1980.
38. **Themes of Indigenous Acculturation in Northwest Mexico**
    Thomas B. Hinton and Phil C. Weigand, eds. 1981.
39. **Sixteenth Century Maiolica Pottery in the Valley of Mexico**
    Florence C. Lister and Robert H. Lister. 1982.
40. **Multidisciplinary Research at Grasshopper Pueblo, Arizona**
    William A. Longacre, Sally J. Holbrook, and Michael W. Graves, eds. 1982.
41. **The Asturian of Cantabria: Early Holocene Hunter-Gatherers in Northern Spain**
    Geoffrey A. Clark. 1983.
42. **The Cochise Cultural Sequence in Southeastern Arizona**
    E. B. Sayles. 1983.
43. **Cultural and Environmental History of Cienega Valley, Southeastern Arizona**
    Frank W. Eddy and Maurice E. Cooley. 1983.
44. **Settlement, Subsistence, and Society in Late Zuni Prehistory**
    Keith W. Kintigh. 1985.

45. **The Geoarchaeology of Whitewater Draw, Arizona**
Michael R. Waters. 1986.

46. **Ejidos and Regions of Refuge in Northwestern Mexico**
N. Ross Crumrine and Phil C. Weigand, eds. 1987.

47. **Preclassic Maya Pottery at Cuello, Belize**
Laura J. Kosakowsky. 1987.

48. **Pre-Hispanic Occupance in the Valley of Sonora, Mexico**
William E. Doolittle. 1988.

49. **Mortuary Practices and Social Differentiation at Casas Grandes, Chihuahua, Mexico**
John C. Ravesloot. 1988.

50. **Point of Pines, Arizona: A History of the University of Arizona Archaeological Field School**
Emil W. Haury. 1989.

51. **Patarata Pottery: Classic Period Ceramics of the South-central Gulf Coast, Veracruz, Mexico**
Barbara L. Stark. 1989.

52. **The Chinese of Early Tucson: Historic Archaeology from the Tucson Urban Renewal Project**
Florence C. Lister and Robert H. Lister. 1989.

53. **Mimbres Archaeology of the Upper Gila, New Mexico**
Stephen H. Lekson. 1990.

54. **Prehistoric Households at Turkey Creek Pueblo, Arizona**
Julie C. Lowell. 1991.

55. **Homol'ovi II: Archaeology of an Ancestral Hopi Village, Arizona**
E. Charles Adams and Kelley Ann Hays, eds. 1991.

56. **The Marana Community in the Hohokam World**
Suzanne K. Fish, Paul R. Fish, and John H. Madsen, eds. 1992.

57. **Between Desert and River: Hohokam Settlement and Land Use in the Los Robles Community**
Christian E. Downum. 1993.

58. **Sourcing Prehistoric Ceramics at Chodistaas Pueblo, Arizona: The Circulation of People and Pots in the Grasshopper Region**
María Nieves Zedeño. 1994.

59. **Of Marshes and Maize: Preceramic Agricultural Settlements in the Cienega Valley, Southeastern Arizona**
Bruce B. Huckell. 1995.

60. **Historic Zuni Architecture and Society: An Archaeological Application of Space Syntax**
T. J. Ferguson. 1996.

61. **Ceramic Commodities and Common Containers: Production and Distribution of White Mountain Red Ware in the Grasshopper Region, Arizona**
Daniela Triadan. 1997.

62. **Prehistoric Sandals from Northeastern Arizona: The Earl H. Morris and Ann Axtell Morris Research**
Kelley Ann Hays-Gilpin, Ann Cordy Deegan, and Elizabeth Ann Morris. 1998.

63. **Expanding the View of Hohokam Platform Mounds: An Ethnographic Perspective**
Mark D. Elson. 1998.

64. **Great House Communities Across the Chacoan Landscape**
John Kantner and Nancy M. Mahoney, eds. 2000.

65. **Tracking Prehistoric Migrations: Pueblo Settlers among the Tonto Basin Hohokam**
Jeffery J. Clark. 2001.

66. **Beyond Chaco: Great Kiva Communities on the Mogollon Rim Frontier**
Sarah A. Herr. 2001.

67. **Salado Archaeology of the Upper Gila, New Mexico**
Stephen H. Lekson. 2002.

68. **Ancestral Hopi Migrations**
Patrick D. Lyons. 2003.

69. **Ancient Maya Life in the Far West Bajo: Social and Environmental Change in the Wetlands of Belize**
Julie L. Kunen. 2004.

70. **The Safford Valley Grids: Prehistoric Cultivation in the Southern Arizona Desert**
William E. Doolittle and James A. Neely, eds. 2004.

71. **Murray Springs: A Clovis Site with Multiple Activity Areas in the San Pedro Valley, Arizona**
C. Vance Haynes, Jr., and Bruce B. Huckell, eds. 2007.

72. **Ancestral Zuni Glaze-Decorated Pottery: Viewing Pueblo IV Regional Organization through Ceramic Production and Exchange**
Deborah L. Huntley. 2008.

73. **In the Aftermath of Migration: Renegotiating Ancient Identity in Southeastern Arizona**
Anna A. Neuzil. 2008.

74. **Burnt Corn Pueblo, Conflict and Conflagration in the Galisteo Basin, A.D. 1250–1325**
James E Snead and Mark W. Allen. 2011.

75. **Potters and Communities of Practice: Glaze Paint and Polychrome Pottery in the American Southwest, A.D. 1250–1700**
Linda S. Cordell and Judith Habicht-Mauche, eds. 2012.

76. **Los Primeros Mexicanos: Late Pleistocene and Early Holocene People of Sonora**
Guadalupe Sánchez. 2015.

77. **The Ceramic Sequence of the Holmul Region, Guatemala**
Michael G. Callaghan and Nina Neivens de Estrada. 2016.

78. **The Winged: An Upper Missouri River Ethno-Ornithology**
Kaitlyn Chandler, Wendi Field Murray, María Nieves Zedeño, Samrat Clements, and Robert James. 2016.

79. **Seventeenth-Century Metallurgy on the Spanish Colonial Frontier: Pueblo and Spanish Interactions**
Noah H. Thomas. 2018.

80. **Reframing the Northern Rio Grande Pueblo Economy**
Scott G. Orman, ed. 2019.

81. **Oysters in the Land of Cacao: Archaeology, Material Culture, and Societies at Islas de los Cerros and the Western Chontalpa, Tabasco, Mexico**
Bradley E. Ensor. 2020.

82. **Households on the Mimbres Horizon: Excavations at La Gila Encantada, Southwestern New Mexico**
Barbara J. Roth. 2022.

83. **Coastal Foragers of the Gran Desierto: Investigations of Prehistoric Shell Middens along the Northern Sonoran Coast**
Douglas R. Mitchell, Jonathan B. Mabry, Gary Huckleberry, and Natalia Martínez-Tagüeña, eds. 2024.

84. **El Fin del Mundo: A Clovis Site in Sonora, Mexico**
Vance T. Holliday, Guadalupe Sánchez, and Ismael Sánchez-Morales, eds. 2024.